Jules J. Berman
Fundamentals of Reality

De Gruyter Series in Mathematics and Life Sciences

Volume 13

Jules J. Berman

Fundamentals of Reality

Emergent Systems Development from Vacuum and Waves

DE GRUYTER

Mathematics Subject Classification 2020
Primary: 97A10, 83-XX, 03B35; Secondary: 81T80, 83A05

Author
Jules J. Berman
USA

ISBN 978-3-11-149814-0
e-ISBN (PDF) 978-3-11-149869-0
e-ISBN (EPUB) 978-3-11-149941-3
ISSN 2195-5530

Library of Congress Control Number: 2025945950

Bibliographic information published by the Deutsche Nationalbibliothek
The Deutsche Nationalbibliothek lists this publication in the Deutsche Nationalbibliografie; detailed bibliographic data are available on the Internet at http://dnb.dnb.de.

Typesetting: VTeX UAB, Lithuania
Printing and binding: CPI books GmbH, Leck

www.degruyterbrill.com
Questions about General Product Safety Regulation:
productsafety@degruyterbrill.com

To my grandchildren: Everett, Luca, Kenzie, and Noah. I hope they come to look favorably upon the reality of their world.

Acknowledgments

Special thanks to Henry Hoffman, who patiently scrutinized every sentence in this book, offering hundreds of helpful suggestions in the process. And to Alex Flournoy, whose videotaped lectures on quantum physics have informed and inspired me.

https://doi.org/10.1515/9783111498690-203

Preface

> Before I speak, I have something important to say.
>
> – Groucho Marx

All things have a beginning, but for the starting points of the fundamental components of our reality (e. g., space, time, motion, energy, matter, light, life, and death), we have no witnesses who can testify to their origins. Should we capitulate to mystery? Must we get on with the daily grind of our lives, ignorant of how our universe came into existence? As it happens, we have a great number of scientific methods by which we can reconstruct the past, based on observations made in the present. We may not have witnessed the emergence of our reality, but we can determine what almost certainly occurred, if we put our minds to the task.

Fundamentals of Reality provides a comprehensive view of the origins of just about everything, drawing from a wide array of scientific advances. In the process of doing so, this book has a set of goals that may be of particular benefit to readers:

1. To vigorously challenge the perception that a general audience will never read a book that contains a serious discussion of mathematical concepts.
   The widely held notion that non-mathematicians cannot grasp mathematical concepts has had a chilling effect on the scientific community and on society as a whole. Mathematics can be understood and enjoyed by non-mathematicians when explained clearly, in narrative text. Also, any conclusions drawn from mathematical analysis must not be defended by insisting that equations do not lie. Equations are, like most things, open to interpretation, and our interpretations are formed with the concepts provided by spoken language.
   History dictates that many of the greatest advances in science come about when we re-examine widely-held assumptions. Our best works may come from scientists working across several disciplines, or from gifted novitiates (e. g., students), or from dedicated amateurs. [1] [2] [3]
2. To develop a coherent narrative that draws no distinction between the various scientific disciplines.
   Nature does not chop the sciences into sovereign disciplines. There is no real difference between chemistry, biology, and physics; they all operate under the same principles and natural laws. To understand reality, it is best to abandon allegiance to any one scientific field so that we may focus on general truths.
3. To highlight recent scientific breakthroughs.
   One of the chief tasks of every scientist is to discern how new findings may be applied to old, unresolved, issues. In *Fundamentals of Reality*, we examine recent findings whose deep significance may have been overlooked by science journalists.
4. To eliminate magical thinking from the natural sciences.
   Throughout our past, mankind has accepted reality as something given to us, as if by magic. We have been taught to believe that whatever may have happened when

https://doi.org/10.1515/9783111498690-204

space, time and life emerged from chaos, the specifics lay beyond our human comprehension. Today, we have finally arrived at the moment when we have the scientific tools to form a credible understanding of our origins. Using powerful telescopes, we can visualize events that occurred about 13.8 billion years ago. We can examine exoplanets orbiting distant stars, and we can make some educated guesses about life outside our solar system. With high-energy particle accelerators, we can come very close to recreating the conditions of our early universe, and capturing the kinds of events that occurred when the first particles interacted. Molecular biologists, using gene sequencing techniques, have learned a great deal about evolution, tracing the lineage of all living things backwards in time, to the first forms of life on our planet. In addition, we now know enough about the behavior of energy and matter to understand how our reality may have developed from a set of initial conditions and laws. Though we have moved closer to truth, today's scientific community understands that all hypotheses must be tested and re-tested. All interpretations of data must be subjected to re-analysis. We now understand that trusted conclusions are tentative, and subject to modification. In *Fundamentals of Reality*, our material skirts the edge of human understanding, and readers are encouraged to shield their sanity with skepticism.

5. To show that knowing how things begin always tells us a great deal about how things develop over time.
   Scientists and engineers are taught how to determine the outcome of events, beginning with an initial set of conditions (e. g., mass, velocity, temperature, composition). The initial conditions are a "given" and are not typically open to much scrutiny. In *Fundamentals of Reality*, we will be asking how those initial conditions came into existence. For example, we would like to know how it is possible for something to be derived from nothing; to discover whether our universe follows a predetermined course, and to isolate the ultimate source of the small collection of fundamental forces and universal constants that account for our universal laws. Initial conditions set the stage for all that follows, and we shall never understand our reality if we cannot understand the first moment of our past.
6. To refute the notion that the fundamental laws and principles that govern our universe make no intuitive sense.
   For physicists, mathematics provides the methods by which one equation can lead to new equations that reveal otherwise unforeseen physical relationships. No one can deny that mathematics has a form of beauty that is unmatched in non-mathematical language. When we use mathematical equations, we can be certain that our conclusions can be unambiguously understood by specialists in all fields who understand the language of mathematics. Regardless, every step in the mathematical process, and the meaning of every inference drawn from the process, relies upon concepts that are understandable to the human brain. Scientists who cannot justify their conclusions without falling back on abstruse mathematical symbolism may easily lose track of their mistakes. The so-called cosmological catastrophe serves as a case in

> point. Starting with measurements of vacuum energy, and building upon relationships described by the theory of general relativity, physicists had concluded that the universe doubles in size every 10^{-43} seconds, a rate that differs from observation by a factor of 10^{120}, accounting for what is widely regarded as the worst prediction in the history of the universe! If their prediction had any physical validity, then all the atoms in the universe would be torn apart in well under a trillionth of a second.
> We can often achieve more by exploring the questions posed by physicists than by understanding the mathematical "answers" to those questions. Accordingly, we will linger on topics that focus on the reasons why various lines of inquiry are important, and how our tentative conclusions may help us grasp a variety of complementary questions raised in all of the natural sciences.

The text is written for intellectually curious generalists. All that is expected of the reader is a high-school level familiarity with the natural sciences (physics, chemistry, mathematics, and biology). For young readers contemplating careers in science, *Fundamentals of Reality* provides an introduction to topics covered in college-level courses. Didactic aids have been added to accommodate readers at different stages of their scientific training. For readers unfamiliar with the subject matter, footnotes are provided that provide short definitions or clarifications of concepts and terminology. For more advanced readers, there are in-text links to notes containing proofs and expanded discussions of difficult topics. There are also in-text citations to relevant bibliographic manuscripts. Serious readers may choose to use this book as a reference source, returning to the notes, footnotes, and citations in moments of renewed curiosity.

Contents

List of Figures

https://doi.org/10.1515/9783111498690-206

1 Introduction: A preview of book contents

The beginnings and endings of all human undertakings are untidy.

– John Galsworthy

Let us look at a hypothetical chronology of the conditions and events that created our current reality.

1. It all begins with a particular kind of nothingness; a nothingness that accommodates wave motion.
2. Wave propagation implies the presence of energy (to move the waves), plus space-time (in which the waves move), plus momentum.
3. Because the bare-bones universe consists of a vacuum through which waves can propagate, the evolution of the universe over time can be described by a wave equation.
4. A universe of waves moving through vacuum places limits upon observability (and hence, limits on the precision of observations).
5. Heisenberg's uncertainty principle sets limits on the precision by which we can simultaneously measure both position and momentum or both energy and time.
6. The Uncertainty Principle tells us that there can never be a measurably zero energy in space. Hence, the energy at every point in space is in constant flux.
7. Conserved physical quantities, and the laws that describe their conservation, arise as a consequence of symmetries (Noether's Law).
8. Symmetries of energy and motion are preserved in different frames of reference. Reference frame symmetries impose a limit upon the speed of light and yield Einstein's laws of special relativity.
9. The laws of special relativity account for the concept of mass, representing that which resists acceleration by energy.
10. Symmetries can "break" to form additional symmetries. A symmetry that arises from another symmetry has its own properties while also preserving the properties of the symmetry from which it arose.
11. Symmetries account for the fields described by the Standard Model of quantum physics.
12. Each of the fields described by the Standard Model of quantum physics accounts for a physical force (i. e., the force distributed through the coordinate space of the field).
13. The interactions between fields described in the Standard Model require fundamental constants.
14. Each field of the Standard Model is characterized by its own particular quantum particle.
15. The identicality of quantum particles of a kind leads to the Pauli's exclusion principle.

https://doi.org/10.1515/9783111498690-001

16. The Pauli's exclusion principle accounts for much of what we recognized as the material universe.
17. The first atoms to form after the Big Bang are the two smallest atoms of the periodic table, hydrogen and helium, with hydrogen predominating.
18. Gravitational attraction produces concentrated collections of hydrogen and helium. At a critical point, dense collections of atoms commence a sustained nuclear fusion. Stars are born and heavy elements are produced.
19. Molecules form from elemental atoms.
20. Additional types of molecules form from reactions between existing molecules.
21. Catalysts facilitate chemical reactions.
22. Chemical reactions lead to chemical pathways.
23. Reactions emerge that create protein catalysts (i. e., enzymes).
24. Chemical pathways yield reproduction of molecular species.
25. Circumstances arise in which self-reproduction proceeds without limit.
26. Self-sustaining chemical systems are established within small, water-filled, lacunae and recesses in volcanic rock (e. g., pumice). These chemically active containments can be described as non-living cells.
27. The chemical machinery within rock cells float between rock lacunae, permitting a crude form of cell reproduction.
28. Specialized enzymes emerge containing a reaction template that can be reproduced exactly.
29. Early forms of virus emerge.
30. Early self-coding templates eventually lead to templates that code for multiple reactions in addition to coding for their own self-replication.
31. New chemical pathways synthesize lipids, and these synthesized lipids self-assemble as an outer membrane enclosing the watery chemical machinery housed within rock lacunae.
32. Lacunae spew their contents into the primordial ocean, in lipid-lined bubbles.
33. Living cellular life begins.
34. DNA emerges as the template molecule of the genome of all cellular life.
35. Gradual alterations in the genetic machinery of cells lead to heritable alterations in cellular organisms (i. e., *evolution*).
36. Fusions of different cells or the capture of parts of different cells lead to an increasing variety of cellular species, with heritable properties obtained directly from other cells of the same or of different cell species.
37. The first eukaryotes emerge from prokaryotic life forms.
38. The mitotic pathway of cellular replication evolves in eukaryotic organisms.
39. Meiosis, the process by which germinal cells replicate, evolves from biochemical machinery of mitosis.
40. Collections of cellular organisms working as a unit led to multicellular organisms.
41. The earliest multicellular organisms arise as clonal clusters of eukaryotic cells dependent upon the multicellular lifestyle for their continued survival.

42. Subclones of cells within the clonal organism demonstrate specialized properties, through a process known as differentiation.
43. Evolving gene pools result in speciation.
44. Meiosis leads to sexual reproduction.
45. An alternating life cycle of haploid and diploid organismal stages develops in the animal, plant, and fungal kingdoms.
46. Two major kingdoms of cellular life (animals and plants) evolve a method of organismal development through progressive stages, wherein each stage is itself a distinct organism.
47. Evolution of the various cell types in animal organisms eventually yields the neuron.
48. Organization of small sets of sensory cells and effector cells yield bi-directional signal pathways that can be described as reflex arcs.
49. Primitive subconscious thought emerges.
50. Evolutionary enhancements of the subconscious produce improved thought processes and inherited instincts.
51. The conscious mind emerges from the subconscious mind.
52. The conscious and subconscious mind produce intellect.
53. A trait emerges in Homo sapiens wherein thought is conducted in words, leading to the development of complex languages.
54. An important intellectual achievement, occurring as an accessory to language, is reading and writing.
55. Additions to the body of written language (i. e., books) lead to continuous advances in science, medicine, engineering, and the arts.
56. Maturation of scientific thought leads to an understanding of the origins of reality.
57. The culmination of all the preceding steps leads us to the publication of *Fundamentals of Reality*.

How credible is this list? Not very. At least, not yet. The bulk of this book is devoted to building a reasonable justification for these assertions. In Chapter 10, titled “Recap”, we will return to this same list and summarize, point-by-point, the evidence supporting each assertion.

2 Beginning with life

If you want to make an apple pie from scratch, you must first create the universe.

– Carl Sagan

2.1 First Earth

It is, in the first place, remarkable that all the planets move nearly in one plane, corresponding with the centre of the sun's body. Next, it is not less worthy of attention, that the motion of the sun on its axis, those of the planets around the sun, and the satellites around their primaries, and the motions of all on their axes, are in one direction – namely, from west to east. Had all these matters been left to accident, the chances against the uniformity would have been, though calculable, inconceivably great. Of the motions of the twenty-three bodies known in the early part of this century, it was found by Laplace, that the adverse chances were as upwards of four millions of millions to one. It is thus powerfully impressed on us that the uniformity of the motions, as well as their general adjustment to one plane, must have been a consequence of a single cause acting throughout the whole system.

– Robert Chambers' Vestiges of the Natural History of Creation, published in 1884 [59]

Robert Chambers, back in 1884, understood that because the sun and the planets lay more or less flat upon an oval plane of space, and because all of the planets move in the same direction around the sun, it is more than likely that they all originated from the same disk-shaped swirl of rotating debris. Chambers did not know when the sun and the planets formed, but he knew that they all formed at about the same time. We are now led, by many different lines of reasoning, to believe that the sun and planets formed about 5 billion years ago. One of the simplest methods of dating our planet's origin involves cataloging the long-lived radioactive isotopes of elements found in the Earth's crust. Elements, much like species of animals, can become extinct. They do so by transforming, over time, into other elements. The elements that transform into other elements are all unstable isotopes, capable of emitting alpha, beta, or gamma rays, and yielding so-called daughter isotopes, in the process. The daughter isotopes, which have an atomic number that is different from the parent isotope, will also decay, until a stable isotope is produced. Every radioactive isotope is an intermediary element, on a transformative journey towards a stable and non-radioactive element. The original unstable isotopes, present at the time that the earth formed, will deplete, over time. Here are 16 elemental isotopes, along with their half-lives (in millions of years). All of these listed elemental isotopes were present when the earth formed, but are now undetectable.

https://doi.org/10.1515/9783111498690-002

Isotope	Half-life, millions of years
Samarium-146	103.5
Niobium-92	34.7
Iodine-129	15.7
Curium-247	15.65
Lead-205	15.3
Hafnium-182	8.9
Palladium-107	6.5
Manganese-53	3.7
Dysprosium-154	3.01
Caesium-135	2.33
Technetium-98	4.23
Iron-60	2.62
Technetium-97	2.6
Gadolinium-150	1.798
Zirconium-93	1.53
Aluminum-26	0.7

We see that the half-lives of these extinct isotopes fall into a range from about zero up to a little over 100 million years. The extinct isotope with the longest half-life is Samarium-146, which has a half-life of 103.5 million years. On Earth today we find numerous radioactive elements with half-lives much longer than 103 million years; half-lives of billions and trillions of years are not particularly unusual. The revealing finding is that all of the original radioactive isotopes, present when the earth formed and having half-lives exceeding 103 million years, are still detectable. What is the significance of the strict 103 million year half-life cut-off that separates extinct elements from extant elements?

The explanation is simple. Unstable isotopes with half-lives under 103 million years have diminished to the point where they can no longer be detected. Unstable isotopes with half-lives greater than 103 million years have not had enough time to diminish to the point of extinction, and can still be detected. If we could compute the length of time for an element with a half-life of 103 million years to decay to the point that it can no longer be detected, then we would know the approximate age of the Earth. As it happens, it takes about 40 half-lives for an element to decay to the point that it can no longer be detected by current monitors. For an isotope having a half-life of 103 million years to reach extinction, Earth must have formed over $103 \times 10^6 \times 40 = 4$ billion years ago. To summarize, Earth must have formed somewhere between 4 and 5 billion years ago in order to produce the distribution of extinct and extant isotopes that we see today.

Earlier in this section, we reasoned that the sun and all of the planets in our solar system formed at the same time. Therefore, we would infer that all the planets and the sun formed about 4–5 billion years ago. If this were true, then we would expect to find roughly the same distribution of extant and extinct isotopes throughout the solar system. A contrary set of findings would send us scurrying to construct a new theory of the origin of the solar system.

2.2 First cellular organism

We can guess that humans have always puzzled over the origin of living things. Most historical religions come bundled with a creation myth explaining how the first living beings came to be. In all such cases, there seems to be an understanding that living things are fundamentally different from non-living things, being endowed with the "spark of life," or the "life force." The source of the life force was generally considered to be of spiritual, mystical, or magical origin, and could not have arisen from inanimate materials in the absence of divine intervention. It was later shown that all of the processes that occur in cells can be duplicated in a test tube, with inanimate chemicals. No "life force" is required for us to witness cellular metabolism. In the twentieth century, the structure of genes and the principles of gene replication, gene transcription (to RNA), gene translation (to proteins via RNA), gene regulation, and protein modifications were elucidated. Again, no "life force" was involved. Because the constituents of life are chemical, and the activities of cells are chemical reactions, we can begin to imagine that life may have had its origin in inanimate matter.[4] [5]

The pathologist Rudolph Virchow famously wrote, "Omnis cellula e cellula" (all cells come from cells), indicating that all living organisms are composed of cells, and all the cells in every organism arose from other living cells. So true! But, if all cells come from cells, then where did the very first cell come from? The earliest signs of cellular life, etched into ancient iron rocks, dates back to between 3.77 billion and possibly 4.28 billion years ago [77]. If we can accept this date, then the first life on Earth appeared as early as a few hundred million years following the formation of planet Earth (estimated to be 4.5 billion years ago) [165].

The word "abiogenesis" describes the creation of life from non-living matter. In this section, we will try to show that abiogenesis provides a credible explanation of how the first cells came into being, if we accept the following three assumptions:

1. All living things are composed of non-living things
   Life on Earth consists of cells, and cells are just bags of chemicals. It happens that these bags of chemical can replicate themselves (i. e., self-replicate), accounting for why there are so many cells on the planet. Because all living matter is composed entirely of non-living matter, we can guess that life may have gotten its start from some process involving non-life.
2. Catalysts are molecules that facilitate interactions between other molecules, sometimes leading to the synthesis of new types of molecules, including new types of catalysts.
   Cells contain all manner of molecules that are exclusively produced by cells (i. e., would not be found in an acellular world). DNA, RNA, enzymes, and many forms of lipids are synthesized in cells and would not be found in an acellular world.

3. The earliest metabolic reactions must have occurred in small connected spaces, containing water.
   Within a small space, the products of reactions might be expected to be available for subsequent reactions. Without connections between spaces, substrates would not be able to float into and out of cells and would not be available for reactions. The spaces must not have been so small as to preclude the acquisition of reactants and must not have been so large as to dilute reactants to an extent that substrates could not encounter one another.[6]

We seem to be hypothesizing that small spaces served as the crucibles of life, but what were these small spaces, and what processes occurred within them? To answer these questions, we need to learn something about catalysts and minerals.

In biology courses, we are taught that catalysts are specialized proteins, known as enzymes, and that these proteins are shaped in such a way that specific substrate molecules can be fitted and held in close physical proximity to chemical reactants. When the substrate and reactant come together, the reaction can proceed, to produce a chemical product. The product, having no special affinity for the catalyst, is released, completing the process. The catalyst does not directly participate in the reaction, and when the reaction is complete, the catalyst is left unchanged.

Catalysts need not be proteins [251, 250, 124]. A familiar example of a non-protein catalyst is the plate upon which we prepare our peanut butter and jelly sandwiches. It is difficult to spread peanut butter and jelly on a slice of bread without something hard and flat to keep the bread in place while the ingredients are spread. In a pinch, you can hold the bread in one hand, using the other hand to spread the ingredients with a knife, but it is very awkward and the bread always tears. Don't we hate it when the bread tears in our hand? The plate is the go-to item when we want to prepare a sandwich with the minimal expenditure of time and energy. The plate facilitates the reaction, but is not consumed therein. The peanut butter and jelly sandwich is our product, and the plate is our catalyst.

Before there were proteins, chemical reactions were assisted by mineral catalysts. These catalysts may have been hard, metallic surfaces, such as we see in rocks, holding substrates steady for the duration of a reaction. Today, cellular catalysts are proteins, but many of these catalysts are metal-protein complexes, such as iron-containing hemoglobin and ferredoxins, zinc finger proteins, and copper co-factored with cytochrome C oxidase [57]. It seems as though we cannot escape our rocky origins, insofar as minerals continue to facilitate the chemical reactions that occur in living cells.

We have just described how the surfaces of rocks may have served as the first metabolic catalysts, but how did rocks provide the spatial structure (i. e., small, interconnected spaces filled with water) in which metabolic reactions could proceed? When volcanic rock forms, a network of bubble-like pores develops, when conditions permit [12]. Examples of porous volcanic rocks include tufa, tuff, travertine, pumice, stromboli

Figure 2.1: Volcanic pumice, a highly porous rock. The pores in pumice, and other seabed rocks may have served as the earliest substrate for the evolution of living cells, providing a hard catalytic substrate for metabolic reactions, while also providing an enclosure for chemical reactions and a path by which reactants and biologically active molecules (e. g., nucleic acids and enzymes) may move from one pore to another. Source, Wikipedia, and entered into the public domain by deltalimatrieste.

basalt and scoria (Figure 2.1). Any of these water-drenched rocks could have been crucibles for catalytic reactions. It may well be that just as the early hominids lived in rock caves, so too did the earliest cells, though their caves were just microns in diameter.

We can begin to imagine rocks containing millions and millions of tiny bubbles lined by catalysts, and filled with floating substrates and products, diffusing from one pore to another. We can imagine that if the product of the reaction participates as a substrate in some other reaction, then we might see an emerging system, producing a range of products. Under these circumstances, natural selection might intervene, increasing the likelihood that particular reactions may prevail over others, depending on the local concentrations of substrates, products, and catalysts. In time, the activity of certain catalysts may be modified (increased or decreased) based on chemical interactions with rock minerals. It all seems a little bit like a living process.[7]

When bubbles in rocks are interconnected, molecules such as RNA virus-like molecules may have moved from bubble to bubble, exchanging genetic materials between bubble chambers. We can begin to imagine that RNA viruses may have existed prior to the evolution of free-living cells. We can also guess that RNA preceded the appearance of DNA. RNA can code for proteins, but DNA can only code for RNA. There would be no reason for DNA to have evolved in the absence of a preexisting mechanism by which RNA translates into protein.

Why did DNA evolve? DNA is a much more stable molecule than RNA, less prone to replication error, and less susceptible to insertions by RNA viruses that commuted between rock bubbles.[i] [ii] [9]

What came next? If we choose a definition of life that reflects the simplest organisms observed today, then we would say that the first form of life was a cell, enclosed by a lipid membrane, and capable of self-replication. We can guess that at some point, cell membranes made primarily of lipids appeared. It is known that phospholipids spontaneously form lipid bilayers in agitated water. It seems possible that rock-dwelling organisms, endowed with an enclosing bilayer membrane assembled from phosphorylated small molecules, synthesized from an RNA template, would eventually float into the open water, encountering new sources of substrates for metabolic reactions.[iii]

An alien being composed entirely of gas and silicon dust, and visiting our planet for the first time, may fail to notice the negligible differences that distinguish politicians from plague bacteria. When we list the properties common to all organisms living on Earth, we find that we are all made of much the same stuff. Namely:

- All cellular organisms contain DNA and RNA.
  The three cellular classes of organisms (bacteria, archaeans, and eukaryotes) have both DNA and RNA. Non-cellular organisms (i. e., viruses) have DNA or RNA, using their host cells to supplement their replicative requirements.[10] [11]
- All cellular organisms contain and require water.
  Every cellular organisms require water to survive; and not just any water. The water in our cells must be salinated (i. e., must contain sodium chloride), simulating the marine milieu from which the first cellular organisms arose. Pure water is actually quite toxic to naked cells, causing osmotic disruption of the cellular membrane.[iv]
  There are cells, even whole organisms, capable of enduring long periods without water. Drought tolerance is a temporizing measures at best, intended only to sustain the organism until such time as water can be acquired. Examples of organisms

---

i The evolution of DNA modifications (adherent proteins and base methylations that regulate the transcription of genes, collectively known as the epigenome) may have originally developed as a defense against infection by RNA viruses.

ii The genome is an organisms's hereditary information, encoded as DNA. For eukaryotic organisms, this would include the chromosomal DNA found in a somatic cell, plus any DNA found in organelles such as mitochondria or chloroplasts.[8]

iii The early divergent evolution of membranes enclosing the first cells, is suggested by the profound differences in membrane structure among the earliest known classes of living cells: Class Bacteria and Class Archaea [138, 243]. If the two earliest classes of organisms had split off from the same, membrane-enclosed ancestor, you might expect them to have similar membranes; but they do not.

iv Ponder the humble erythrocyte (red blood cell). Erythrocytes are extremely rugged, withstanding multiple collisions and deformations as they tumble through the bloodstream and squeeze through tiny capillaries. But if you drop a red blood cell into pure water, it will instantly lyse (break open).

capable of surviving long stretches in a dormant and dehydrated state would include: spore-producing bacteria (e. g., clostridia), fungal spores, and tardigrades (microscopic animals commonly known as water bears). Some plant seeds have been known to survive in a dormant and desiccated state for hundreds or possibly thousands of years. Nonetheless, all organisms on Earth require water to actively conduct metabolic processes (such as cell division and replication).[v]

Deinococcus radiodurans (from the Latin radiation plus surviving) is a mostly harmless bacteria whose remarkable resistance to radiation came to light when food biologists were experimenting with irradiation as a method for sterilizing and preserving food (Figure 2.2). Deinococcus radiodurans survived radiation levels expected to kill all life forms.[vi] Under natural conditions, no terrestrial organisms are subjected to the radiation levels that biologists directed against D. radiodurans; hence evolution could not have selected for a defense mechanism against high levels of radiation. If D. radiodurans developed defenses against radiation, those defenses must have evolved to satisfy some other purpose entirely. It seems that D. radiodurans is subjected to conditions of extreme dryness. Unlike several other families of bacteria, D. radiodurans does not survive dry conditions by undergoing spore formation. D. radiodurans endures dryness by protecting its genome with elaborate DNA repair processes. The same repair mechanisms that come in so handy under extreme conditions of dryness happen to confer resistance to gamma radiation.[12]

- All organisms contain a similar ribosomal RNA sub-unit.

  All organisms on Earth use ribosomes to translate messenger RNA to proteins. Archaeans, bacteria, and eukaryotes have their own characteristic ribosomal structures, but the general design of the ribosome is equivalent for all organisms.
- All organisms on Earth, with the exception of viruses, are composed of cells consisting of a lipid bilayer bag filled with protoplasm (i. e., the fluid containing the metabolic chemicals that sustain life).

  We take for granted today that organisms are cellular, but we should find it remarkable that no other forms of life have evolved a purely non-cellular mode of existence. There are no talking rocks, no fire demons, and no pure energy beings. To

---

v Aside from its role as the medium through which chemical interactions occur (e. g., the transport of ions into and out of cells), water participates directly as a metabolic reactant. In photosynthesis, glucose is produced by carbon dioxide, water and light, while energy is released by oxidizing glucose to water and CO2. The synthesis of proteins, nucleic acids and carbohydrates involves water elimination. Catabolism of these same substances involves hydrolysis (i. e., breaking water bonds.)

vi Deinococcus radiodurans, sometimes called the world's toughest bacterium, has evolved multiple mechanisms to protect against prolonged dryness. Most surprising is its ability to modify the configuration of its DNA, and to apply novel techniques to preserve genetic information and to repair damaged DNA [154]. The desiccated organism can survive 5,000 Gray units of radiation (5 Gray units is sufficient to kill a human).

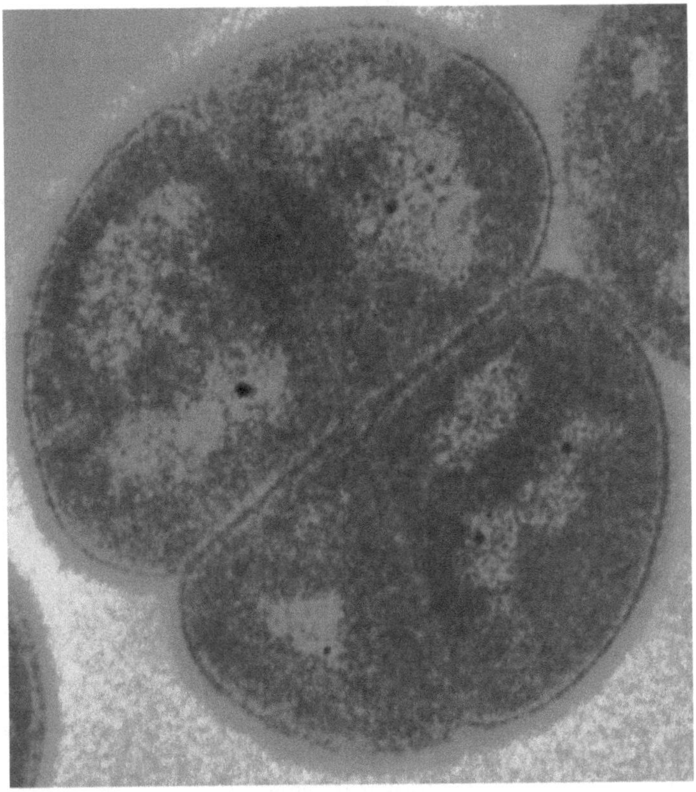

Figure 2.2: Photomicrograph of the bacteria Deinococcus radiodurans. Source, U.S. Government, micrograph prepared TEM of D. radiodurans in the laboratory of Michael Daly, Uniformed Services University, Bethesda, MD.

the best of our knowledge, we live in a living world inhabited by cells and viruses and nothing else.[vii]

– All cells have similar enzymes, with each enzyme operating at about the same rate. Regardless of which organism we inspect, we will find enzymes, and the enzymes in one organism will look and behave much like the enzymes in other organisms. The reason for this universal sameness among enzymes and metabolic pathways has much to do with the common ancestries of all living organisms. Nearly all of the genes, and the enzymes they encode, were inherited over the ages. The enzymes that are widely shared among highly diverse classes of organisms tend to be the housekeeping enzymes (e. g., enzymes involved in cellular replication, regulation,

---

vii Ammonia and methane may also support life, though methane becomes liquid only at minus 296° Fahrenheit (minus 182° Celsius). There are large lakes of liquid methane and ethane on Titan, one of the moons of Saturn. Who knows what forms of life may exist on Titan?

respiration, and construction) that are equally vital to every cell. We also encounter, among distantly related classes of organisms, some of the same or similar gene families; groups of genes that evolved over long periods of time, to serve similar cellular activities in diverse types of organisms.

Most of the enzymes on Earth operate within the same narrow temperature range. In humans, when the body temperature rises to 104° Fahrenheit, we call this hyperthermia. Unrelieved hyperthermia will often lead to multi-organ dysfunction and death. The reason why hyperthermia is so very toxic to cells (human and nonhuman) is that when the temperature rises just a few degrees, enzymes fail to function properly. Mammals have evolved complex homeostatic pathways to maintain a nearly-constant body temperature, despite wide seasonal and diurnal variations in the temperature of our atmosphere.

If enzymes operate efficiently in a narrow range of body temperature, then how can animals lacking endothermic maintenance of a constant body temperature (i. e., poikilotherms) survive? Some poikilotherms simply slow down when the temperature drops. Such organisms have adapted themselves to a low level of metabolism and respiration. Other poikilotherms survive by having multiple enzyme systems, each operating at different optimal temperatures, to fulfill the same set of biochemical pathways [58]. A poikilotherm may have up to ten enzyme systems, each attuned to a different temperature range. Poikilothermy may appear to be an overly complex way to run an organism, but it makes sense under low energy conditions. We eutherian endotherms (i. e., warm-blooded placental mammals) generate our own heat, expending up to 80 % of our energy just to maintain an optimal body temperature [10]. Poikilotherms neither heat nor cool their bodies, hence saving energy and lowering their food requirement.[viii]

All told, cellular metabolism, regardless of species, is a slow phenomenon, as compared with events occurring in the nonliving physical world. The speed of light, the speed of oscillation of electromagnetic fields, the speed of electrons orbiting a nucleus, and the speed of planets orbiting the sun, all occur much faster than anything we experience in our cellular lives. Even when we consider the rapid conduction of a nerve impulses (50–60 meters per second), cellular organisms are excrutiatingly slow. We can blame our enzymes.

---

viii The evolution of enzymes that function within specific ranges of environmental temperatures has been of great benefit to geneticists. The PCR (Polymerase Chain Reaction) technique, wherein sequences of DNA are highly amplified, involves incubating DNA at high temperatures that would denature human enzymes. The thermophilic bacterium, Thermus aquaticus, produces a thermostable DNA polymerase that functions within the required temperature range. The availability of the Thermus aquaticus DNA polymerase (abbreviated as Taq) was essential to the development of PCR (Polymerase Chain Reaction), an important technique used in molecular biology laboratories.

2.3 First oxygenic photosynthesis

It is not easy to release the oxygen from water. If it were, our planet would have a highly explosive atmosphere and not very much water. Accordingly, there is no single enzyme that can do the job. The hydrolysis of water into oxygen and hydrogen uses a highly complex pathway. The first occurrence of oxygenic photosynthesis on Earth required the independent evolution of two distinct metabolic pathways that eventually combined to complete the full biosynthetic cycle [151]. The first pathway involves the capture of photons to hydrolyze water, yielding oxygen and two key energy transfer molecules: NADPH (nicotinamide adenine dinucleotide phosphate) and ATP (adenosine triphosphate). The second, light-independent pathway, is the Calvin cycle, driven by NADPH and ATP to produce simple sugars from carbon dioxide.

The overall reaction describing oxygenic photosynthesis is:

$$\textit{carbon dioxide} + \textit{water} + \textit{light energy} \rightarrow \textit{carbohydrate} + \textit{oxygen}$$

The first and only organism to achieve oxygenic photosynthesis was a cyanobacterium, which mastered the metabolic pathway about 2.5 billion years ago. All organisms in which oxygenic photsynthesis occurs have inherited the pathway directly from the cyanobacteria.[ix]

Until the cyanobacteria evolved oxygenic photosynthesis, the atmosphere of the planet contained very little Oxygen gas, and the minerals inside the earth were unoxidized. No one knows with certainty the timeline of terrestrial oxygenation events, but evidence suggests the following scenario:

1. About 3.5 billion years ago, a prokaryote (bacteria) evolved that could produce oxygen from water, through a rather inefficient pathway that did not involve photosynthesis. All cellular life on Earth up to this time was prokaryotic (bacteria and archaeans), and virtually all living organisms were anaerobes (i. e., organisms that live without air) to which oxygen was toxic. A relatively small amount of oxygen was produced, all of which was rapidly trapped by substances within the Earth, particularly oxidizing minerals such as iron.[x]
2. About 2.7 billion years ago, cyanobacteria evolved oxygenic photosynthesis, producing oxygen using the energy of photons. No organism since has independently evolved oxygenic photosynthesis. Early eukaryotes (cells that had evolved a nucleus) of Class Archaeplastida captured cyanobacteria and adapted them as or-

---

ix Chloroplasts, the organelle that conducts oxygenic photosynthesis in plant cells, are the evolutionary byproduct of cyanobacteria, conscripted to live symbiotically within the earliest members of the plant kingdom.

x As the result of the oxygenation of our planet, all the iron ore mined on Earth is oxidized. Pure metallic iron is produced in foundries, but does not occur naturally. The only natural source of unoxidized iron on the Earth's surface is found in meteorites, that have traveled to us through airless space.

ganelles (chloroplasts) much later, probably no earlier than 2.1 billion years ago. These chloroplast-laden eukaryotes became the first plant organisms. At this moment (i. e., about 2.7 billion years ago), a great deal of oxygen was being synthesized, but this oxygen was absorbed by the oceans and the emerging land masses, and very little oxygen made its way into the atmosphere [119].

3. About 2.5 billion years ago, the oceans and the Earth's crust had become saturated with oxygen, and the excess gas bubbled its way into the atmosphere.
4. About 2.3 billion years ago, atmospheric oxygen reached a level that was toxic to most of the existing anaerobic organisms, producing the so-called oxygen crisis or oxygen catastrophe (Figure 2.3). The harmful effects of reactive oxygen species on cells (including all Earthly cells) include:
   – Damage to DNA or RNA
   – Oxidations of polyunsaturated fatty acids in lipids (lipid peroxidation)

Figure 2.3: From the moment when cyanobacteria improbably evolved oxygenic photosynthesis, some 2.7 billion years ago, unchecked cycles of cyanobacterial replication eventually led to an enormous rise in atmospheric oxygen levels. The process calls to mind the tale of the sorcerer's apprentice (made famous in the Disney's movie *Fantasia*). Once the magic begins, it cannot stop. Source, Illustration of Der Zauberlehrling, from "Goether's Werke" by Künstler(1842–1892), now in the public domain.

- Oxidations of amino acids in proteins
- Oxidative deactivation of specific enzymes by oxidation of co-factors

The relatively few anaerobic species that survived the oxygen catastrophe were those living in relatively low-oxygen locations (e. g., deep in soil) and those with efficient scavenging enzymes that absorbed and removed oxygen molecules.[xi] [13] [14]

5. The really large atmospheric oxygen concentrations, comparable to those we see today, did not come until about 1 billion years ago.
6. Until about 500 million years ago, just prior to the Cambrian explosion, animals had persistently adapted to low oxygen level, low energy, sluggish existences. Passively feeding off of dead organisms that sank to the ocean floor, pre-Cambrian animals were largely bottom-feeding detrivores. About half a billion years ago, there evolved a profusion of new animal species that were capable of exploiting a high-energy metabolism fueled by ample oxygen levels. These Cambrian species had fast muscle fibers enabling them to chase, seize, kill and devour other animals. It was during the Cambrian explosion (a short period spanning about 50 million years) that the vicious predator/prey lifestyle took hold, and all of the body plans characterizing today's predatory animals had evolved [226].

2.4 First virus

> The question whether or not "viruses are alive" has caused considerable debate over many years. Yet, the question is effectively without substance because the answer depends entirely on the definition of life or the state of "being alive" that is bound to be arbitrary.
>
> – EV Koonin and Petro Starokadomskyy [142]

Viruses have been on Earth for a very long time, possibly preceding the first cellular forms of life [97, 96]. The evidence suggesting that viruses preceded the appearance of other classes of living organisms is based in part on the observation that some archaeans, bacteria, and eukaryotes have homologous genes (i. e., genes similar in DNA sequence by virtue of descent from a common ancestor) that code for capsid proteins, such as would be derived from a class of viruses.[xii] A simple explanation for homology to a viral-type gene, present in all extant cellular forms of life is that an ancient gene came from a viral species that lived at a time that preceded the emergence of other forms of life [15]. Over the eons, retroviruses have had a chance to insert lots of DNA fragments

xi The ancient ancestor of our oxygen-delivering hemoglobin molecule was an oxygen scavenging enzyme that served to protect anaerobic organisms from highly toxic oxygen molecules.

xii The capsid is the primarily protein shell of a virus that encloses the genetic material of the virus when the virus is outside its host cell. The capsid aids the virus with its attachment to the target host cell, and with the subsequent penetration of the viral genome into the host cell.

into eukaryotic species. About 8 % of the human genome is composed of fragments of RNA viruses [121, 81, 110].[15] [16]

The earliest cellular organisms on Earth seem to have existed no later than 3.7 to 3.4 billion years ago, and this would indicate that viruses, which preceded cellular organisms, were here earlier still. We can imagine that viruses may have existed in the first billion years of Earth's formation, when the planet was a molten cauldron. But wait. We seem to have trapped ourselves in another origin paradox. We know that viruses are basically packages of nucleic acid that exist in a dormant state until they infect cells (either bacteria, archaea, or eukaryotes). After entry to the cell, the virus hijacks the cellular machinery and replicates. This being the case, then how could there be viruses before there were host cells to support viral replication?

The solution to the paradox requires that we revisit our proposed origin of cellular life, discussed earlier in this chapter. We hypothesized that cellular life began as water-filled connected bubbles within porous rock of volcanic origin (e. g., pumice). These bubbles served as crucibles for cellular reactions, in which diverse products such as RNA, proteins, lipids, and glycolipids were first created. Within one of these rocky protocells, a virus particle may have been produced, and this tiny particle may have traveled through the microscopic connections between bubbles. It was just a matter of time before one of these traveling nucleic acid chanced upon a lacunae containing the enzymes necessary for viral replication. Before long, replicating viruses may have traveled through all of the rock communities on Earth, protected in their journeys by membranes (i. e., viral capsids) collected within lacunae. The earliest viruses may have originated in porous rocks, without requiring the pre-existence of free-living cells.

It is ironic that viruses, putatively the oldest form of life on planet Earth, were only recently discovered. No scientist had actually seen a virus until 1939, when Kausche and his colleagues examined fluid containing the filterable infectious agent of tobacco mosaic disease under the electron microscope (Figure 2.4). We should keep in mind that many scientists alive today were born at a time when viruses had not been morphologically identified [4]. The short history of ancient viruses should cause us to wonder whether there are other entities living among us that humans have not countenanced.[17]

Because viruses are a relatively new idea to humans, we should not be surprised that our thoughts about viruses are blunt and unsophisticated. To many scientists, viruses are nothing more than non-living packaged genetic elements that pass from cell to cell.[xiii] Unlike most living organisms, viruses do not partake in metabolism and do not yield energy. For the most part, viruses do not adjust to changes in their environ-

---

xiii A virus particle is the genome (RNA or DNA), plus a protein capsid, and, in some species, a lipoprotein envelope covering the capsid. Envelopes are acquired from the host cell when the virus buds off through the host cell membrane. The infectious virus package is called the virion. The simplest viruses contain only enough RNA or DNA to encode as few as four proteins [157].

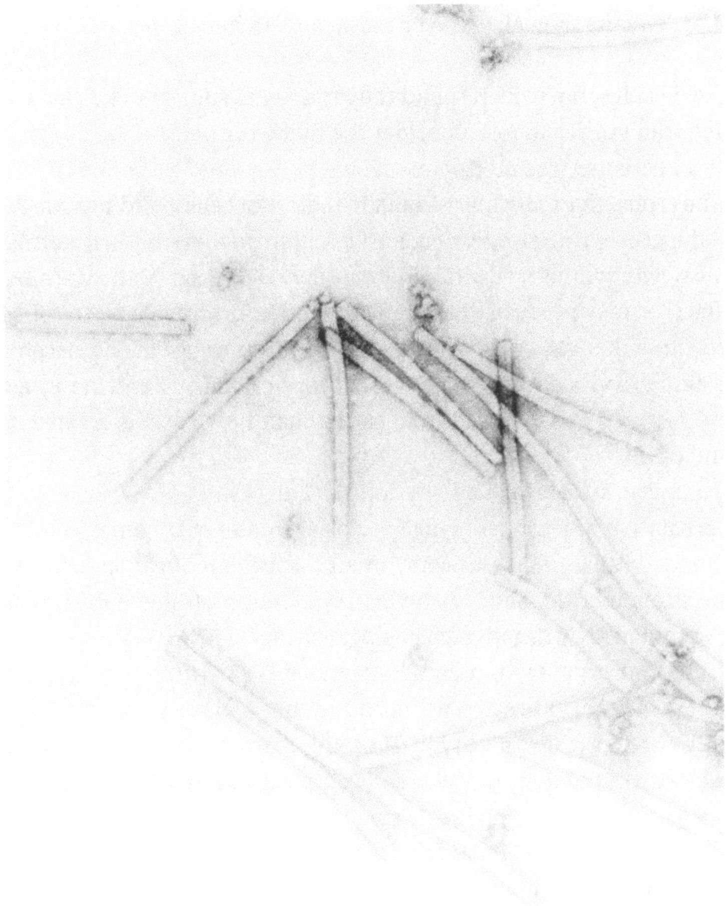

Figure 2.4: Transmission electron microscopic image of Tobacco Mosaic Virus (magnification approximately 160,000), appearing as rod-shaped organisms. These were the first viruses ever to be identified. Source, Prepared for the public domain as a U.S. government work.

ment (i. e., no homeostasis), and do not respond to stimuli. It seems that viruses depend entirely upon their host cells for every activity that bears semblance to life. For these reasons, viruses are often considered to be nonliving organisms.

In fairness, all organisms depend on other organisms for life; viruses do it better. For "nonliving" organisms, viruses have done extremely well for themselves [144]. We observe that every class of cellular organism hosts viruses [157]. Viruses are literally everywhere in our environment, and are easily the must abundant life form in the oceans, in terms of numbers of organisms [9]. The number of different species of viruses seems to far exceeds that of the eukaryotes, bacteria, and archaea [230, 229]. A single class of newly characterized aquatic viruses, the Megaviridae, appears to have more species and more phylogenetic diversity than all of the bacterial and archaean species found in ocean water [180].[xiv]

Over the past two decades, the evidence and the arguments supporting viruses as living organisms, have won much support. Consider the following points:

- Viruses replicate and create virocells.
  Viruses employ the replicative machinery found in their host cells, building viral assembly plants in the process. After infection, host cells forego many of their normal functions, and the newly synthesized viral products may cause cell death (so-called viral cytolytic effect) or may produce disturbances in cellular physiology (so-called cytopathic effects). In either case, the host cells, after infection by virions, become something more akin to a viral factory than to eukaryotic organisms, and these new living entities are referred to as virocells, indicating that the virus has created its own form of cellular life.[xv]
- Viruses exhibit chemical, structural, and physiological diversity.
  Viruses display great diversity in their range of hosts, habitat, size, genes, mechanisms of infection, and capsids. Viruses come in an assortment of genomic forms, including: double-stranded DNA; single-stranded DNA; double-stranded RNA; positive sense single-stranded RNA; negative sense single-stranded RNA, single-stranded RNA with a reverse transcriptase, and double-stranded DNA with a reverse transcriptase. Viruses of each form breed true, meaning that a virus of any type will replicate to produce viruses of the same type. Thus, the types of viruses, determined by their genomic structure and replicative modes, can be organized as biological classes.
- Viruses do not lead a totally intracellular life.
  It has been assumed that extracellular viral existence is relegated to long-term storage: a static viral genome double-wrapped in a protein capsid, and shrouded in an envelope purloined from the cell membranes of its abandoned host cell.

---

xiv Phylogenetics is method of classification based on ancestral lineage. The classification of living organisms is the prototypical phylogenetic classification. Though the term "phylogenetics" is most often applied to classes of organisms, its usage can be extended to any hierarchical classification in which each subclass derives directly from its immediate ancestral class (the so-called parent class). For example, the *Nomina Embryologica Veterinaria* classifies tissues under the anlagen from which they develop [202]. This classification lists the hierarchical lineage of embryonic blood islands in the following way: Gastrula → Mesoderma → Mesenchyma → Textus hemopoeticus → Insulae sanguineae. Setting Latin nomenclature aside, we can see the classification of embryonic structures is phylogenetic.

xv The virion is the form of a virus existing outside the host cell. Viruses can be thought of as having an alternating life cycle: the virion and relatively dormant form consisting of nucleic acids and a few proteins encased in a capsid and existing outside of any host cell; and the virocell, the infected host cell containing swarms of replicating viruses. If the virus is cytopathic, the virocell dies, releasing manufactured virions into the immediate environment. In some instances (e. g., several types of Herpes viruses), viruses may live within its host cell for extended periods without leading to cell death.

It has been recently demonstrated that some viruses enjoy life outside of their hosts. The Acidianus Tailed Virus can be cold-stored at room temperature for long periods, without changing their morphology. When the temperature rises, they undergo a structural transformation, forming bipolar tails [113]. This indicates that viruses can react to external stimuli and perform biological activities, much as free-living organisms do.

- Viruses evolve and speciate.
  We accept that there are distinct species of viruses. The definition of a viral species, as suggested in 1985, is essentially the same definition as any other species of living organism: an evolving gene pool [135]. Despite the high mutation rate of viral genomes, there is sufficient stability in their genomes, particularly in genes that code for constitutive proteins, to establish a phylogenetic lineage for classes of viruses [128].[xvi] [18] [19]
  Presumably, much of the stability of viral species is imposed by their host organisms. How so? All viruses must replicate within a host, and the process by which a eukaryotic cell is infected and transformed into a virocell is very complex. Viruses are forced to adapt some level of host preference, and must fine-tune all their functions to the available conditions within the host [16]. Thus, for viruses to successfully replicate, they must evolve into relatively stable species.
- Virus speciation yields phylogenetic lineages (i. e., descendant families).
  In 2017, the ICTV (International Committee on Taxonomy of Viruses) recognized 9 orders, 131 families, 46 subfamilies, 803 genera, and 4,853 different viral species [252]. The number of viral species known at the time (4,853) is thought to represent a tiny fraction of the total number.
- Viruses participate in the evolution of eukaryotic organisms.
  Some of the greatest evolutionary leaps among eukaryotic organisms can be attributed to genes prised from non-eukaryotes (e. g., mitochondria and chloroplasts acquired from bacteria). Viruses have also contributed to eukaryotic evolution, and it is fair to say that if it were not for viral genes, you would not be reading this book (it would be left to some octopus or insect to finish the chapter). Specifically, at least four major advancements in Class Metazoa can be attributed to the acquisition of viral genes in eukaryotic cells.
  1. The acquisition of adaptive immunity, in Class Gnathostomata (the jawed vertebrates), and every descendant class [132].[20]
  2. The acquisition of the placenta in Class Theria (i. e., the placental mammals plus the marsupials, which have a rudimentary placenta); and

---

xvi Mutations are alterations in the nucleotide sequences of genomic DNA. These would include point mutations (changes in a single nucleotide) and segment mutations (e. g., losses of a string of nucleotides) and structural variations that delete, amplify, or move segments of DNA.

3. The maintenance of pluripotent stem cells, important in embryogenesis, is achieved, in part, with transcripts of stem cell-specific long terminal repeat retrotransposons, themselves derived from retroviral sequences [94, 80, 254, 159].[21]
4. In Class Mus (containing the common house mouse), an MuERV-L retrovirus capsid gene, was domesticated to code for an antiviral factor [195, 37].

One of many competing theories is that the first eukaryotic nucleus (i. e., the organelle containing the chromosomes, found in non-bacterial cells) was a giant virus that was not totally successful in transforming its proto-eukaryotic host into a virocell [88, 98, 97]. A hybrid giant virus/virocell/proto-eukaryote may have stabilized and replicated to form an early, nucleated cell; the first eukaryote. Participating as it does in eukaryotic evolution, and accounting in no small way for the generation of new eukaryotic species, viruses have demonstrated a range of biological functions indicative of genuine viral life.

For a virus, extracellular existence must be akin to a state of suspended animation. A virion (the viral genome encased in a capsid) animates when it attaches to a cell and intrudes its genome into the cytoplasm to create the hybrid virocell; a living organism consisting of the hijacked eukaryotic cell whose nuclear machinery is redirected to synthesize viral progeny. If every living cell is conceptualized as a potential virocell, then every living species is a potential incubator, just waiting to be enslaved by invaders from the Kingdom Viridae. If viruses could think, they would certainly believe that they were doing us metazoans (animals) a favor. Frankly, most of the cells in a metazoan body lead a vegetative existence, doomed to a fully differentiated, post-mitotic, and short existence. Viruses re-animate post-mitotic cells, and create a thriving center for viral life from a lackluster population of eukaryotic cells, as virocells.[22] [23] [24]

The long debate over the validity of viral life can be reduced to a definition [142, 143]. How we define life is up to us, and we can certainly construct a definition that accommodates the viruses, if that is what we choose. There is no urgency to the matter. In short order, when our robots become self-replicating and our computers attain consciousness, the definition of life will be addressed anew. When that time comes, perhaps a non-cellular life form will take the lead and create for itself a definition for life that encompasses viruses of the organic type and viruses of the software type. If we are lucky, the new definition of life will include humans.

2.5 First eukaryotes

At the simplest level, all cellular life can be divided into two forms: the eukaryotes, which contain a membrane-bound organelle, known as the nucleus, that holds the cell's genetic material; and the prokaryotic life forms (bacteria and archaeans) that have no nucleus.

No one knows when the first eukaryotes appeared on Earth. One school of thought estimates that eukaryotes were here 2.7 billion years ago, about one billion years following the first appearance of prokaryotic life forms. This theory is based on finding sterane molecules in shale rocks, dating back nearly 3 billion years. Eukaryotic cells are the only known source of naturally occurring sterane molecules. Their presence in shale is held as evidence that the first eukaryotic organisms must have existed no later than 2.7 billion years ago [50]. Other biologists tie their estimate of the beginning or eukaryotic life to the epoch in which the first eukaryotic fossils were found, about 1.7 billion years ago. This leaves a gap of one billion year between estimates for the origin of the eukaryotes (i. e., 1.7 to 2.7 billion years ago).

Among the smallest eukaryotes are picoplankton that have a diameter as small as 0.2 to 2 microns in diameter [20]. There are many eukaryotic species that have never been adequately studied, and these include pico- and nano-sized heterokonts [13]. Despite the wide range in eukaryotic sizes, it is worthy to note that the majority of eukaryotic organisms are populated by cells of about the same size (i. e., 25–50 microns in diameter). The outliers at the high end of size and weight tend to be specialized, non-dividing cells (e. g., oocytes, multinucleated, platelet-producing megakaryocytes).[xvii] There seems to be an upper limit to the size of dividing eukaryotic cells, beyond which the complex and tightly controlled process of cellular division cannot proceed. We should not be surprised to learn that the cells of a whale are no larger than the cells of a mouse.[25]

All eukaryotes have a number of chemical and metabolic features in common. For example, actin and related filamentous cytoskeletal molecules are components of every eukaryote, and are only found in eukaryotes [271]. The earliest eukaryotes, and all of their living descendants, came equipped with three structures that separate them from prokaryotes: mitochondria (at least one per cell), undulipodium, and a nucleus.

1. Mitochondria.
   Mitochondria are membrane-delimited organelles, having their own genome, that proliferate within the eukaryotic cell (Figure 2.5). Current theory holds that the first mitochondria originated as an obligate intracellular endosymbiont, from an ancient member of Class Rickettsia.[xviii] This theory is based upon the finding of gene sequence similarities between the close relatives of modern rickettsia and that of eukaryotic mitochondria. Since all eukaryotic organisms contain mitochondria, we infer that the close relationship between Class Alpha Proteobacteria (of which the rickettsia are members) and Class Eukaryota extends to the first mitochondria in the first eukaryotic cell.

---

xvii The heaviest eukaryotic cell is the egg of the ostrich (Struthio camelus), which typically weighs between 3.5 and 5 pounds.

xviii Class Alpha Proteobacteria, from which Class Rickettsia descends, are characterized by their small size, their Gram negativity, and their intimate associations with eukaryotic cells. The Alpha Proteobacteria live as sybionts, endosymbionts or as intracellular parasites..

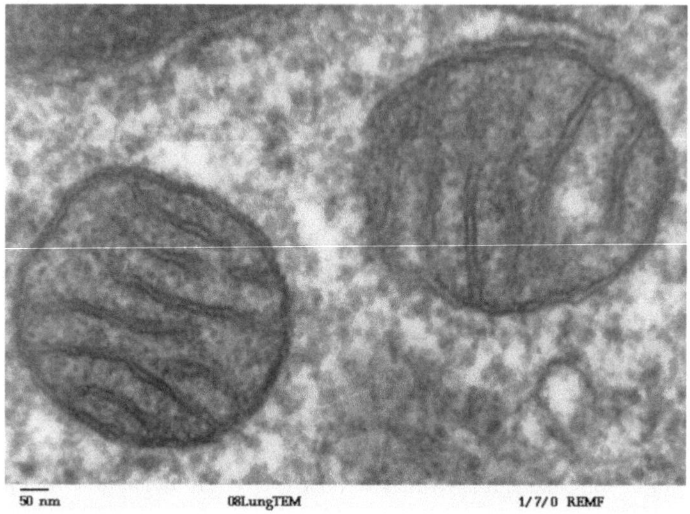

Figure 2.5: An electron micrograph of mammalian lung cells, featuring two mitochondria. The two-layered parallel lines traversing the organelles are cristae. Cristae are infoldings of the mitochondrial inner membrane, providing a large surface upon which cellular respiration may proceed. Source, Wikipedia, from a work that was released into the public domain by its author, Louisa Howard.

All existing eukaryotic organisms, even the so-called amitochondriate classes (i. e., organisms without mitochondria), contain vestigial forms of mitochondria (i. e., hydrogenosomes and mitosomes) [227, 239, 240, 55]. The mitochondria provide eukaryotes with a source of internal energy, via oxidative phosphorylation. If fuel is abundant, then the amount of energy produced by a cell is proportionate to the number of mitochondria, and the cells that utilize the most energy contain the greatest number of mitochondria. For example, nearly half the cytoplasmic volume of continuously beating heart muscle cells is composed of mitochondria, with several thousand mitochondria per cell.

The mitochondria provide eukaryotic cells with many evolutionary options. Eukaryotic cells can grow much larger than bacteria, relying upon their mitochondria to provide sufficient energy to maintain the complex set of physiologic processes necessary for a large organism. Eukaryotic cells may have very large genomes, and each cell cycle of a dividing eukaryotic cell requires enough energy to replicate the entire genome.[xix] The energy costs of replicating a genome containing billions of base nucleotides must be paid in full by mitochondria. Energy-rich cells, containing

xix For comparison, the genome of a typical bacteria, such as Escherichia coli, is about 5 million base pairs in length. The genome size of a human is about 3 billion base pairs (on the order of a thousand times larger than that of a bacteria). Plants tend to have larger genome sizes than mammals, for example wheat has a genome of 16 billion base pairs. Accordingly, plant genomes tend to contain more protein-coding

mitochondria, can afford to carry large genomes and perform all manner of strenuous tasks.[xx]

With the acquisition of mitochondria came the onset of mitochondriopathies; diseases whose underlying cause is mitochondrial pathology (i. e., dysfunctional mitochondria). Mitochondriopathies can be genetic or acquired. Most of the genetic mitochondriopathies are caused by nuclear, not mitochondrial, gene mutations. Though mitochondria live outside the nucleus and have their own genomes, mitochondrial DNA codes for only 13 proteins of the respiratory chain. All other proteins and structural components of mitochondria are coded in the nucleus.

As we might expect, mitochondriopathies affect the cells that are most dependent on their mitochondria for their functionality. It is not surprising that most of the mitochondriopathies produce multisystem disorders that are characterized by muscle weakness, cardiomyopathy (disorders of the heart muscle), ataxia (movement disorders), and deafness [139].[xxi]

Deafness is present in some inherited mitochondriopathies. Deafness may also occur as an acquired condition, such as seen with deafness following antibiotic usage (e. g., aminoglycosides). About a quarter of individuals receiving aminoglycoside therapy have some loss in hearing, as measured by audiometry testing. Why would an antibiotic produce a mitochondriopathy? We must remember that our mitochondria evolved as captured bacteria that adapted to an intracellular existence within the early eukaryotes. Our mitochondria, true to their bacterial origins, are susceptible to toxicity induced by some antibacterial agents [131].[xxii]

2. Undulipodia.

   Prokaryotes and eukaryotes have flagella, rods that protrude from the organism; their back-and-forth motion propelling cells forward through water. Aside from a superficial resemblance, the flagella of eukaryotes have no relationship to the flagella of prokaryotes [13]. Eukaryotic flagella are orders of magnitude larger than the prokaryotic flagella, contain hundreds of species of proteins not present in the flag-

---

genes than we find in the human genome; with rice having 46–56 thousand genes, and humans having 20–25 thousand genes [274].

xx A single cell eukaryotic cell such as a hepatocyte, or a renal tubular cell, or a myocyte may contain hundreds of mitochondria. Alternately, a eukaryotic cells may contain as few as one mitochondria, as is seen in Prasinophytic algae.

xxi Additional features of mitochondriopathies may include: pigmentary retinopathy, ocular atrophy, deafness, gut motility disorder, and sideroblastic anemia, among many others. A mitochondriopathy is something to consider when seeking the cause of baffling multi-system disorders, especially those arising in childhood [89].

xxii All dividing cells in the body contain mitochondria, so shouldn't antimicrobial antibiotics produce systemic damage in all organs (not just deafness)? Presumably, all cells are somewhat affected, but among the cells involved in hearing is a sub-population that requires high levels of mitochondrial activity, and these seem to be the cells that are most affected by mitochondrial deficits.

ella of prokaryotes, have a completely different internal structure, anchor to a different cellular location, and do not descend phylogenetically from prokaryotic flagella [166].[xxiii] Specifically, prokaryotic flagella are composed primarily of flagellin. Eukaryotic flagella, which are better known as undulipodia, are composed primarily of tubulin and contain more than 100 other identified proteins, including dynein. Most biologists continue to apply the misleading term "flagellum" (plural, "flagella") to eukaryotes. Regardless, every existing eukaryote descended from an organism with an undulipodium. The undulipodium is a highly conserved feature of eukaryotes, and all descendant eukaryotic classes contain undulipodia or structures that evolved as modified forms of undulipodia. For example, we humans have cilia on the surface of mucosal lining cells, which are a shortened type of undulipodium, and we have long, undulating undulipodial tails on spermatocytes, facilitating locomotion.[xxiv] [26]

3. Nucleus.
   As far as anyone knows, the very first eukaryote came fully equipped with a nucleus. Theories abound, but no one really knows where the nucleus came from. It has been suggested that the eukaryotic nucleus was formed from a captured archaean cell. Indeed, there are many commonalities between the eukaryotic nucleus and archaean cells, in terms of the structure and organization of DNA, RNA, and ribosomes.
   Here are few observations supporting an archaean origin for the eukaryotic nucleus:
   - Only eukaryotes and archaeans have a TATA box (a sequence of thymidine-adenine-thymidine-adenine that specifies where RNA trascription can begin). Bacteria have the so-called Pribnow box, consisting of a TATAAT sequence.
   - The nuclear DNA of eukaryotes and the DNA of archaeans both have attached histone proteins [145].
   - The RNA polymerase and ribosomes of eukaryotes and archaeans are similar. [50].

   A competing theory, follows the recent discovery of giant viruses, such as the mimivirus, that rival eukaryotes in terms of genome size and complexity. Genes of giant virus origin have been found in eukaryotes, prompting the suggestion that a viral origin of the eukaryotic nucleus is possible [88, 98, 97].[27] [28]

---

xxiii Prokaryotic flagella have a diameter of 0.01 to 0.025 microns. Eukaryotic flagella, or undulipodia, have a much larger diameter (0.25 microns).

xxiv Several structures in eukaryotic organisms have evolved as homologous derivatives of undulipodia, all of which contain tubulins [167]. These include pericentriolar bodies, centrioles, kinetids, specialized receptors, haptonemes of coccolithophorids, and undulating membranes of trypanosomes. The primary cilium, a derivative of the undulipodium, is found exclusively in vertebrates. Disorders of the primary cilium are a recently characterized family of human diseases known as the ciliopathies.

2.6 First sex

Sex in Eukaryotic Microorganisms: More Voyeurs Needed

– from a manuscript written by Dave Speijera, Julius Lukes, and Marek Elias [225]

Sexual reproduction is the fusion of two haploid cells, contributed by different individuals of a species, to produce a new diploid organism of the same species whose genome is composed of contributions from both haploid cells. Haploid organisms arise from diploid organisms through a process known as meiosis, in which the diploid organism, having two sets of chromosomes in each diploid cell, divides a nucleus and produces cells that each have one set of chromosomes. Defined this way, sex can be enjoyed by any eukaryotic organism, even single-celled eukaryotes commonly known as protists [168].[xxv] [29] [30]

Sex becomes bewildering when we try to find a general way to express the mechanics of the process, which varies tremendously among species. Let us keep it simple by restricting our attention to meiosis, the process by which diploid cells (cells with a double complement of chromosomes) produce haploid cells (cells with a single complement of chromosomes). In the case of multicellular organisms (e. g., animals and plants), haploid cells can have a complement of genes that can be identified as either male or female, based on genetic attributes. In the case of single cell eukaryotic organisms (commonly known as protists), where gender does not apply, haploid cell fusion may occur. Regardless of how they manage to meet one another, when two haploid cells fuse, they yield a genetically unique diploid organism having genes derived from two different parent organisms of the same species.

In the case of mammals, a haploid cell produced by a male organism (e. g., the spermatocyte) fuses with a haploid cell produced by a female organism (e. g., the oocyte). The spermatocyte may contain a male complement of genes or a female complement of genes. The oocyte contains two female complements of genes. The fusion of the sperm and the oocyte produces either a male-complement/female-complement diploid organism (yielding a genotypically male offspring) or a female-complement/female-complement diploid organism (yielding a genotypically female offspring). Even in cases where there is no sexual distinction among cells of an organism (e. g., in the case of protists), the ability of diploid cells to produce haploid cells will raise the opportunity for two haploid cells (each one of the two produced by different organisms of the same protist species) to fuse. The process by which haploid cells from two different organisms fuse to form a new and unique diploid organism is known as sex.

---

xxv Strict taxonomists should deprecate using "protists" to describe all classes of single-celled eukaryotes. The term includes classes of organisms of unrelated lineage and excludes classes of closely related lineages that happen to be multicellular. For a discussion of taxonomic principles, you may want to read my book, *Taxonomic Guide to Infectious Diseases: Understanding the Biologic Classes of Pathogenic Organisms, Second Edition* [32].

Mitosis is the process by which somatic cells (i. e., diploid cells of the body comprising all cells other than the haploid germ cells) reproduce by division to form two genetically identical diploid cells. The enzymatic pathways required for meiosis are largely identical to the pathways involved in mitosis, suggesting that the two processes may have evolved together. Furthermore, when organisms are sampled from the major classes of eukaryotes, most exhibit sexual competence, as evidenced by meiosis, or genetic recombination (the exchange of genetic material among paired chromosomes, as occurs in meiosis), or nuclear fusions [225]. We also observe that the genes involved in meiosis are conserved in divergent classes of eukaryotes, even in those classes that no longer indulge in sexual activity, such as Trichomonas [164]. These findings indicate that sexual reproduction, by way of haploid fusions, was available to all eukaryotes, beginning with the root eukaryotic organisms that appeared about 2.7 billion years ago [50].

Sex is not the only way for organisms to reproduce. Single-celled eukaryotes can reproduce by mitosis (division of diploid cells). Fungi can reproduce by mitosis or by fusion of haploid spores. Many plants can reproduce by cloning or by hermaphroditic reproduction, in which one organism produces both male and female haploid cells that can fuse to yield a new organism without the assistance of a partner of the opposite sex (Figure 2.6).[31]

Asexual reproduction occurs in the animal kingdom. The bdelloid rotifers, tiny animals best observed under a microscope, famously decline the benefits of sex, preferring to reproduce exclusively by parthenogenesis [190]. The all-female league of bdelloids has managed to diversify, evolving over 500 species, thus defying generations of evolutionary biologists who would insist that speciation is impossible without sexual reproduction. Apparently, the bdelloids are able to acquire new genes directly from other organisms, a process known as horizontal gene transfer. The bdelloids bypassed the messiness of sex, preferring to expand their gene pool (and the opportunity for speciation) by stealing genes from other organisms. It is estimated that about 10 % of the bdelloid gene pool is derived from non-bdelloid organisms [190]. All eukaryotes, including the bdelloids, apparently have the genetic machinery for meiosis. We have no way to exclude the possibility that the bdelloid rotifers may indulge in sexual reproduction, under some circumstances, but we can at least say that the bdelloid rotifers, being uniformly female, must have defended their chastity for a very long time [190].[32] [33]

The bdelloids are not the only animals that acquire genes through horizontal gene transfer. Tardigrades, also known as water bears, are tiny animals that have obtained about one-sixth of their genome from an eclectic assortment of organisms, including bacteria, archaeans, plants, and fungi [42]. Their mode of gene acquisition seems to be more than adequate for their survival, insofar as tardigrades are widely held to be the hardiest of animals. Aside from enduring a wide range of environmental conditions here on Earth, the tardigrades are capable of surviving the extreme cold, and the blistering cosmic radiations of outer space.

Figure 2.6: Drawing of a branch of an English walnut tree (Juglans regia) with prominent male catkins (right, with enlarged view of the same on the left), and a small cluster of female flowers (center, at top bifurcations). Source, The Natural History of Plants, by AK von Marilaun, 1902 [248].

What are the conditions that might sway a species to follow a life of crime, perpetually stealing genes from other species, rather than earning their genes through the hard work of evolution? The answer is simple. Evolution works through the selection of naturally occurring gene variants that are available within the species' gene pool. Evolution is not a very efficient way to achieve metabolic breakthroughs that involve two or more independently evolved pathways that work together to perform a new function. If two pathways are required to achieve one metabolic process, then each pathway must evolve without the benefit of selective pressure for the dual pathway (which would require rare beneficial mutational events to occur in both pathways to achieve survival benefit). Achieving a complex, combined pathway through natural selection is possible, but it is highly improbable. Such pathways may require many millions of years to evolve in any given species. There is no guarantee that the feat will ever be repeated. In this case, the mechanism through which a complex pathway will be attained by unrelated classes of organisms is most likely going to be through genetic acquisition of the complete pathway from an endowed organism (e. g., through horizontal gene transfer);

not through the natural selection of pre-existing gene variants. The process by which one organism steals components of another organism is known as kleptoplasty. We can imagine that an organism that is adept at kleptoplasty may not need to evolve through natural selection.

When we are taught evolution (if we are taught evolution), the dual concepts of natural selection and mutation are locked together, as though the process of evolution depends upon the occurrence of mutations. New mutations have a role in evolution, but most of what we see in terms of the diversity of individual organisms with a species, and in terms of the creation of new species is achieved through the natural selection of favorable gene assortments that exist within the available gene pool of a species. Species achieve fitness when they produce a genetically diverse population, from which some individuals will have the ability to survive unfavorable conditions (e. g., extremes of temperature, lack of food or water, infections, toxins introduced to the environment). Sexual production provides a diversity of the types of individuals in the general population of a species. The diversification of the genetic expression in a population is accoumplished largely through gene shufflings, not through gene mutations. There are several ways in which gene shuffling is achieved in sexual reproduction:

1. Gene contributions from two parent organisms.
   Each parent contributes a haploid set of chromosomes. Together, they form the paired diploid set of chromosomes.
2. Multiple chromosomes.
   A diploid cell has two sets of chromosomes, but the two chromosomes in a matched chromosome pair are not identical to one another (one chromosome of each pair is of maternal origin and the other of paternal origin). Meiosis proceeds after the two chromosomes of each pair are shuffled.
3. Homologous recombination during meiosis.
   Homologous recombination is a shuffling trick that assorts alleles (matched genes in the paired chromosomes of diploid cell wherein one of the pair had a maternal origin and the other had a paternal origin). As a the chromosome of a haploid will have a mixture of genes, with some having paternal origin and others having maternal origin. This shuffling process assures that the paired chromosomes of the offspring will have allele-allele combinations that were not present in either the father or the mother [56].[34]

As a consequence of shuffling, the species population becomes diversified and populated with genetically unique individuals. When an environmental catastrophe occurs (e. g., a pandemic, a climate change, a crop failure), some individuals are likely to have a set of genes that maximize their chances of survival. A subpopulation that thrives despite adversity will have a diverse gene pool that is different from that of the general population. If such a subpopulation were isolated (e. g., through migration), then a new species would emerge; and it could all happen without the occurrence of any new gene mutations.[xxvi]

2.7 First plant

As mentioned earlier in this chapter, about 2.7 billion years ago, cyanobacteria evolved oxygenic photosynthesis, to produce oxygen from water and photons (Figure 2.7). Later, no earlier than 2.1 billion years ago, early eukaryotes of Class Archaeplastida captured cyanobacteria and adapted them as organelles (chloroplasts). Today, all life forms participating in oxygenic photosynthesis are either the descendants of Class Archaeplastida (plants and algae living in symbiosis with internal chloroplasts) or they are the cyanobacteria themselves.[xxvii] There is no evidence that organisms other than the cyanobacteria were successful at evolving the pathways required for oxygenic photosynthesis.

What evidence do we have that the chloroplasts in descendants of Class Archaeplastida arose from cyanobacteria? Aside from both chloroplasts and cyanobacteria having much of the same enzymatic apparatus, inspection of the membrane enclosing chloroplasts reveals two distinct layers. All other plant organelles are wrapped by a single membranous layer. The double-layered chloroplast is presumed to arise from the original membrane enclosing the captured cyanobacteria plus the membrane derived from the engulfing eukaryote. Over time, the chloroplast established a symbiotic relationship with its host eukaryote.[xxviii]

The Archaeplastida, like every other organism at the time, lived in the sea, and it should be no surprise that the first multicellular descendants of Class Archaeplastida were the pelagic algae and, later, the seaweeds. Having achieved multi-cellularity, the ancestors of modern plants needed to wait for the evolution of the embryo, before they could produce the complex organisms that we know today as seed plants (Class Embryophyta).[35] [36]

---

xxvi The mechanism by which new species emerge from old species is well understood, but seldom explained. We will return to the topic of speciation in Section 8.3, "New species; old species lost".

xxvii It seems that plant chloroplasts can be acquired temporarily by animals through a process called kleptoplasty. The kleptoplastic cell captures a chloroplast from an algae and uses the captured chloroplast for a short period (a few days to a few months) until the chloroplast degenerates. These chloroplasts are not self-replicating. When new chloroplasts are needed, the kleptoplastic organism simply ravages another colony of algae. The Sacoglossan sea slug, has achieved a photosynthetic life-style, all thanks to kleptoplasty.

xxviii In the case of kleptoplastic organisms that simply engulf plant cells, harnessing their captured chloroplasts for their own purposes, we see chloroplasts that are lined by three or four membrane layers, derived from the algae or plant cell that was captured plus the membranes of the chloroplast, plus the membrane of the engulfing organism.

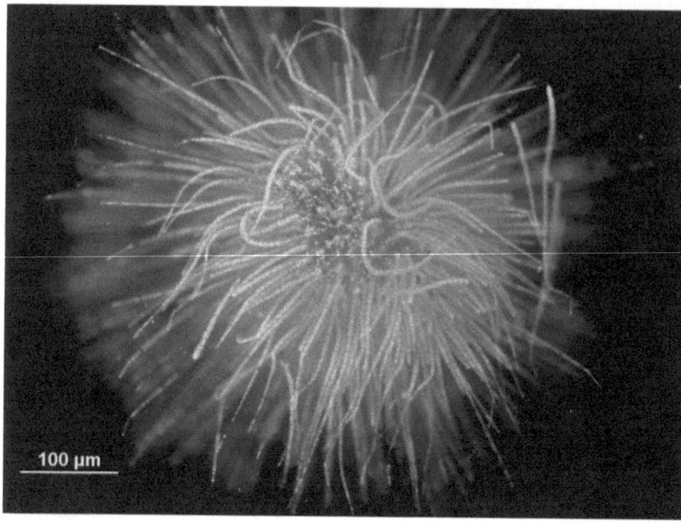

Figure 2.7: Gloeotrichia echinulata, a member of Class Cyanobacteria, the first and only class of terrestrial organisms to evolve oxygenic photosynthesis. Source, Wikipedia, from a public domain image commissioned by the U.S. Geological Survey, and photographed by Barry Rosen.

2.8 First embryo

With the advent of the embryo, multicellular organisms acquired structural complexity in which individual organs (e. g. brain, liver, kidney, lungs) mutually cooperated for the benefit of the "self".

Well before the first embryo arrived, there were multicellular organisms with complex structural parts. For example fungi, then and now, produce specialized anatomic parts for the purpose of spreading spores and facilitating sexual unions (e. g., sporocarps of mushrooms). This being the case, then what is the advantage of having an embryonic stage of development in some organisms?

When organisms without embryos develop, all specialized tissues derive from any of a large population of stem cells, all of which are capable of producing many different kinds of tissues. This process places a huge developmental burden on the cells of an organism. As a result, the complexity of organisms lacking an embryo is limited.

In the case of organisms that develop from an embryo, the strategy for development involves stages of progressive levels of cellular differentiation, in which each stage benefits from the prior development of its preceding stage. For example, in the development of the mammalian kidney, the embryo must first produce a mesonephros, which is then replaced by a metanephros, which then develops into the kidney. Along the way, there are scheduled events that must occur in concert with other embryonic structures, so that the kidney is connected via its ureters with the bladder, and the bladder can be emptied via the urethra.

The embryo is itself a living organism whose function is to organize the anlagen of organs and structural components that will be present in the fully formed organism.[37] Because the embryo follows a programmed sequence wherein each step of the sequence is dependent upon the successful completion of its prior step, the embryo essentially replaces itself with a fetus that eventually grows into a mature organism having complex and coordinated parts.

The embryo is found in only two classes of living organisms: Class Metazoa (animals), and Class Embryophyta (containing most of the species that we commonly know as plants). We infer that embryos evolved independently in animals and in plants, as there are profound differences in the embryonic forms of both classes, and because both classes descended from separate lineages, lacking embryos, and the first animal embryos appeared several hundred million years prior to the first plant embryos.[xxix]

Among the earliest animals on Earth were the sponges (Figure 2.8). A major class of sponges is Class Demospongiae, and it happens that species of Class Demospongiae are the only organisms known to methylate sterols at the 26-position, a fact used to place the earliest demosponges to a date preceding the age of the first observed fossils [51, 158]. Current evidence based on correlating the presence of metabolized sterols with their presence or absence in dated rock strata would suggest that sponges were present on Earth earlier than 635 million years ago, but not before 713 million years ago. Because

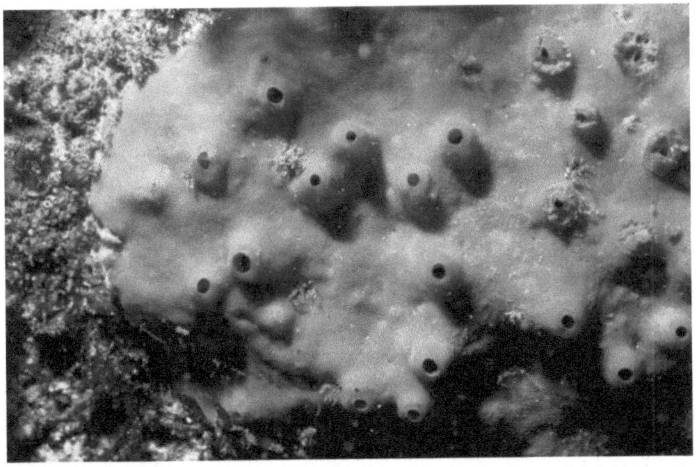

Figure 2.8: Acarnus erithacus, the red volcano sponge. There are at least 8,800 known species of Class Demospongiae. Source, a public domain work of the U.S. National Oceanic and Atmospheric Administration, photographed by SIMoN/MBNMS.

xxix The superclass of Class Embryophyta (plants developing from an embryo) is Class Streptophyta, and the superclass of Class Metazoa (animals developing from embryos) is Class Opisthokonta, neither of which had evolved embryos, leading us to infer that embryos in plants and embryos in animals must have evolved separately.

sponges were among the very earliest metazoans (i. e., animals formed from blastulated embryos), we can tentatively estimate that the first complex animal embryos evolved a bit after 713 million years ago.

The embryophytes (plants that develop from embryos) comprise the land plants, and they evolved somewhat later than the animals, possibly because the continental land masses were inhospitable to land plant growth until the coming of the streptophytes (the ancestors of embryophytes) about 490 million years ago. The plant embryo evolved some millions of years afterward. In hindsight, we can guess that the animal embryo evolved much earlier than the plant embryo because the first animals evolved in ocean waters, before the continents were well formed.

All embryos, whether animal or plant, employ the same biological trick to achieve structural diversity. They start with one totipotent cell (i. e., the diploid cell resulting from the fusion of a haploid egg and a haploid sperm). Totipotency starts with stripping the genome of all or most of its epigenetic baggage.[xxx] [xxxi] A cell equipped with a genome stripped of regulatory constraints is capable of expressing all of the genes of the genome. Having a cell that can express every gene at once would be an impediment to the development of an organism composed of carefully arranged organs, each containing a defined set of cell types?[38]

The key to embryonic development involves timing and cellular commitment. The totipotent zygote, and the embryonic cells thereafter, regulate the timing of the expression of genes, establishing a specific sequence of developmental events. Each milestone of embryonic development sets the stage for the next milestone. As these developmental steps continue, cells are locked into a pattern of constrained behavior that is appropriate for some specific biological role. Eventually, a fully developed organism emerges that is largely composed of committed cells that perform specific functions, in specific anatomic locations.[xxxii] [39]

Animals are thought to have evolved from simple, spherical marine organisms called gallertoids. This living sphere was lined by a single layer of cells enclosing a

---

xxx The epigenome is the collection of chemical and structural modifications of DNA that regulate cellular differentiation.

xxxi All diploid cells of an animal organism have the same genome. The epigenome controls which genes can be expressed, and at what level of activity. In doing so, the epigenome accounts for the different cell types that compose the tissues of the animal organism. A simple way of thinking about the respective roles of the genome and the epigenome is that the genome establishes the unique identity of an organism within its species. The epigenome establishes the identity of the different types of cells (e. g., liver cells, red blood cells, neurons) that compose the organism.

xxxii Stem cells are the undifferentiated cells of the embryo. A stem cell divides to produce two different types of cells: another stem cell plus a cell that is more differentiated than the original stem cell. According to the stem cell theory of development, all of the differentiated cells of the body derive from an ancestral lineage of stem cells, leading back to the zygote. In many cases, the fully differentiated cells (coming at the end of the lineage) are incapable of further division, and are sometimes referred to as post-mitotic cells.

soft center in which fibrous cells floated in an extracellular matrix. The earliest fossil remnants (seabed burrows and tracks) of these early animals, date back to about one billion years ago.

As the gallertoids evolved to extract food from the seabed floor, they flattened out. The modern animals most like the gallertoids are the placozoans, discovered in 1833 found plastered against the wall of a seawater aquarium. These organisms are just under a millimeter in length and each is composed of about 1000 epithelial cells. With the exception of being flat, rather than round, the today's placozoans closely resemble our notion of the early gallertoids, having an outer lining of cuboidal cells, and an inner gelatinous matrix holding a suspension of fibrous cells.

What is the single property that distinguishes the animal embryo from the plant embryo? The answer is the "blastula", present only in the animal embryo. A blastula is a cyst that forms in the early embryo from which the three germinal layers (ectoderm, endoderm and mesoderm) of the embryo arise. For now, let us not concern ourselves with the details of embryonic development. Let us focus on the significance of a fluid-filled cyst in the human embryo. Later, we will explain why such cysts are absent in plant embryos.[40 41 42 43]

We think of epithelial cells, such as those that line ducts and glands, as being rigid, polygonal cells that fit together, like Lego pieces. This is absolutely untrue. Each animal cell that appears to be shaped like a polygon is actually a fluid-filled sphere (Figure 2.9).[xxxiii]

When two soft spheres are pushed together, the conjunction is a flattened surface (Figure 2.10). As multiple cells of the same size are pushed together, each cell takes the shape of a regular polygon, and the net effect is a honeycomb-shaped network of cells called an epithelium.

An epithelial structure built from pushing together a group of soft spheres quickly disassembles into individual, round cells, when the pressure is removed. This being the case, how did animals manage to evolve the stable embryonic blastocyst?

The integrity of cellular, water-tight cystic structures (including the prototypical blastocyst) was achieved by the evolution of two specialized structures that reside on the surface of animal cells, exclusively: desmosomes and tight junctions (Figure 2.11). In the earliest stage of embryogenesis (the morula stage), the embryo consists of a small ball of cells. The cells of the early animal embryo, held together by desmosomes and tight junctions, soon thereafter begin to secrete fluid. Because the desmosomes and tight junc-

---

xxxiii Not all animal cells are floppy, fluid-filled bags. Specialized animal cells may produce intracellular bands of protein that force a spindled or elongated shape (seen in muscle cells producing actin, myosin, or fibrous cells producing desmin). Insect and crustaceans use chitin or other extracellular materials to construct elaborate exoskeletons and shells. Nonetheless, when we speak of the early cells of the embryo, and the fully developed cells of epithelial organs (e. g., liver, gut, pancreas), we are dealing with cells best described as spherical bags filled with a watery protoplasm.

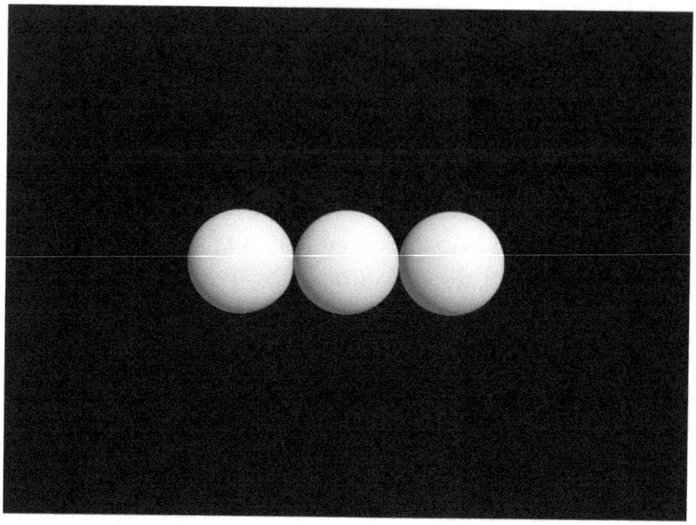

Figure 2.9: Animal cells are essentially soft bags filled with water. When they are suspended in fluid, they assume their natural spherical shape. The illustration depicts how three metazoan (i. e., animal) cells might appear, if they simply touched one another. Source, from author, with image created with Pov-ray rendering software.

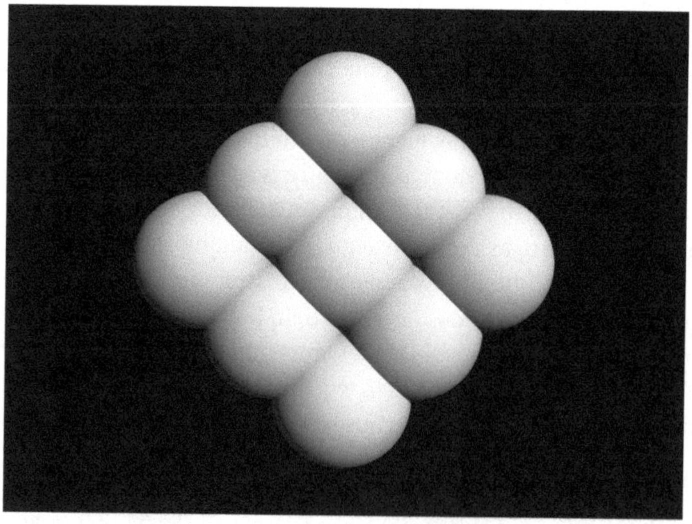

Figure 2.10: Round metazoan cells, when pushed together, form flat surfaces at their intersections. The effect of many spheres pushing together is a polyhedral network, simulating an epithelium. Source, from author, with image created with Pov-ray rendering software.

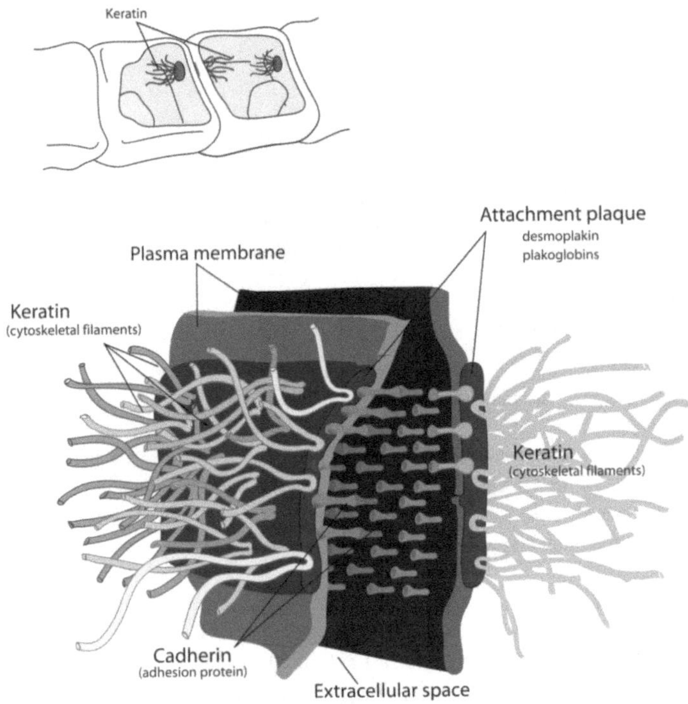

Figure 2.11: Graphic of a desmosome. Desmosomes act like buttons that hold together the flat-surfaced areas where cells touch one another. The net effect is to produce a water-tight epithelial membrane (i. e., composed of polyhedral cells). Desmosomes are characteristic of cells that develop from blastocysts (i. e., all metazoan organisms). Source, Wikipedia, created and released into the public domain by Mariana Ruiz, LadyofHats.

tions holding cells together are water-tight, fluids secreted by the cells accumulate in a cavity (Figure 2.12).

It is through the process of cyst formation that animals form blastocysts, stomachs, ducts, glands, and body cavities.[xxxiv] [xxxv] [44]

The evolution of the desmosome and the tight junction permitted animals to connect their cells into water-tight epithelial surfaces. This being the case, how do fungi and plants, lacking desmosomes, manage to build structural components? Fungi and plants prop up their cells with rigid extracellular materials: chitin in the case of fungi and cel-

---

xxxiv The stomach is credited as the first metazoan organ, inasmuch as all metazoans, other than the first simple metazoans (e. g., placozoans), are equipped with a stomach. The placozoans digest food by flopping over particles and using desmosome junctions to create a temporary enclosure into which digestive enzymes are secreted and nutrients are absorbed.

xxxv There is a downside to every embryonic innovation. With the advent of the animal desmosome came a collection of desmosomal disorders that only affect members of the animal kingdom, and these include blistering diseases as well as several forms of arrhythmogenic cardiomyopathies.

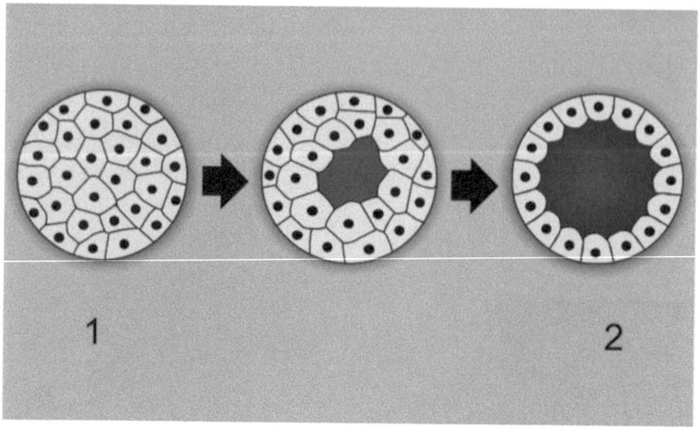

Figure 2.12: Graphic of blastulation. The early, solid, embryo secretes fluid into a central viscus, the developing blastocyst. Blastulation is accomplished with specialized membrane channels that transport ions and water, and with desmosomes, that provide a water-tight boundary between adjacent cells. Blastocysts are only seen in members of Class Metazoa (animals). Source, Wikipedia, created and released into the public domain by Pidalka.

lulose in the case of plants. It is the extracellular matrix in plants that determines the shape of plant cells. When plant cells are stripped of their extracellular matrix, they revert to their original spherical shape. Because plant cells cannot form tight junctions, they cannot form a blastocyst (the earliest fluid-filled chamber to form in the embryo). Consequently, plant embryos are solid, without the epithelial-lined glands, cysts, and cavities that characterize animal embryos (Figure 2.13).

Aside from the inability to form water-tight glands lined by soft epithelial cells, the process of embryonic development is much the same in plant and animal. One of the

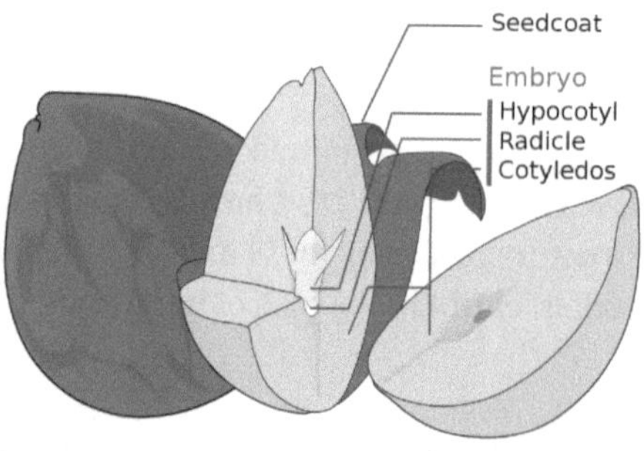

Figure 2.13: Diagram of avocado seed. The small avocado embryo is situated at the base, along the central axis of the seed. The embryo is solid and hard. Peanuts have an embryo much like that of the avocado. To find it, split the shelled peanut in half and look for a small hard node at one end. The peanut embryo, like the avocado embryo shown here, has two epicotyls that extend from the embryo giving the embryo the appearance of a face with a beard. Source, Wikipedia, and entered into the public domain by its author, Sarang.

most striking similarities involves the production of extraembryonic tissue by the early embryo. In mammals, the embryo produces three germ layers from which the adult organism develops (endoderm, mesoderm, and ectoderm). The embryo of plant and animal also produces a layer of cells that are intended to support the development of the embryo only for the duration of gestation, at which time they are ejected from the organism. The fourth germ layer of the mammalian embryo, the trophectoderm, produces all of the so-called extra-embryonic tissues (e. g., the placenta and umbilical cord). The suspensor of the plant is analogous to the trophectoderm in mammals. The extra-embryonic suspensor pushes the plant embryo into proper position and attaches the embryo to maternal cells for nutrient transfer, serving a function that is strikingly similar to that of the mammalian placenta [156].

2.9 Where do babies come from?

One of the most commonly held misconceptions is that embryos are miniature babies. This belief may have arisen from the fully discredited theory that the early zygote (the cell resulting from a fusion of a spermatocyte and an oocyte, from which the embryo develops) is a preformed human that has only to expand in size during the gestation period. We know this not to be the case, but we also know that embryonic cells have the same genes as those of the infant and of the adult. Furthermore, every baby develops from a fetus, which, in turn, developed from an embryo. The fetus and the adult organism, though derived in stages from the embryo, are anatomically, physiologically, metabolically, and cytologically distinct from one another.

When we observe various species of organisms, we find that the embryonic forms of plants and insects are nothing like the adult forms into which they develop. For example, an acorn is not a miniature oak tree. A caterpillar is not a miniature butterfly. With these species, we can say without hesitation that the embryo is not a miniature version of the adult. Can we generalize from the tree and the insect and apply our knowledge directly to humans? Let us see.

Human development is not fundamentally different from the development of any other animal. All animals (i. e., all members of Class Metazoa) develop from an embryo, and all embryos develop in a similar fashion, even the embryos of insects. Insects develop through a series of distinctive morphologic stages, called instars. Growing instars molt, to become the next instar. The development of the fly Drosophila melanogaster begins very soon after fertilization, with the first instar at 24 hrs. The second instar is seen at 48 hrs, and the fifth at 118 hrs. The pupa appears on the sixth day. When the pupa molts, on the ninth day, the adult insect emerges. During each stage, structures from the preceding stage may be reorganized or otherwise eliminated. The adult is the only stage that exhibits the familiar body structures that we commonly associate with insects: antenna, legs, wings, and so on. These adult structures arise from undifferentiated organizing cells that form imaginal discs. Each of the imaginal discs is committed to form a specific structure (i. e., the wing disc produces a wing, and the antenna disc produces an antenna). Because insects develop through a succession of morphologically distinctive living forms, all of which lack the external structural features characteristic of the adult form, it is easy to see that the instar is not a small version of the adult insect.

Similarly, in the human embryo, we do not find the organs that are present in the adult organism. Instead, we see committed tissues that undergo a series of developmental steps that eventually lead to the creation of a small organ, in the fetus. For example, the human kidney develops from a focus of mesoderm that eventually forms a primitive mesonephros, which is the anlage tissue for the subsequent development of the metanephros, which is the anlage for the development of the kidney. The embryo does not contain two, small, formed kidneys. As a general rule, miniature, anatomically complete versions of adult organs do not emerge until the fetal period.

In plants, which also develop from embryos (called seeds), the distinction between the embryo and the developed organism is stark. Acorns and oak trees are the same genetic species and both are members of Class Quercus. No one can mistake an acorn (i. e., containing the oak embryo) for an oak tree. The acorn and the oak tree are different stages of the same organism, each with its own biological function.

Because embryonic development is crucial to the development of the adult organism, and to the evolution of the species, we might expect that natural selection would favor organisms capable of producing an abundant number of embryos. In every species that develops from an embryo (all metazoans and all seed plants (angiosperms and gymnosperms)), there are many more embryos produced than fully developed organisms. This is most evident in the plant kingdom, wherein every seed contains an embryo. We see examples of single plants that annually produce billions of embryos (i. e., seeds). To understate the situation, every adult organism develops from an embryo, but not every embryo will develop into an adult organism.

Adult males are phenotypically different from adult females in many ways, but the most significant differences pertain to organs. Women have a uterus, a vagina and a clitoris. Males do not. Males have a prostate and a penis. Women do not. What accounts for these differences? Are there separate anlagen for female organs and for male organs? Actually, no. Female organs develop from the exact same anlagen as do the male organ. The gender-neutral anlagen respond to hormones and other gender-specified regulators to produce female organs in women and male organs in men.

Examples in which a single embryologic anlage, common to all human embryos, produces different organs in males and in females include:

- The embryonic gonad develops into the testis in men, the ovary in women.
- The paramesonephric duct develops into the appendix testis and prostatic utricle in men, and the Fallopian tubes, uterus, cervix and vagina in women.
- The labioscrotal folds develop into the scrotum in men, and the labia majora in women.
- The genital tubercle develops into the penis in men, and the clitoris in women.

Each embryonic anlage develops into some particular adult structure, even if it is a non-functional tissue remnant, in men or in women (e. g., nipples in men, prostate in women). To press a point, embryonic structures are not tiny versions of adult male or female organs. Embryonic structures are composed of precursor cells that are programmed to develop into organs, in the fetus. Because men and women share the same embryonic anlagen, they are both at-risk for developing the same types of neoplasms, and these tumors can resemble the anlagen or the tissues derived from the anlagen.[xxxvi] Thus men may develop breast cancer. Women may develop Leydig cell tumors of the ovary (exhibiting cytologic features of the Leydig cells in testes). Men and women may develop germinomas (called seminomas in men and called dysgerminomas in women).[xxxvii] There are a host of rare tumors of males and females that are presumed to derive from the equivalent male and female embryologic anlagen [265].

We should not succumb to the belief that embryonic cells are just somewhat less differentiated versions of the cells that populate the fully developed human. Abundant evidence indicates that embryonic cells are highly specialized to fulfill their predetermined biological functions. As such, embryonic cells are different, in many ways, from adult cells. As examples, embryonic cells respond very differently to toxins, and other injuries than cells in adults, and embryonic cells interact with their adjacent tissues in a coordinated manner that is not seen in the tissues of adult organisms.[xxxviii]

A fundamental difference between the embryonic organism and the adult organism is observed in a particular fruit fly mutation. Mutation of the gene Sas-4, coding for a centriolar protein in flies, and inhibits the growth of cilia in the fly embryo.[xxxix] Remarkably, these mutant embryos mature normally, into morphologically unremarkable adults. Soon after birth, these flies die, because adult flies need their cilia for a wide variety of vital physiologic functions [19]. In this case, a mutation that is lethal to the adult organism had no effect on embryonic development.

As another example, mutated oncogenes tend to produce developmental disorders (not cancers) when they are present in embryonic cells. Oncogenes produce cancers

---

xxxvi Neoplasm translates from the Greek as "new growth", and is nearly synonymous with "tumor". Neoplasms can be benign or malignant. Leukemias grow as a population of circulating blood cells and do not generally produce a visible mass (i. e., do not produce a tumor), but are included under the general term "neoplasm". Hamartomas, benign overgrowths of tissue, are typically included among the neoplasms, as are the precancers (cancer precursor lesions), which are often small and scarcely visible.

xxxvii A seminoma is a tumor of male germ cells, almost always arising in the testis. The female homoplasy for seminoma (i. e., the embryologically analogous tumor arising from female germ cells) is the ovarian dysgerminoma.

xxxviii Teeth are formed through a recprocal induction process involving odontogenic ectoderm and odontogenic mesoderm. Nothing similar occurs in adult tissues, leastwise not in adult human tissues.

xxxix Cilia are the slender, waving structures that are found on cells lining ducts and bronchi. In the developed human lung, the cilia cleanse the bronchi of mucus and debris by wafting material upwards. Cilia move ductal contents from exocrine glands (e. g., salivary gland, pancreas, gallbladder) into the gut.

(not developmental disorders) when they arise in adult cells.[xl] In addition, neonates and children develop totally different types of tumors than those observed in adult humans, indicating that the first cells involved in prenatal carcinogenesis are fundamentally different from the cells involved in adult carcinogenesis. These examples provide us with some notion of the biological differences between an embryo and a fully developed organism, both of which perform their own functions, respond in their own way to mutations, and have their own requirements for survival. The tumors that arise from embryonic cells have been mistakenly referred to as "primitive" to indicate that they lack the features of the cells we encounter in the tissues of adult organisms (Primitive neuroectodermal tumor). This is just wrong. The various cells of the embryo have their own highly specialized embryonic functions and the tumors derived from embryonic cells are not primitive versions of the cells found in adult organisms.[45]

Let us return to our earlier question. When, after conception, does human life begin? This is not a question that science can answer. We can say, however, with absolute certainty, that an embryo is not a baby in miniature.[46]

2.10 First organism to die of old age

> Only accidental, externally caused death, existed at the origin of life.
>
> – Lynn Margulis and David Sagan in their book, *What is Life?*, published in 2000 [168]

When we contemplate the grim facts of life, the inevitability of aging and death intrudes upon the psyche and darkens the soul. It seems to us that we all must grow old and die. Yet, when we think objectively about aging, we find it to be a biological process, much like any other biological process. Humans get old and die, as do dogs, and horses, but the process proceeds at a species-specific mode and pace. Something about the genetic machinery of each species determines when the aging process takes its toll; why some species have a longer life-span than others. If aging is an inherited biological process that follows a species-specific tempo, then we can safely infer that aging, like all other biological attributes of animals, must have arisen through evolution. If so, then there must have been a time, before the aging process had evolved, when organisms did not age. Before there was aging, death occurred largely by accident; not from the aging process [168]. This being so, we should expect to find evidence of the evolution of aging, somewhere in our ancestral lineage.

When we survey all the classes of organisms living today, we see that the process of aging, leading to death, is something that applies to only a small minority. Aging seems

---

xl In particular, germline BRAF mutations are associated with developmental syndromes such as Noonan, LEOPARD, and cardiofaciocutaneous syndromes [213]. BRAF mutations in somatic cells are associated with a variety of cancers, notably melanoma.

to be a trait exclusive to animal classes descended from a few subclasses of bilaterians. The remainder of the classes of life, including single-celled eukaryotes, bacteria, archaeans, many species of plants, and various branches of metazoans (i. e., animals), simply do not age. They can live thousands of years. When they die, it is typically due to some misadventure (e. g., predation, starvation, suffocation, infection, dessication, etc.). Even among the animals for which aging and death come naturally, there are wide variations in life-span, as though the trait of aging is more fully expressed in some classes of animals than in others.

What is aging? We ought to have a precise definition for something that affects all of us, but words fail. Superficially, aging is a sequence of events that occur in an organism, over time, that reduce the ability of the organism to survive, and which eventually result in the death of the organism. We infer that the rate of the aging is genetically determined because all of the members of a species will undergo that same aging process, more or less, and will reach the same end-point (i. e., death) at roughly the same age. In the case of humans, we can all expect to die from the aging process before the age of 120 years, unless we die earlier, from intervening events.

Perhaps aging is best understood on the cellular level. During life, most of the tissues of our body are constantly renewing, with old cells being replaced by new cells. This process of constant renewal is most evident in our blood cells, which account for most of the cellular turnover in our bodies. In point of fact, each of us contains 20–30 trillion red blood cells, representing more than 70 % of all the cells in the human body [219]. A red blood cell circulates for about 120 days, accumulating bumps and scratches and various internal derangements along the way. Old and damaged red blood cells are scavenged (by macrophages) and replaced by freshly minted cells. Because the number of circulating red blood cells is constant, we can assume that each day, about 1/120 or a bit over 1 % of the total population of red blood cells are replaced every day of our lives (i. e., 0.2 to 0.3 trillion red blood cells). As we grow old, the red blood cell replacement process never stops. Many superannuated individuals have a normal red blood cell count, with no obvious alteration in red blood cell morphology. Our blood-forming capacity never seems to age.

Likewise, we each enjoy an eclectic assortment of tissues and organs that, like our blood cells, undergo constant renewal throughout our lives, with no signs of slacking with age. For example, the uppermost levels of our gut lining (i. e., the epithelial layer in esophagus, stomach, small and large intestines) are constantly shedding. The shedded cellular debris is mixed with our digesting food and eventually exits the body as feces. Every cell that is shed into the gut is replaced by cells from the lowest layer of the gut lining. These lining cells are renewed without any apparent age-related deficit. No one dies from an old colon.

Whereas some of our organs chug along through the decades, seemingly unfazed by the aging process, others seem to be affected by aging at the earliest moments of development. Cartilage cells (lining joints), neurons (of the brain and the nervous system), cardiac muscle cells, oocytes, the enamel-forming cells of our teeth, all seem to have a limited ability to be replaced by progenitor cells. Cells that lack the ability to divide are called post-mitotic, indicating that they can no longer replicate.[xli] As time progresses, and as more and more of these cells die without being replaced, we begin to see clinical deficits in the affected tissues. Our joints lose their cartilage cushions, and we may suffer arthritis, particularly in weight-bearing joints. As our heart suffers minor or major attacks, the involved muscle fails to regenerate, and our cardiac function deteriorates. Our skeletal muscle cells diminish in number and size, a process known as sarcopenia, and we become frail. The brain atrophies, and we may experience age-related cognitive impairment. It does not bode well.

We can describe aging as the process that occurs in tissues when particular cell types cease to divide, resulting in functional deficits. In contrast, we can describe cancer as the process that occurs when particular cells in a tissue cannot stop dividing.

When does the aging process begin in humans? It is tempting to believe that aging begins after we have stopped growing, and after we have attained our peak in physical and mental abilities. We often imagine that after we've reached age 40 or 50 or so, the rest of our lives will be spent in continuous decline. To be young, we are told, means being free from the ravages of the aging process. This concept is easily disproven. As previously mentioned, oocytes form in utero, ceasing replication well before the birth of the female organism.[xlii] By the time a woman is born, the majority of her eggs have aged out of the gene pool. The prenatal death of oocytes indicates that aging, a process that always involves the inability of a cell population to regenerate itself, must begin before birth.

In fact, all of the inherited diseases of aging are first evident in infants and children. These disorders of aging do not wait until middle age to exert their effects. This tells us that the aging process is something that begins very early in human development; not late in life.[xliii] [47]

Society devotes a great deal of attention to diseases that occur in the elderly. Indeed, the bulk of research funding in the field of aging is devoted to geriatric conditions. Does it make much sense to devote all of our attention to the final stage in the aging process that

---

xli Scientists apply a variety of terms to describe the cells that replace dying cells. Variously known as reserve cells or stem cells, the process of cell replacement seems to be more or less the same. The reserve cell, divides to produce a reserve cell (replacing itself) and a cell that will become fully differentiated, and serve as a replacement for the cell that has died. In this way, the number of differentiated cells and reserve cells remains static.

xlii The number of oocytes reaches a peak of about 7 million cells at 5 months gestation. After the peak is reached, about three months before birth, the oocytes begin to die; they are not replaced. The number of live oocytes declines until the number falls below a threshold of 1,000, triggering menopause [93].

xliii We will not describe all of the rare congenital aging disorders, but the list includes Bloom syndrome, Cockayne syndrome, Dyskeratosis congenita, Fanconi's anemia, Hutchinson-Gilford progeria, Werner syndrome, Wolfram syndrome, and Xeroderma pigmentosum [28].

directly precedes death, when we know that the first step of aging must have occurred in utero?

How do aging cells die? Individual cells die through a carefully choreographed process known variously as apoptosis, programmed cell death, programmed senescence, and cell suicide. Apoptosis begins with a triggering event (such as a toxic change in cellular homeostasis or a developmental signal to reduce the number of cells in a population) and is followed by a series of biochemical processes involving the shrinkage of cellular compartments and the excision of DNA into small pieces, all accompanied by a cascade of lytic enzymes that reduce the cell to detritus that can be engulfed by macrophages. The characteristic morphologic phases of apoptosis are: chromosomes break into small fragments; the nucleus shrinks (i. e., karyopyknosis); and the cytoplasmic membrane protrudes outward, as blebs.

Whether we examine aging at the level of the species, the individual, the organs and tissues within the individual, or the biochemical process of the death of single cells, we see evidence of a carefully controlled biological process that evolved in our ancestral genes. Was there a time when aging did not exist? Was there a moment in evolution when aging appeared? Can we look back through time and find the origin of aging?

The key to the origin of aging lies in the classification of life, which contains the chronologic ancestry of every organism that has ever lived on earth. Let us take a step back to the first forms of cellular life on Earth. Cellular organisms fall into one of two broad classes: the prokaryotes (cells with no nucleus, comprising Class Bacteria and Class Archaea); and the eukaryotes (cellular organisms containing a nucleus). When we think of aging, we generally confine ourselves to processes that occur in multicellular organisms because single-celled organisms (i. e., all of the prokaryotes and most of the eukaryotes) continue to divide so long as they are well-nourished, uncrowded, and protected from environmental pathogens. The prokaryotic organisms do not age.

Nearly all of the child classes of Class Eukaryota are single-celled organisms or simple colonies of single-celled organisms. None of the single-celled organisms undergo anything akin to aging. When one cell divides to become two daughter cells, nothing remains of the original cell other than its genome replicated in the daughters. The two daughter cells emerge as brand-new individuals, not as aged replicates of the parent cell. Hence, the continuously dividing single-cell eukaryotes and all of the prokaryotes have a life-style that evades our concept of aging.

Only two of the original Eukaryotic classes became ancestors of complex multicellular organisms: Class Viridiplantae, the ancestral class of all plants; and Class Opisthokonta, the ancestral class of fungi and animals (Figure 2.14). Some species from both of these major eukaryotic classes achieved the dubious ability to age and divide. However, most of the branches of Class Viridiplantae and Class Opisthokonta lack the ability to age.

In the plant kingdom, aging, when it occurs, follows a different biological pathway than what we see in the animal kingdom. For example, many individual trees can live

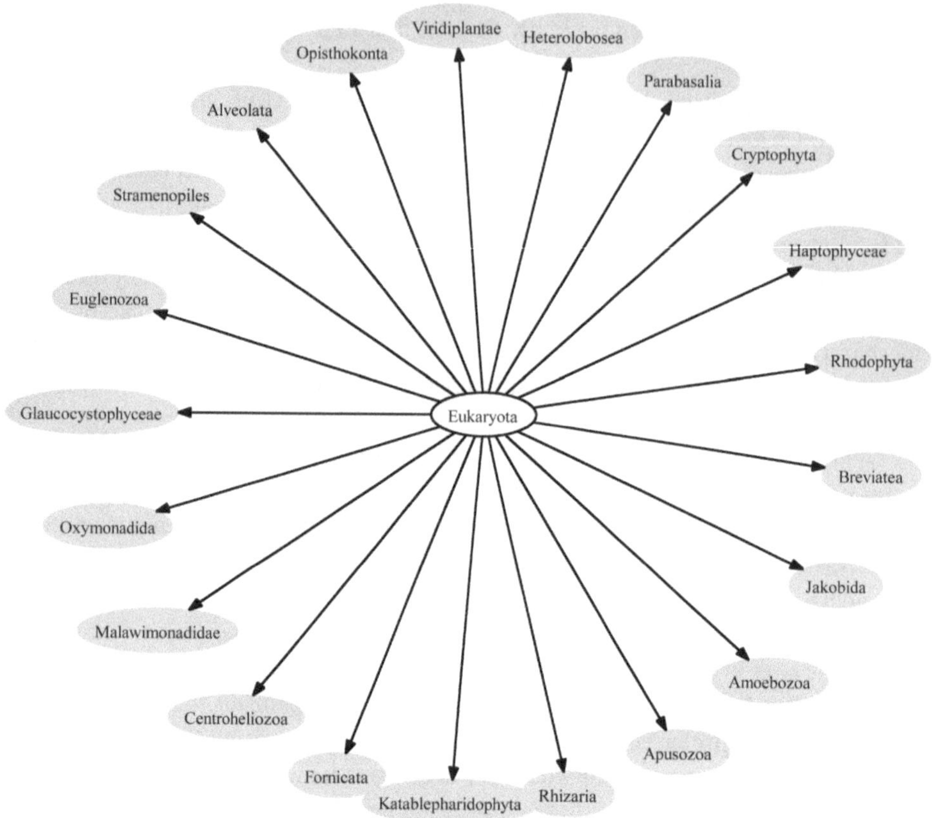

Figure 2.14: Class Eukaryota has 21 child classes. At the top of the chart is Class Viridiplantae, from which all plants descend. From Class Opisthokonta (just left of Class Viridiplantae) come the fungi and the animals. Only the descendants of Class Opisthokonta contain subclasses exhibiting what we have come to think of as aging. Source, author.

hundreds or even thousands of years. Methuselah, a Great Basin bristlecone pine residing in Inyo County, California, is reputed to be about 5,000 years old, which seems to be about the observed limit for the life-span of any individual standing tree [28].

Some species of trees self-clone. Self-cloning occurs within a copse, producing a group of trees all having the same genetic identity, with new clonal growths replacing dying trees. It is not unreasonable to consider the copse itself as a single biological organism, characterized by one genome and by a stable collection of growing and dying cells. Such clonal organisms are virtually immortal. A copse of Quaking Aspen (Populus tremuloides), living in Fishlake National Forest, Utah, is thought to be hundreds of thousands of years old [182]. Considered as a single organism, the Quaking Aspen colony occupies 106 acres and has an aggregate weight of about 6 million kilograms (Figure 2.15). Farmers in ancient and modern times have benefited from the self-cloning nature of

Figure 2.15: A grove of quaking aspen trees, all belonging to the same clone. These trees are referred to, collectively, as Pando. Source, Wikipedia, produced for the U.S. Department of Agriculture.

plants by developing the agricultural technique known as coppicing. Young trees are cut to near-ground level, and new, clonal trees re-shoot from the stump. By repeated cuttings, the trees are maintained as juveniles. Regularly coppiced trees never seem to age or die; they just spread out from the center. Individual coppiced trees have been maintained for centuries [28].

Unlike most animal species, individual plants contain totipotent cells that can divide indefinitely and from which portions of plants, or fully formed plants, can regenerate. Not so for most metazoans (animals) who, in the natural course of development, lose their totipotent cells as they differentiate to develop mature organs. Because all of Class Viridiplantae is characterized by species that maintain a totipotent cell population, we can infer that every member of the class (i. e., all plants) has a mechanism for achieving continuous clonal growth. Hence, death in the kingdom of plants is much more like a biological option than it is an inevitability.

Other than Class Viridiplantae, the only class of eukaryotes that experience anything akin to aging is found in Class Opisthokonta. Class Opisthokonta has child classes: Class Fungi, Class Choanoflagellida, Class Metazoa, and Class Aphelidea. Organisms of Class Fungi have no upper bound in terms of size or age. The Malheur National Forest in Oregon is host to a fungus (Armillarea ostoyae, the honey mushroom) that covers 2,200 acres of land, to an average depth of about 3 feet. It is estimated, based on growth rates, to be about 2,400 years old. Class Aphelidea are basically fungi and do not experience aging, as we recognize it. Likewise, members of Class Choanoflagellida are single-celled organisms that grow in close-packed social colonies, but have not attained status as a multicellular organism and do not exhibit aging in the usual sense. Consequently, among the child classes of opisthokonts, it is only Class Metazoa that contains aging organisms (Figure 2.16).

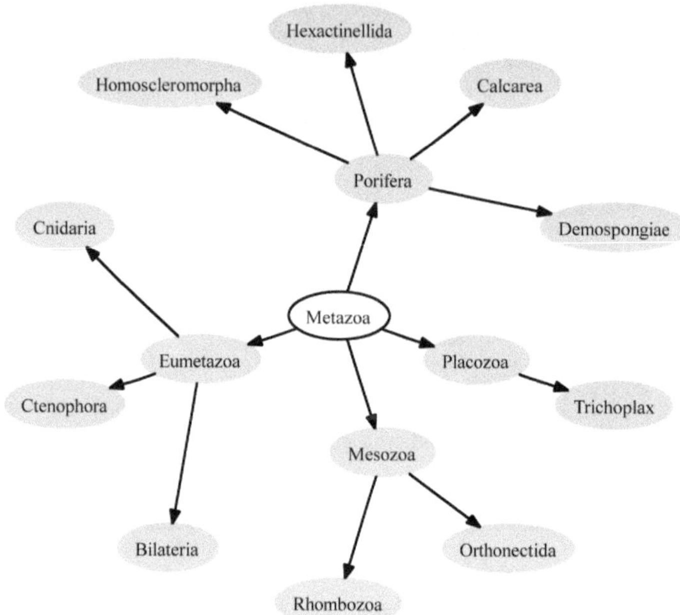

Figure 2.16: Class MetazoA, with its child classes and its grandchild classes. These are the ancestral classes of all animals on earth. Source, author.

The child classes of Class Metazoa are: Class Porifera (the sponges); Class Placozoa (with one extant species, Trichoplax); Class Mesozoa; and Class Eumetazoa (so-called true animals, of which we humans belong).

Sponges, among which are some of the longest-living animals, belong to Class Porifera. Estimates based on growth rates of Antarctic sponges suggest a very long lifespan in these animals. A two-meter-high sponge, discovered in the Ross Sea, is estimated to be 23,000 years old. Due to fluctuation in the levels of the Ross Sea, skeptical scientists doubt that the specimen could have lived for much more than 15,000 years. Still, a low-estimate 15,000 year lifespan is impressive enough.

Class Placozoa contains one extant species, Trichoplax, a flat organism, about 1 mm in diameter, that lacks internal organs. Trichoplax can regenerate their entire bodies from small groups of cells. As such, they are essentially immortal.

Class Mesozoa is an informal class containing simple, miniscule animals. There is currently some controversy over the legitimacy of Class Mesozoa, with some evidence suggesting that the monophyletic mesozoan species are subclasses of the lophotrochozoans. In any case, the accepted mesozoan species do not exhibit anything akin to the aging process observed in most eumetazoans.

It is only within Class Eumetazoa that the aging process is found, so let us look at the subclasses of Class Eumetazoa to see which subclasses exhibit aging (Figure 2.17). Class Eumatazoa has three subclasses: Class Cnidaria, Class Ctenophora, and Class Bilateria.

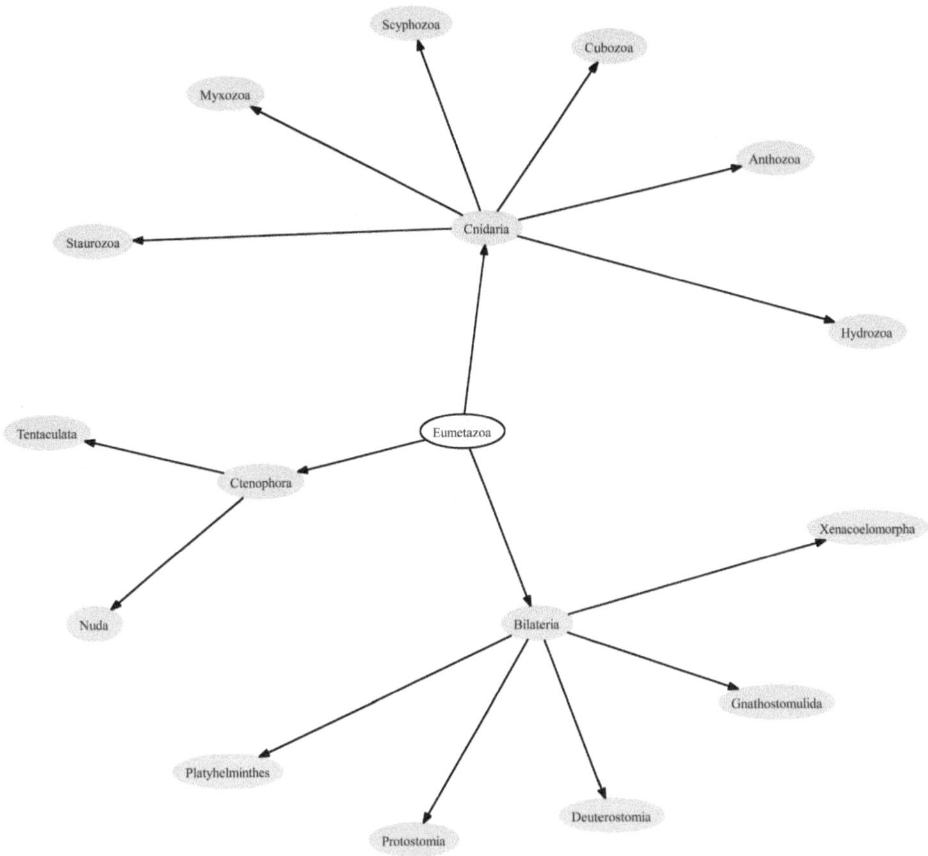

Figure 2.17: Subclasses of Class Eumetazoa (animals). The first branching gives us the cnidarians (also known as the coelenterates), the ctenophorans (the comb jellies), and the bilaterians (animals whose embryos have bilaterial symmetry along a central axis). Source, author.

Some members of Class Cnidaria seem to be immortal. The medusae (fully mature form) of the jellyfish Turritopsis nutricula has the rare ability to revert back to its immature polyp stage, basically reversing its life cycle. The immature polyp can reproduce by budding. There seems to be no limit to the number of times in which a medusa may form from a budding polyp and revert back to a polyp stage. Hence, the jellyfish Turritopsis nutricula, and several other species, including Turritopsis dohrnii and Hydractinia carnea, may all be immortal [215, 170, 171].[xliv]

xliv See also the article by Nathaniel Rich, "Can a Jellyfish Unlock the Secret of Immortality?" in The New York Times, November 28, 2012.

Class Ctenophora (the comb jellies) contains the bdelloid rotifers, tiny creatures whose females can clone themselves. Although the body of the short-lived female dies, its identical genome persists clonally, basically side-stepping the entire issue of aging.

Within Class Bilateria, we have the following subclasses: Platyhelminthes, Deuterostomia, Protostomia, Gnathostomulida, and Xenacoelomorpha.

Class Platyhelmithes (the flatworms) contain species that regenerate their bodies from small pieces. In particular, the planarian flatworms can be cut into pieces, and an entire organism can be regenerated from each of the pieces. Hence, researchers quip that planarians are "immortal under the edge of a knife." Observations of blastemal whole-body regeneration indicate that planarian cells are totipotent and able to form blastemas. We do not know how planaria manage to maintain a population of totipotent stem cells, but it has been suggested that a key step involves the maintenance of normal telomere length in planaria [233]. Telomeres are sequences of non-coding DNA attached to the termini of chromosomes. Without telomeres, cells cannot divide. Telomeres are typically lost from non-dividing somatic cells in animals that age.[48]

Another subclass of Class Platyhelminthes, Class Turbellaria, contains species that clone themselves by transverse or longitudinal division (Figure 2.18). Still other turbellarians reproduce by budding. The term "aging" does not easily apply to species that can regenerate whole organisms from single slices, or that divide themselves in two parts to produce two clones of themselves, or that self-reproduce by simple budding. Clearly, members of Class Platyhelminthes did not get the memo instructing them to curl up and die, like most of their bilaterian sister classes.

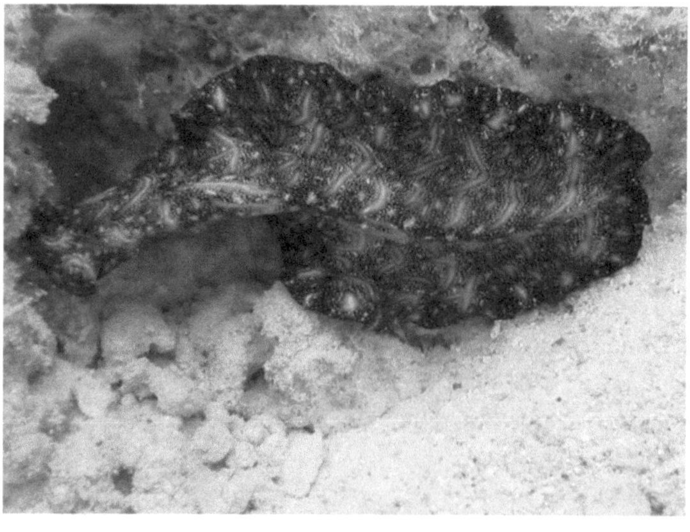

Figure 2.18: Pseudobiceros bedfordi, a species of Turbellarian flatworm (Class Platyhelminthes). Source, Wikipedia, and contributed to the public domain by its author, Jan Derk.

Class Gnathostomulida consists of tiny aquatic organism that are self-reproducing hermaphrodites. Much like the comb jellies, their reproductive life-style essentially bypasses the issue of aging.

Class Xenacoelomorpha is closely related to both Class Deuterostomia and Class Protostomia and has been included within the deuterstomes by some taxonomists [199]. The xenacoelomorphs are tiny aquatic hermaphroditic animals that engage in sexual reproduction. Just to simplify things, we will tentatively include Class Xenacoelomorpha with the deuterostomes and protostomes in our discussion of the aging classes of animals.

Without prolonging our suspense, we can say that with the evolution of Class Protostomia and Class Deuterostomia, the process of aging first evolved. As far as anyone can tell, all of the species, extant or extinct, that descended from either of these two classes exhibit some form of the aging process. All of the classes that are not descended from either Class Deuterostomia and Class Protostomia contain species that do not age; at least, they do not age in a manner that we would consider equivalent to aging as seen in deuterostomes or protostomes.

Let us summarize. Single-celled eukaryotes are ageless. Although we haven't discussed them, the prokaryotes and the viruses are ageless in the same sense as the single-celled eukaryotes are ageless: they replicate to produce exact copies of themselves. Of the three major classes of multicellular organisms (i. e., plants, fungi and animals), the plants and the fungi contain ageless species. Of the remaining class of multicellular organisms (i. e., Class Metazoa, the animals), the classes produced by the first several phylogenetic branchings are ageless. All of these ageless organisms are characterized by continuous growth and or continuous cell renewal. Aging is an evolved trait that arose when Class Bilateria branched and produced the protostomes and the deuterostomes about 550 million years ago; and the aging process is restricted to their descendants.

The notion that aging is an evolutionary innovation seems a bit absurd to those of us who stare in the mirror to see our thinning hair and our turkey necks, and contemplate our mortality. It's best to shirk these depressing thoughts and to think of aging as a trait that benefits the species (at the expense of individual members of the species). Our species does not care that individuals grow old and die. Individuals can be replaced. Species achieve immortality by speciating (producing new species), not by simply persisting.[xlv]

xlv All evolutionary innovations develop from the embryo. Accordingly, medical scientists who study aging should devote some of their attention to the embryo; and maybe a little less time to observing superannuated adults. The rare genetic diseases of aging (i. e., the various progerias) are manifested in very young children, indicating that aging, as a biologic process, begins at the earliest stages of life.

2.11 On first thought

> I used to think that the brain was the most wonderful organ in my body. Then I realized who was telling me this.
>
> – Emo Philips

Thinking about how we think is a messy business. When we depend on our own minds to provide objective information about our thought processes, we can expect to be disappointed. We use our conscious faculties to analyze thoughts, but we cannot account for the origin of those thoughts. When we find ourselves whistling a tune while we walk, we might stop for a moment to ask, "How did that tune enter my head?", or, more broadly, "Why am I thinking about what it is that I'm thinking about?" The tune, or some inchoate semblance of the tune, was passed into consciousness by the subconscious mind. The conscious mind embellished the tune as a whistled opus. Much the same process accounts for human speech. Words, or something akin to words, pop into our minds, and we use our intelligence to form polished sentences. It is the subconscious delivery of words that forms the basis for the popular word association test. The psychologist instructs the patient to say the very first word that comes to mind when some prompt is said or shown. The patient's response originates in the subconscious mind and, therefore, serves as a clue to the patient's psyche. As a working definition, we can say that thoughts that do not originate in the conscious mind must have come from the subconscious. Beyond that, we really have very little understanding of how the subconscious mind operates.[xlvi] [49]

Philosophers, through the ages, have come to think of the topic of consciousness as the "hard" problem; something that no one has approached satisfactorily.[xlvii] We foolishly continue to beg our brains to explain to us how our brains work, with nothing much to show for our efforts. In the past half century, we have witnessed incredible advances in the field of brain imaging, including the introduction of computed tomography and nuclear magnetic resonance imaging.[xlviii] Scientists can now determine the areas of the

---

xlvi Ultimately, we cannot accept our own assessments of our thought processes, because all such self-descriptive statements are self-referential, and there is simply no way by which self-referential statements can be trusted. Self-referential statements lead us into nonsensical paradoxes. An often repeated example is the Liar's Paradox. A liar says: "I am lying". If the statement "I am lying" is true, then the liar is telling the truth (and not lying at all), which means that statement "I am lying" is false. There are dozens of popular paradoxes of self-reference. These exercises in futility all demonstrate that logical inferences cannot be drawn from self-referential statements. Hence, we cannot trust ourselves to know the truth of our thoughts about our thoughts.

xlvii On the subject of consciousness, Stuart Sutherland wrote, "Nothing worth reading has been written about it." In "Consciousness", Macmillan Dictionary of Psychology, Macmillan press, 1989.

xlviii Additional brain imaging techniques are: positron emission tomography, functional magnetic resonance imaging, multichannel electroencephalography, magnetoencephalography, near-infrared spectroscopic imaging, and single-photon emission computed tomography.

brain that are selectively activated during specific physiologic functions. Encouraged by their new-found technology, brain scientists were confident that mankind would soon unlock the mystery of consciousness. Furthermore, anticipated breakthroughs in brain research should have uncovered cures for psychiatric illnesses. Alas, none of our early expectations have led to an understanding of how we might understand and fix disorders of consciousness. Excluding neurodegenerative and traumatic brain conditions (e. g., Alzheimer disease, Parkinson disease, chronic traumatic encephalopathy), there is, at present, no psychiatric condition that is consistently associated with anything visualized by advanced imaging modalities [43].

Before capitulating to our ignorance on the subject, here are a few questions pertaining to consciousness that we might contemplate:

1. How, and when, did consciousness first occur on our planet?
   Some philosophers entertain the notion that consciousness can only arise from consciousness. That is to say that a conscious being could not be assembled from unconscious components. If this were the case, then the physical components of consciousness (i. e., brains, neurons, organelles, and their composite molecules, atoms and subatomic particles) should have some form of consciousness. Therefore, consciousness was present from the moment the universe began. Really? Because we lack any real understanding of consciousness, all manner of wild speculation can be entertained.
2. When did consciousness evolve in living organisms?
   Specifically, what were the evolutionary steps that led to the first occurrence of consciousness? This question would require us to know which organisms have consciousness and which organisms do not. Unfortunately, we have no way of making any such determination. For example, we can generally agree that humans have conscious thoughts, and that our dogs and cats have conscious thoughts. Do worms have consciousness? How about trees? We haven't a clue.
3. How is consciousness restored in individuals awakening from sleep, or rousing after anesthesia, or recovering from coma?
   During sleep, and anesthesia, and even coma, the mind does not turn off. There is brain activity during states of unconsciousness. What is our mind doing while our consciousness is gone?
4. Is there anything we can do to elevate our level of consciousness?
   It would be wonderful if we could increase the level of our consciousness ten-fold. Or would it? What would we do with all that consciousness? What does it mean to have a heightened consciousness?
5. Will there come a time when we can pour our consciousness into a computer, achieving an immortalization of our thoughts, in silicon?
   Is consciousness something that can be extracted and preserved?
6. Most importantly, what is consciousness?

A glib and circular definition of consciousness is that consciousness is our ability to claim consciousness. The Cartesion version of this definition would be: "I think I am conscious; therefore I am conscious." Does a claim of consciousness have any merit? Imagine a scenario in which an individual is under deep hypnosis. The hypnotist asks the subject whether he is conscious. The subject answers "Yes, I am conscious." Soon thereafter, the hypnotist wakens the subject and asks whether he has any memory of their session. The subject answers, "No," indicating that an unconscious state had prevailed at the moment when the subject claimed consciousness.

Let us examine the issue of consciousness by turning our answer around. Let us imagine that we ask the subject under hypnosis whether they are conscious, and they answer "No, I am not conscious." Can a subject hear a question and form a cogent reply in the absence of consciousness?

Perhaps we should fall back to a minimalist definition: "Consciousness is how the organizing properties of the mind are expressed." Through a process that we do not understand, our collective neurons manage to organize their signals into something that we associate with our own consciousness.

2.12 The subconscious, master of the conscious

> What is mind?
> Doesn't matter.
> What is matter?
> Never mind.
>
> – Nihilistic folk wisdom

Of course, nothing we experience is quite as it seems. Our thoughts, our speech, our visions, and all of our perceptions are fed to us by a part of our minds that evades awareness: the subconscious.[xlix] As we live through each day, our minds buzz with thoughts. Despite our best efforts, most of us cannot clear our heads of incoming thoughts. But where do those thoughts come from? We do not consciously create those thoughts because, if we did, we would be aware of how those thoughts came into being. Our conscious minds are preoccupied with thoughts that arose from our subconscious minds.

Let us take a look at some of the thoughts that are manufactured for us by our subconscious mind:

- **Vision.**
  We fool ourselves into believing that our vision represents an image of the world that is focused, by the lenses of our eyes, onto our retinas. Of course, this is not the

---

xlix Here is a simple and functional way to define the "subconscious". When we do something with our minds, but we are not conscious of what we are doing, then we can say that it must have been the product of our "subconscious" mind.

case, inasmuch as the retinal projection is just the first of many internal processing steps that precede vision. Let us review. Our retinas register the received light on an immense array of specialized cells, excitable by photons. Electrochemical impulse are distributed though a complex sequence of neural processes, to the visual cortex of the brain, and to various other neural pathways and destinations, where further processing ensues. The product of all this neural activity is somehow assembled, without any conscious awareness, into an animated movie, in real time, that our gullible mind accepts as "reality". In other words, the vision that reaches our conscious awareness is prepared for us by our inscrutable subconscious director.
Two different phenomena testify to the handiwork of the subconscious in the creation of vision. One is the occurrence of optical illusion. When we look at an image, and we know that what we see cannot possibly represent what is actually there, then we can be certain that our subconscious must have botched the job (Figures 2.19 and 2.20).

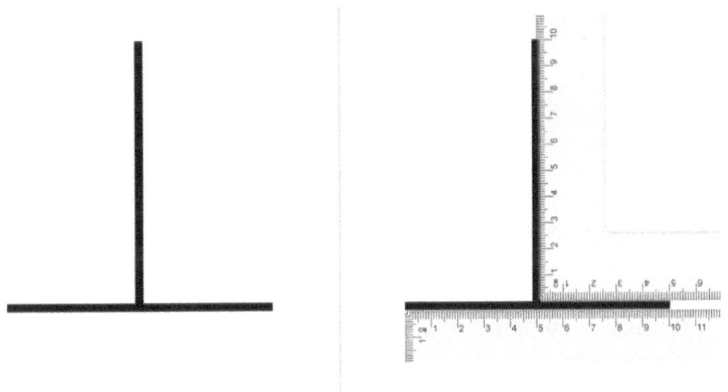

Figure 2.19: The vertical bar appears to be longer than the horizontal bar, but when the two bars are measured, we find that they have the same length. A libertine interpretation of the image, served up by the subconscious mind, demands a conscious misinterpretation of reality.

Another distortion provided by the subconscious is the erasure of scotomas from our perception of visual fields. Scotomas are blind spots caused by imperfections of the retina. Imagine what we would expect to see if there were a "dead zone" in the retina of our left eye involving half of its visual field. Such scotomas can occur following a stroke, or following a focal area of retinal detachment, or following some direct injury to the internal eye. In such a case, we might expect to see a visual field obscured by a dark area corresponding to the missing retinal surface. This is what we would find when we take a photograph using film with a focal defect in the surface emulsion, or when we use a digital camera with a missing portion of its sensor array. This is not what we find in humans with retinal scotomas. Instead, we see a

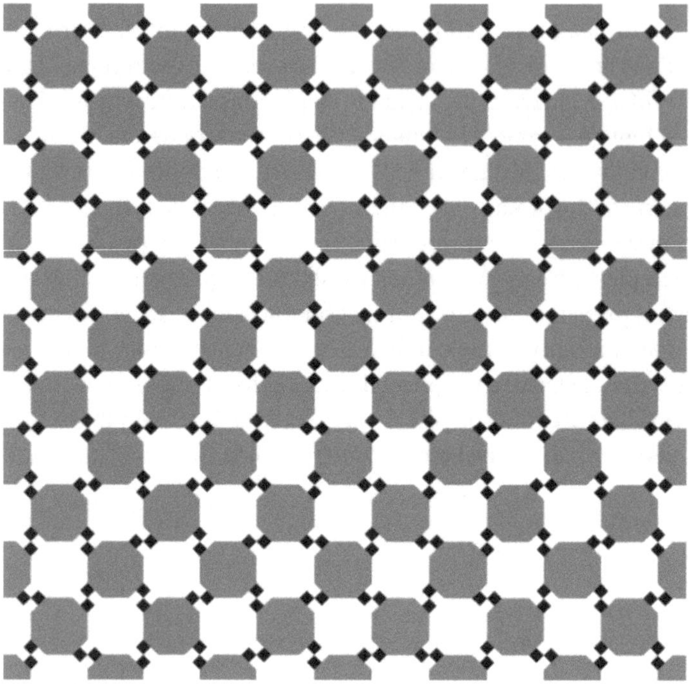

Figure 2.20: Small angled bars at the corners of checkers appear to twist the checkerboard, but the rows and columns are straight. A subconscious interpretation of the image imposes a conscious misinterpretation of reality. Source, image entered into the public domain by AnonMoos.

visual field that seems to us complete and normal. The patient with a scotoma is unaware of anything missing in his or her vision (until such time as they bump against an object that seems to have materialized from nowhere). In a valiant effort to create a panoramic view of reality, our subconscious creates the illusion of unimpaired vision that our conscious mind cannot dispell by act of will.

– **Dreams.**
  Dreams are movies prepared for our viewing pleasure by a furtive director who we may call the subconscious. Our fulsome dreams serve as testimony to the power of the subconscious. In sleep, we exist in a dream world, without any awareness of how that world was created. Because the visions in our dreams, and the visions of our woken reality are both prepared for us by our subconscious minds, it can be difficult to distinguish one from the other. Anyone who lives with a dog knows that slumbering dogs enjoy a rich inner life. The dog moves his legs furiously, as he runs through the dreamscape, possibly chasing a rabbit or a ball, or racing against another dog. My dog Bailey, on rare occasion, would waken from a nightmare. Unsure whether his dark terror persisted into the light of day, he would refuse to leave his bed. Fearful at first, he would refrain from eating his breakfast until enough time

had passed that he could once more distinguish the conscious world from the unconscious dream-state. Only then would breakfast commence. So far as we know, every eutherian species (i. e., the placental mammals) is issued with about the same brain anatomy and the same neural processes.[1] It should not surprise us that all eutherians have a subconscious mind capable of producing pleasant dreams or terrifying nightmares.

- **Proprioception.**
  When we descend a stairway, we do not look down, at our feet. We look straight ahead. Most of us don't bother to hold onto the bannister. We are confident that we will not trip and fall. The ability to know where we are, and how we shall proceed is a feature of the subconscious, known as proprioception. Proprioception, like most types of subconscious thought, plays out quickly. If we were to suppress proprioception and opted for consciously guiding our feet, we would not be able to run through a forest, or catch a high fly ball, or type a manuscript before its deadline.

- **Nascent thoughts.**
  Where do thoughts come from? They seem to just pop into our heads. For the most part, thoughts begin in the subconscious mind. The conscious mind imposes something akin to a logical discourse upon the nascent thoughts received from the subconscious.
  How can we be certain our thoughts arise from the subconscious? If our conscious mind created our thoughts, then we would be able to explain how those thoughts came into being (that's basically what it means to be conscious of something happening). But we cannot do so. If thoughts do not originate in our conscious mind, then they must come from the subconscious.
  As an experiment, try not to think for a while. It can not be done. As long as we are conscious, the subconscious mind feeds us thoughts. At day's end, when we surrender to sleep, the subconscious mind sets the agenda for our dreams.

- **Instinct and collective intelligence.**
  In animals (dogs, cats, birds, ants, spiders) there is a species-specific set of instructions, known as instincts, that drive much of their behavior. Without instincts, animals could not survive in a harsh and indifferent environment, and could not cooperate with other organisms to establish social systems that would include mutual protections, mating, and raising offspring.
  Are humans, like other animals, driven by instinct? We like to think not, preferring to believe that Homo sapiens have replaced instinct with reason. We believe that

1 Differences in brain anatomy among the different eutherians are typically minor. For example, it seems as though only mammals have evolved a corpus callosum, that curved sheet of white matter slung like a hammock between the right and left cerebri, connecting neurons across either hemisphere. All other vertebrates rely on the relatively small anterior commissure for their inter-hemispheric connections.

> we are fundamentally different from all other creatures. Still, there are modes of thinking that are common to all humans, such as fear, anger, love, anxiety, empathy, and curiosity. These behaviors are not taught to us, and they cannot have originated through conscious reasoning. We have come to recognize the assortment of inborn human ideas as our "collective intelligence", a term roughly synonymous with human instinct.

To understand the significance of collective intelligence, we must re-examine the differences between the conscious mind and the unconscious mind. Our conscious minds are molded by personal experiences and memories. Our subconscious minds are waiting for us, from the moment of birth, or earlier. Without the subconscious mind, we could not understand what we see, and we could not use our senses in any useful manner. We could never learn to speak, and we could never be human.

Presumably, the subconscious minds of humans are more or less standard issue, much like our legs and livers. There are differences among individuals, but all of us have the same functioning parts, drawn from the human gene pool, which evolved incrementally through the lineage of our ancestral species, from the moment that life began on Earth. Assuming that it is true that every human subconscious is roughly equivalent, we might expect to uncover identical driving forces that characterize the human race.

Carl Jung (1875–1961) promoted the idea of a collective intelligence, derived from the subconscious, and acting upon the conscious mind. The inherited intelligence of our species is claimed to provide us with cross-cultural symbols, myths, personality archetypes, dreams, and even memories; all of which act to shape the human psyche. Jung taught that these thoughts must be properly integrated into our conscious psyches; otherwise, all manner of psychiatric disorders may arise [130].

It is tempting to dismiss Jungian psychoanalysis as a pseudoscience, insofar as its theories are built on clinical and personal observations; not upon controlled laboratory experiments. Jung's philosophy, and his approach to psychoanalysis and dream interpretation, is neither proven nor disproven. Perhaps, when we validate examples of equivalent myths and symbols influencing diverse cultures, separated by time and geography, we will temper our skepticism. By asserting the existence of a collective human subconscious, and then populating that domain with visions (i. e., dreams and hallucinations), defined objects (e. g., symbols, myths, archetypes), and alchemies (observed as induced transformations towards a more perfect state of existence), Jung had found a credible method by which the conscious mind can examine the subconscious mind.[50]

2.13 The “self” and its intelligence

What’s in your head, in your head?
Zombie, zombie, zombie-ie-ie-ie, oh

– lyrics from Zombie, by Dolores Mary O’Riordan

What is the relationship between intelligence and consciousness? If we say that intelligence is the ability to achieve a goal through directed thought, then our definition implies that there is a formed intention, which suggests that some consciousness serves as the source of the intention. In this manner, we derive a circular definition wherein intelligence requires consciousness, in the form of intention, and must include a sense of “self” from which the attention arose. Consciousness, in turn, requires intelligence with which to draw inferences from the information collected by the “self”. It is all very confusing.

At some point in the near future, the related fields of robotics and artificial intelligence will have matured to the point that androids will function autonomously, with their own form of intelligence. If androids achieve consciousness while capable of self-replication, then the human race will have a problem on its hands. Will we be obsolete? On the bright side, we can ask our androids to explain the meaning of consciousness. At long last, we may hear an objective solution to our “hard problem”.

We must not rule out the futuristic option of a hybrid consciousness, combining human and android minds. Perhaps we shall evolve into a largely bionic creature retaining a small cellular component. Perhaps our descendants will change their bionic forms, whenever needed, to suit the changing environments on extraterrestrial planets. These creatures may retain some level of human consciousness in the form of regenerative brain tissue that can be transplanted to succeeding generations of android/human hybrids. Will one of the two components (android and human) of the chimeric brain achieve dominant consciousness? Stay tuned.

We leave the topic of the origin of consciousness with one question. Is there a limit to what an entity that has intelligence but lacks self-awareness (e. g., a computer, a robot, or a massive hive of buzzing neurons) can do? For example, could an unaware computer create a universe? If so, is it possible that our universe was created by an unaware intelligence (i. e., an intelligence that had no sense of self and did not know what it was doing)? The mind boggles.

2.14 Intelligence in the animal kingdom

Let us put the definition of human intelligence aside, for the moment, and proceed to search for intelligence in the animal kingdom. We humans like to believe that we know which animals have advanced reasoning skills. Class Mammalia, which includes humans, whales, dogs, and cats is generally regarded to be a fairly smart class. Birds, we know, are also capable of a high level of reasoning and self-awareness. The closest ancestral class containing both birds (Class Aves) and mammals (Class Mammalia) is Class Amniota. All of the amniotes have the same basic brain structure (i. e., bihemispheric cerebrum and cerebellum), and it is not at all surprising that intelligence in one amniotic species would be found in the other amniotes.[li]

Are there any smart animals outside of Class Amniota? Apparently so, and these intelligent animals emerge from a class that is commonly considered to be among the dumbest of animals, the molluscs. We are referring here to the members of Class Cephalopoda (i. e., octopi, cuttlefish, and squids), a descendant class of Class Mollusca, and a fairly close cousin to clams (Class Bivalvia) and snails (Class Gastropoda) (Figures 2.21, 2.22, and 2.23). The cephalopods, particularly octopi and cuttlefish, are credited with being highly intelligent and self-aware animals [118, 1]. Functionally, the octopus brain supports short and long-term memories, sleep, and the ability to explore objects and learn their shapes.[lii] For example, octopi easily open jars of food shut with a screw-cap lid. Captive octopi are known to respond with different behaviors based upon their

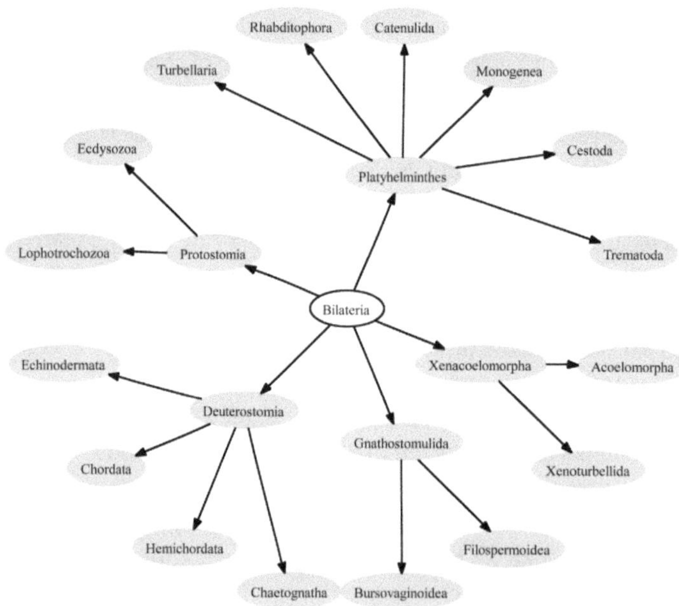

Figure 2.21: Class Lophotrochozoa (ancestral to octopi) is the child class of Class Protostomia. Class Chordata (ancestral to humans) is the child class of Class Deuterostomia. Source, author.

li The amniotes are a subclass of tetrapods characterized by the presence of a specialized membrane that covers and protects their embryos. We humans are proud members of Class Amniota.

lii See article by Peter Godfrey-Smith titled "The mind of an octopus," in the January 1, 2017 edition of Scientific American Mind.

Figure 2.22: Class Mollusca is a child class of Class Lophotrochozoa. Source, author.

established relationships with individual human caretakers and visitors. They seem to resent captivity and have been known to contrive ingenious strategies to escape confinement and to seek a more hospitable environment. Octopi have a social life and some female octopi exhibit some of the most protective brooding behavior observed in the animal kingdom.[liii]

Observational evidence suggests that all three subclasses of the cephalopods are intelligent. Considering that we humans have no seriously creditable way of measuring our own intelligence, it is unlikely that we will soon fathom the full potential of the octopus mind.

On the anatomic and physiologic level, the octopus brain and nervous system has virtually no resemblance to that of humans [118]. Molluscs, like all invertebrates, have a distributed nervous system. Octopi, like other molluscs, have multiple brains, one for each of the 8 arms, plus a central brain. How all of these anatomically isolated brains coordinate thought and action is something of a mystery, but we can see that the brains of the octopus are organized much differently than the centralized amniote brain (such as our own). Aside from gross anatomy, we find fundamental differences between the brains of octopi and amniotes at the cellular level. The axons of amniotes are wrapped

liii See Jonathan Webb's article, "Broody octopus keeps record-breaking four-year vigil," in BBC News Science and Environment, July 30, 2014.

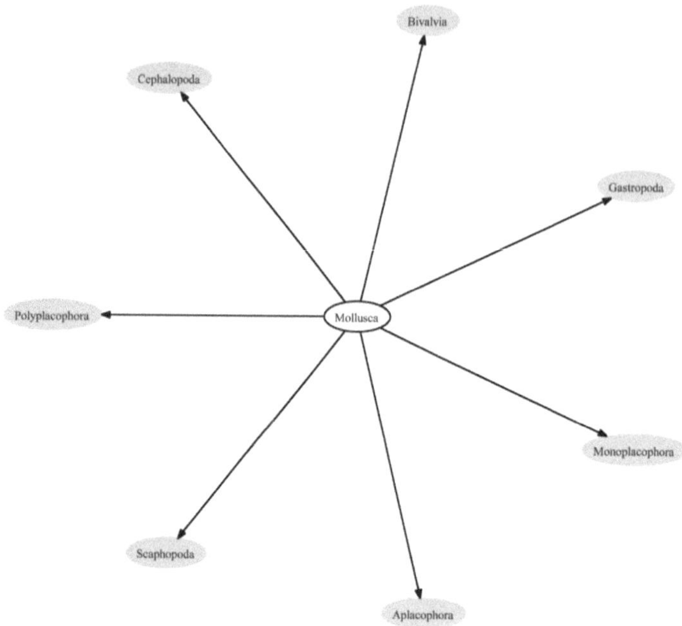

Figure 2.23: Class Cephalopoda (containing octopi, squid and cuttlefish) is a child class of class Mollusca and a sibling class of Class Bivalva (including clams) and Class Gastropoda (including snails and slugs). Because neither the sibling classes or the ancestral classes of the cephalopods seem to exhibit much intelligence, we can infer that the cephalopods evolved their high level of intelligence from within their own class. *De novo*. Source, author.

by cells that produce myelin (glial cells in the brain and Schwann cells in the peripheral nervous system). Myelin greatly increases the conduction velocity of neural impulses. The axons of neurons in octopi are unmyelinated, slowing the rate at which impulses are conducted [1]. Hence, we can infer that quick nervous system responses in the octopus must be achieved through a mechanism that does not require fast conduction through long axons.

When we examine the ancestors of molluscs and mammals, in the Classification of Life, we find that the two classes have no closely shared ancestors.

When we follow the lineage of the major classes of organisms, we find that Class Mollusca (including the cephalopods, such as octopi and squids) are related to Class Chordata (including the amniotes, such as humans and birds) only by the most distant of ancestral animal classes, Class Bilateria. As such, humans and octopi have almost no common evolution. In the case of the cephalopods, high intelligence arose in a class whose sister classes seem to be totally devoid of thought. These findings merit a few inferences:

1. The intelligences on Earth arose independently of one another, using dissimilar anatomic and physiologic strategies.

The early classes in the two lineages leading to humans and to octopi were, so far as we can tell, brainless. Therefore, the high intelligence in the cephalopods and in the tetrapods (including humans) must have evolved separately. Consistent with this conclusion is the observation that nervous systems in humans and octopi have radically different anatomic and cellular organizations.

One additional supportive observation is that the first cephalopods may have appeared as early as 522 million years ago [117]. The first amniotes seem to have appeared about 312 million years ago [22]. The asynchronous appearances of cephalopods and amniotes supports the inference that the two classes evolved independently of one another.

2. High intelligence may have evolved more than twice in the history of Earth.
   The history of evolution is marked by occurrences that are so unlikely that they seem to have occurred once and then never again. The prototypical example of a one-hit wonder is the evolution of oxygenic photosynthesis. So far as we can tell, oxygenic photosynthesis evolved once, in an ancient cyanobacteria, and all oxygenic photosynthesis occurring today involves the acquisition of ancient genes coding for the cyanobacterial pathway. Traits that evolve more than once in Earth's long history (such as intelligence) are considered to be repeatable. Because the evolution of higher intelligence evolved independently in Class Amniota and Class Cephalopoda, we can infer that intelligence evolved at least twice. Therefore, there may be additional species of intelligent animal life (extant or extinct) that are unknown to us.
3. In the case of the octopus, intelligence may have arisen de novo, without preceding small steps achieved throughout its ancestral lineage.
   The ancestral classes and the sister classes of Class Cephalopoda exhibit no signs of much intelligence. Hence we suspect that intelligence evolved de novo in the cephalopods.
4. Intelligence does not appear to confer any great evolutionary advantage, in terms of speciation.
   Neither Class Cephalopoda nor Class Amniota are particularly successful classes, in terms of the number of descendant species.[liv] There are about 800 living species of cephalopods, representing well under 1 % of the molluscan species. The gastropods (e. g., snails and slugs) account for about 80 % of the approximately 100,000 known species of mollusc. Among Class Amniota, the extant subclasses are reptiles, birds

liv We tend to think that a successful species is one that is the most fit for long-term survival. This is simply wrong. There have been about 5–50 billion species of organisms on Earth (conservatively), and more than 99 % of them have met with extinction, leaving a relatively scant 10–100 million living species [208]. Extinction is expected for all evolving species. The destiny of every successful species is speciation (the production of subspecies), and the success of a species is determined by the number of descendant species that emerge and survive over time.

and mammals, which represent a small portion of living animals. The smartest subclass of the amniotes is thought to be Class Primata (the primates), most species of which are barely managing to survive. The self-appointed smartest of the smartest classes, Class Homo, contains merely one extant species, Homo sapiens. These facts would all suggest that intelligence barely registers as a factor in the success of species.

5. Intelligence may evolve again, in some unpromising class of animals. For that matter, the existence of two independent paths toward high intelligence auger the development of intelligent life elsewhere in the universe.[lv]

Why do we obsess over the hard questions of intelligence, consciousness and self-awareness? Well, despite our best efforts, we do not have a clear understanding of who we are. Our ignorance of self seems to be part of the human condition, and this upsets us.

2.15 The emergence of human language

> I personally think we developed language because of our deep need to complain.
>
> – Lily Tomlin

There is no way to know, with any certainty, when humans first acquired language. Because language appeared before writing, we can hardly expect to find any record of the event. We can guess that human language arose between 100,000 and 40,000 years ago because most of the intellectual advances of humans (e. g., primitive religions, burial rituals, jewelry making, wall painting) seemed to have begun in this period, and we like to think that language skills may have spurred these societal achievements.

A few casual observations about language may help us to understand something about its origin, its uniqueness in the animal kingdom, its relevance to our understanding of human consciousness, and the role it plays in the human condition.

– There are many thousands of languages (extinct and extant) and no human societies lack a language.

  Infants will learn the language to which they are exposed. An Indian child raised in Argentina will acquire fluency in Spanish just as quickly as a native born child. Humans can learn several languages, if they so wish. Language immersion accelerates the process. In other words, the choice of language is immaterial to the role of language in the human condition.

---

lv Regarding intelligence in the universe outside our planet, it is a chilling fact that the most intelligent animals on Earth are predatory carnivores (e. g., humans, hawks, tigers, and octopi). For our own sakes, we can only hope that intelligent extraterrestrial life has not evolved with the rapacious tendencies exhibited here on Earth.

- Speech is not a prerequisite for language.
  Signed (i. e. nonvocal) languages are an effective form of communication. Therefore, the impressive ability of humans to produce many different sounds with their mouths, vocal cords, and well-controlled breaths, is not a requirement for language acquisition.
- Infants are born without the ability to speak, but typically acquire the ability to understand language in under two years. Passable speech follows soon thereafter. Babies understand the spoken words of their parents well before their own bablings transform into meaningful speech. Verbalization requires fine motor coordination of lips, teeth, tongue, nose, vocal chords, and breath. Language comprehension operates through the baby's developing mind, not its mouth.
- Chimpanzees, dogs, and pigs are often credited with a level of intelligence that exceeds that of infants. The "higher" intelligence of humans seems to appear soon after infancy, at about the same time as language skills are developed.
  We should not rule out the possibility that animal intelligence is, in the absence of language, on par with human intelligence.
- No other animal on earth seems to have the linguistic aptitude of humans, and no animal species seems to use language the way that humans do.
  Nature offers many examples of animals that communicate with sounds. It is common to encounter animals that use distinctive vocalizations to warn one another of danger, or to indicate their willingness to mate, or to convey any number of socially relevant signals (e. g., mating calls), through gestures or sounds. As we study primates, birds, and octopi, we are amazed at their at communication. Yet, there is little evidence to suggest that non-humans have anything equivalent to human language.
- The ability to understand language and to express oneself through language is a defining characteristic of the human species.
  Amnesiacs who cannot remember their own name are typically able to speak normally. Similarly, individuals suffering with dementia, whose memory is greatly diminished, typically retain the ability to speak until the very last stages of their clinical course.
- The paradox of language seems to require that thought and language are one and the same thing.
  Humans cannot express their thoughts in the form of language until there is a preexisting language. But a language cannot exist until humans create thoughts in the form of language. This is a paradox that can only be resolved by accepting that humans have an instinctive aptitude for expressing thought in the form of language.

Noam Chomsky, the great philosopher in the field of human linguistics, has emphasized that the primary purpose of language is to conduct thought. Humans have a freewheeling inner monolog in which thoughts are conveyed as conscious words. The inner monolog of language is much more important to us that our reliance on language to communicate with one another. A monk may live under a vow of silence, but his inner voice continues unabated.

Those of us who have dogs or other pets, are likely to say that they can always tell what their pets are thinking. But surely, these pets are not thinking in language. More than likely, those dogs are thinking in terms of images (e. g., images of food, or of affectionate embraces, or encounters with other animals) or remembered impressions (e. g., what smelled good, what tasted bad). Thinking without language is something that humans do during infancy. We also think without language in response to immediate threats. For example, when someone swings their fist at your head, you are not likely to think: "I'd better duck now." More than likely, the image of the fist approaching your face is enough to warrant a quick duck, without the intervention of language.

If all animals in the world conduct their lives without composing their thoughts in the form of language, then why is language a tightly conserved trait in all humans? Just as we see some humans who are tall and others who are short, why do we not see some people whose thoughts arrive as images and others whose thoughts arrive as words? Let us consider an answer that seems to fit our general observations.

1. Words are much like index items.
   A word is basically a pointer to a trove of trivia. For example, the word "dog" points to images of dogs stored in memory, along with a collection of memories associated with dogs. When we say "grass", we do not need to describe a blade of grass, and we do not need to specify the height of the grass or the variety of grass. The word "grass" conveys a thought in our own minds and in the mind of others who speak the same language. When we construct a sentence using words, we are using a form of shorthand that frees us from the burden of describing every detail of our thoughts. It is enough to use common "pointers" to collected examples of concepts and images.
2. Index items can be classified.
   Aside from being index pointers, words fit into our general classification of the world. A "dog" is a subclass of "animal", which is a subclass of "living being". Therefore, a "dog" inherits the general attributes of a living being. Every item in a classification inherites all of the properties of its ascendant classes. Hence, whenever we use a word, we can apply its inherited class properties. A classification simplifies reality and helps us to understand the relationships between items belonging to different classes [34]. Luckily for us, classifying words in language is something that our minds do for us, without much conscious input.
3. The use of language helps us to associate different items that have been stored in memory. By tracing the associations between remembered items, we can recall related thoughts, quickly.
   Mnemonists, who can recall a long string of digits after just a few moments of study, tell us that they achieve their results by associating numbers with words, often in the form of stories. The memory of digits, we are told, is solidified by word associations.

It is interesting that if you ask an eight year old child to recall events from the first few years of life, they always draw a blank. It is tempting to imagine that the infant, who has not yet acquired a working vocabulary, cannot form memories that are associated with words. The eight-year-old-child, who relies upon her language skills to build memories, cannot access the "wordless" memories created during her own infancy.

To summarize, all humans use language to conduct the continuous inner monolog of the conscious mind. Long after language developed in humans, writing was invented. With the advent of the written word, stories, news, opinions, agricultural methods, mercantile transactions, new inventions, and everything required to build and advance a civilization could be codified and transmitted across generations. With the written word, humanity had hit its stride.

2.16 First contact with others

Starry-eyed science fiction enthusiasts dream of humanity's first encounter with extraterrestrial life. The thinking here is that something good will come of it. Optimists may expect that an advanced alien culture will gift us with a cure for cancer, as a peace offering. It seems far-fetched to think that aliens have developed the cure for a disease from which they do not suffer and of which they are unaware, but hope springs eternal.

Much more likely, an alien civilization capable of traveling to Earth would be industrially-inclined, and accustomed to using force to accomplish goals. Far more intelligent than ourselves, they may see no need to cooperate with inferior life forms, such as humans and mosquitoes. Should the opportunity arise to encounter alien life forms, it seems that we Earthlings should approach with an abundance of caution.

Be that as it may, we humans have a poor track record of noticing diverse forms of life, be they terrestrial or extraterrestrial. We are only now beginning to recognize that there are countless species of organisms living here on earth, to which we have been oblivious. Some of these organisms dwell in heretofore inaccessible regions of the planet (e. g., in ocean depths), while others have shared our familiar environment, exerting profound effects on our lives, while effortlessly concealing their existences. For example, we have the giant viruses, whose existence was not even suspected until well into the twentieth century. As previously mentioned, a single class of newly characterized giant aquatic viruses, the Megaviridae, appears to include more species than all of the prokaryotic species found in ocean water [180]. These organisms were not recognized as viruses until 2003. The number of newly discovered viruses, bacteria, fungi, single-celled eukaryotic orgnanisms, animals, and plants increases at a rate that surpasses our ability to properly describe and classify them [32].

This being so, how will we know when we have encountered extraterrestrial life forms? The clue is complexity, a condition that seems to accompany all forms of life. Simply put, living things are more complex than non-living things. If this is so, then life can be identified by its contribution to local complexity.

Let us consider the following case in point. There are fewer than 100 naturally occurring elements accounting for several thousand cataloged species of naturally occurring compounds observed on Earth. We may have thousands additional natural compounds that appear in small quantities that have not been cataloged. There may be some thousands more that are unstable and will never be counted. Nonetheless, we can estimate that, at the very most, the number of naturally occurring compounds on Earth-like planets is on the order of tens of thousands. When we try to count the number of compounds synthesized by living organisms, we quickly run into the trillions. The aggregate number of unique genomes (one for each organism), plus species of RNA molecules, plus assorted proteins, plus different glycolipids that are synthesized at every moment, from every species of organism on the planet, is so large as to defy any rational estimate. For all practical purposes, the number of distinctive molecular species produced by living organisms is virtually limitless, while the number of molecular species resulting from simply mixing atoms, under various combinations of temperature, pressure and concentration, is limited to the thousands. Reasons for the vast array of chemicals found in living organisms are many; not the least of which is that natural selection thrives on diversity.

In theory, the intrepid explorer of distant planets would take a sample of soil (or gas, or liquid) and evaluate the number of different chemicals found in the sample. There is a practical limit to the number of naturally occurring compounds in any environment. When that number is far exceeded, we can infer the likely presence of life. An Earthly example of how this might work is found when we look for sterols in rocks. Species of Class Demospongiae (a type of sponge) are the only terrestrial species known to methylate sterols at the 26-position. The finding of sterols methylated at the 26-position, in any sample of Earth, indicates the present or past presence of demospongiae [51, 158].

Let us not assume that alien life forms are carbon-based (as they are on earth). It makes no difference whether the alien is composed of silicon crystals, or complex gases, or electromagnetic waves. Regardless of composition, life forms will impose complexities upon their environment. Such complexities might take the form of localized temperatures variations, electromagnetic impulses, pressure shifts, sounds, or any non-random fluctuations of the environmental measurements. Life always leaves a trace.

On September 3, 1976, the Viking Lander 2 arrived intact upon the planet Mars, where it remained operational for the next 3 years, 7 months and 8 days. Soon after landing, it performed an interesting remote-controlled experiment. Using samples of martian dirt, astrobiologists measured the conversion of radioactively-labeled precursors into more complex carbon-based molecules; the so-called Labeled-Release study. For this study, control samples of dirt were heated to a high temperature (i. e., sterilized), and likewise were provided with radioactively-labeled precursors, but did not yield complex carbon-containing molecules. The tentative conclusion, published soon thereafter, was that living Martian organisms residing within unsterilized samples had built carbon-based molecules through a metabolic pathway [35]. As we might expect, the conclusion was immediately challenged, and it remains controversial to this day.

Long after the initial conclusions were published, the original data from the Labeled-Release study was made available to scientists, for reanalysis. New analytic techniques have been applied to the data, and new interpretations have been published [35]. As additional missions have reached mars, more data has emerged (i. e., the detection of water and methane), also supporting the conclusion that there is life on Mars. None of the data is conclusive, because Martian organisms have not been isolated. The search for extraterrestrial life continues.

2.17 The first and only?

The first piece of luggage on the carousel never belongs to anyone.

– George Roberts

In the summer of 1943, a strange fog visited the customarily sun-drenched municipality of Los Angeles, California. A smothering blanket of dirty-orange industrial-strength fetid pollution upon them assailed; responsive only to the gravity of that moment. Eighty years later, we find it comical that Angelinos, startled in that moment, believed that the smog was a one-time only event. Reality proved them wrong.

Humans have a tendency to believe in the stability of existence; that what we observe today will be what we shall observe tomorrow. The sun shall always rise. When some freakish event occurs, that we have never before experienced, we can expect such anomalies will never be repeated. All evidence points us to the contrary. When we search the night sky, we see clear evidence that stars like the sun live for a time, then are gone, leaving with bangs (supernovaa) and whimpers (dwarf stars).

Unlikely events that occurred in the past will almost certainly recur if enough time is allowed, and if there are no subsequent events that block the recurrence. As an example of the latter condition, consider the aforementioned first smog in Los Angeles. If the first smog had wiped out the city's population, then there would be no Angelinos left to drive the cars and the trucks necessary to generate a second instance of LA smog.

Earlier in this chapter, we claimed that oxygenic photosynthesis, by ancient cyanobacteria, was a first and only achievement. The successful evolution of two independent pathways evolving to produce a coordinated complex pathway devoted to producing a molecule that is toxic to all living organisms (i. e., Oxygen), seems so contrary to nature that it simply could not happen twice. Maybe not. Paulinella chromatophora, a member of the eukaryotic class Rhizaria, seems to have captured and adopted its own permanent chloroplast-like organelle [189]. This conclusion is based, in part, on the dissimilarities between the photosynthesizing organelles of Paulinella chromatophora, and those of all other chloroplast-containing eukaryotes. If the finding holds, we would have evidence that oxygenic photosynthesis evolved more than once, and was captured in a eukaryotic species to produce a second instance wherein a eukaryote achieved oxygenic photosynthesis through a permanent symbiotic relationship with a cyanobacteria.

Let us examine another example where the recurrence of an event is dependent upon the initial event. Immediately following the invention of nuclear weapons, atomic bombs were dropped on human populations (Hiroshima and Nagasaki). The aftermath was so devastating that steps were taken to ensure that nuclear weapons would never again be dropped in fear or hatred or greed. The first event suppressed the occurrence of a second event, for a while. At this point in history, it seems as though our world is poised for another nuclear catastrophe. The "First of many rule" may yet exert its authority.

When we began this chapter, we asked what generalizations we might make about the properties of "firsts." As a very general rule, each "first" resets reality and extends the lineage of descendant "firsts", all endowed with properties inherited through their predecessors. We will revisit the concept of new realities emerging from old realities when we discuss symmetry breaking in Section 6.3, "All fields are derived from one root field".

3 Everything from nothing

As I was sitting in my chair
I saw the bottom wasn't there;
Nor legs, nor back;
But I just sat,
Ignoring little things like that.

– Anonymous limerick

3.1 What is nothing?

We can define creation as the act of producing something from nothing. We may think that the universe came into existence at one particular moment in time, before which there was nothingness, just biding time until a creator steps in to get things moving. Frankly, few of us stop and wonder about the properties of nothingness, although nothingness is the putative initial ingredient of our reality.

Suppose, in an effort to settle the matter, we define nothingness as the state wherein there is absolutely nothing: no spacetime geometry, no matter, no energy, no fields, no constants, no properties whatsoever. Does that clarify the meaning of nothingness? Not really. It only serves to produce a paradox that we will call the *Paradox of Nothingness*. If nothingness has no properties, then it has the property of having no properties, which is itself a property. Therefore, when we insist that nothingness has no properties, we are in fact transforming nothing into something.

Our questions about nothingness may seem absurd to us, but we should admit to ourselves that none of us has first-hand experience with the subject. The closest that we can come to a concept of nothingness, in our observable world, is achieved when we remove all of the matter and energy from a defined space. When we do so, we encounter several startling observations indicating that "nothingness" has a particular set of properties that relate directly to our experience of "somethingness".

1. If we have a volume of space with zero energy and zero matter, then we would know the exact measurement of energy at every possible spacetime position within the volume (i. e., we would have zero of everything, everywhere).[51]
   However,
2. The Heisenberg's uncertainty principle applies everywhere in our universe, even in a total vacuum.[i] The Heisenberg's uncertainty principle tells us that at any exact location in space, there is always the probability of a non-zero energy level. Consequently, every point in a vacuum is subject to energy fluctuations. Some points in empty space would have a tiny bit of positive energy. Other positions would have

i We will discuss the Heisenberg's uncertainty principle later in this chapter. For now, we need only know that there are strict limits to what is known about the physical properties associated with exact locations in space.

https://doi.org/10.1515/9783111498690-003

a tiny bit of negative energy, and the amount of energy at any point in spacetime must fluctuate over time.
Therefore,
3. When we try to create nothing by removing all the energy and all the matter in a volume, we can be certain that energy fluctuations contaminate our nothingness. This is the Casimir effect, which we will be discussing later in this chapter, in the Section 3.7, "Nothing, the key to everything".
Also,
4. The law of conservation of energy tells us that the total energy of the system must be zero. However, if the space of "nothingness" is large (or infinite), we might expect to have, at any given moment, a very large amount of positive energy (summed over all of nothingness space) at some location, matched by a very large amount of negative energy (summed over all of nothingness space) at some other location.[ii]
Furthermore,
5. We know nothing about the size of "nothingness", and there is no reason to assign any upper limit to the size of localized energy fluctuations.
Therefore,
6. Being unlimited in magnitude and just by the vicissitudes of chance, there may be localities of "nothingness" that contain huge amounts of energy and matter, for brief moments.
Finally,
7. In such a scenario, we can imagine that with a sufficient amount of "nothingness", a high energy locality might arise called a singularity point, resulting in a Big Bang.

The preceding line of reasoning draws forth a bold new universe, created entirely from nothing. This weird line of reasoning begs for some rational validation. The remainder of this chapter is devoted to providing a plausible framework supporting the notion of a universe of waves having the properties of matter, emanating from nothing.

3.2 Nothing but waves

> The career of a young theoretical physicist consists of treating the harmonic oscillator in ever-increasing levels of abstraction.
>
> – Sidney Coleman

As a bare-bones minimal universe, let us suppose that our reality emerged from a type of unbounded nothingness that had, as its only property, the ability to propagate harmonic oscillations (i. e., waves). Somehow, we must explain how waves propagating through nothingness, gave rise to matter and light.

---

ii When I contemplate a universe made of nothing and having a total of zero energy (positive energy being exactly balanced by negative energy), Shakespeare's play *Much Ado About Nothing*, written in 1599, comes to mind. The play's title summarizes modern quantum field theory nicely.

Because fundamental particles are made of nothing (i. e., they have no substructure and no size; they have only a set of properties specific to their particle types), and because all matter is composed of fundamental particles made of nothing, then everything in our universe can be described by these properties (i. e., the properties of nothing). Moreover, it would seem that the properties of the aforementioned fundamental particles can be adequately described as immaterial waves (i. e., oscillations in fields), with each field bestowing its properties to its quanta (i. e., each field has a field-specific quantum particle).[iii] All together, the universe consists of nothing (i. e., the nothingness of particles), and the best way we have of expressing moving particles made of nothing is as traveling waves endowed with the properties of their respective fields.

To explain how this is all done (i. e., how we magically turn a particle into a wave), we will work our way up a chain of logic, starting with stationary waves, then moving to traveling waves, then endowing waves with energy, via the Schrödinger wave equation. A derivation of the Schrödinger wave equation is provided as a note in Chapter 11,[52] along with two supplementary notes supporting some of the mathematical steps in the derivation.[53] [54]

3.3 Wave properties

Physics is the universe's operating system.

– Steven R. Garman

Now that we have defined reality as a system consisting of waves (e. g., matter waves, electromagnetic waves, gravity waves) traveling (i. e., oscillating over time) through fields (arrays of points having properties), what can we do with our grand conception? It turns out when waves conform to the rules of linear systems, we can understand a great deal about the behavior of our universe, using a bit of high school level mathematics (i. e., linear algebra and introductory calculus). let us take a few moments to discuss linear systems and how they may apply to our wavy world.

Linear systems are mathematical representations in which the interaction of the elements of the system can be described with a linear operator, such as addition. In physics, the elements of the linear system are typically waves. All linear systems have a few features in common.

1. Additivity.
   The components that add up to an input signal, when sent individually through the system, produce a set of output signals that, when added together, produce the same

iii As with everything in physics, the concept of fields has mathematical nuance. The development of the Standard Model (which predicts the existence of every known fundamental particle) requires an understanding of field properties.

output signal as we would get if we sent the aggregate input signal directly through the system.

For instance, if we know that input x produces an output y, and we know that $x = x_1 + x_2$ and that an input of x_1 produces an output of y_1, and an input of x_2 produces an output of y_2, then we can be certain that $y_2 + y_1$ equals y.

2. Homogeneity.

   If a system takes an input x and has an output, y, then homogeneity implies that $k \times x = k \times y$, where k is any constant value.

3. Time-shift invariance.

   In most physical systems, time-shift invariance also applies to linearity, meaning that a time-shift of the input will produce the output that we would expect to find when the output had the same time-shift. This is an important feature when applied to waves.

4. Superpositioning.

   When a system has additivity and homogeneity, it is linear, and it comes endowed with all of the convenient physical and mathematical features that facilitate system analysis, including the superposition property; multiple waves can exist at the same place and time.

If we think about homogeneity and additivity, we can see that the additivity property always confers the homogeneity property insofar as multiplication of a function by a constant is equivalent to addition of the function to itself by the constant's value (e. g., $4x = 1x + 1x + 1x + 1x$. We can think of adding two signals to one another as being equivalent to superpositioning one signal on the other (e. g., superpositioned light waves passing through one another). Hence, additive systems have the properties of homogeneity and of superposition. Similarly, in additive systems, all signals (including waves) can be represented by additive (superpositioned) components. Finally, most linear systems have the additional property of time-shift invariance. That is to say that shifting the signal in time produces an output with the same time shift.

Much of what we do with signal processing comes courtesy of the linearity of signals; and signals are linear because they have the property of superpositioning, a feature of light and of all other waves. "Digitalizing" involves sampling (and measuring) a wave signal at short intervals. When waves meet, they sum (as you would expect from a linear system that supports superpositioning). If two equal waves are in phase, the sum of the two waves at the point of intersection is double the magnitude of either wave. If they are 90 degrees out of phase, the sum of the two waves is zero at that point, a phenomenon known as interference. But each of the two waves are in no way diminished by interference, emerging just as they were before the point of intersection.

All wave signals, regardless of how complex they may seem, are composed of simple sinusoidal signals. One of the convenient features of linear systems is something called "sinusoidal fidelity". If the input of a system is a sinusoidal wave, then the output will be a sinusoidal wave having the same frequency as the input. The input and the output waves may have different amplitude and phase, but they will both be analyzable as waves of the same frequency. Inputs other than sinusoidal waves (e. g., square waves) lack output fidelity. The superpositioning property of waves allows us to extract the components of a complex signal (composed of many different sinusoidal waves), using either Fourier analysis or signal convolution.

3.4 Converting waves into digital signals

A digital signal is just an analog signal that has been sampled and its frequency measured at short intervals of time. This produces a data file of time/frequency pairs. Digital signal data is commonly represented as a graph, with sample times displayed in chronologic order on the x-axis and the corresponding frequencies on the y-axis. The same features that characterize electromagnetic waves (i. e., linearity and superpositioning) apply to the digital signals that represent those waves.

The importance of digital signals is that they serve as the *Lingua Franca* that can represent every system characterized by input/output signals. For example, if you were to speak into a microphone, the auditory signal could be converted into a digital signal that would be analyzed or recorded or that could be transformed to an electric signal displayed on an oscilloscope or broadcasted as an electromagnetic wave or heard through an electromechanical transducer, such as an audio speaker. Creating the intermediary digital signal means that our input device and input signal need not be of the same type as our output device and output signal. Any output device that can receive and interpret the intermediary digital signal is sufficient.

For the most part, digital signal processing (DSP) is based on one simple concept, widely known as the fundamental concept of DSP:

- When we pass such a signal through a linear system (e. g., a microphone, a light sensor), the output of signal is the same whether we choose to send the original signal, or whether we choose to send through each of its components separately, and combining them later.

  Therefore,
- The primary tools of digital signal processing are the mathematical techniques by which signals are decomposed and recomposed [223].

Waves (e. g., light waves) moving through space and time constitute linear systems. In particular sinusoidal waves have the property of superpositioning (i. e., two waves can occupy the same place and time). We can combine any number of waves together to create a summed wave, and we can decompose the summed wave into component parts. Therefore, the fundamental concept of DSP can be directly applied to quantum physics, wherein everything consists of waves oscillating through fields. The "systems" in physics are the mathematical transformations that preserve invariances and are equivalent to the “systems” analyzed in DSP that preserve linearity. Not surprisingly, the analytic tools of DSP can be directly applied to physics.[iv]

It is all very neat and simple, but do we have any right to expect reality to be a linear system? Not exactly. Much of our understanding of the universe is a purely “linear” conception, but it is easy to envision instances in which linearity does not apply. For example, if the period of a wave changes with the amplitude of the wave, then a wave may dampen, and the strict rules of linear wave equations do not apply. In the macro world, friction and heat and other emergent properties can detract from the predictability of linear systems. A practical example of reality’s stubborn non-linearity is found in the “Mpemba paradox”. Under certain experimental conditions, we observe that hot water will freeze faster than cold water, even though hot water must pass through the same temperature as the cold water as its temperature is lowered [277]. This apparent violation of thermodynamics and of common sense, results from non-linear events that occur during the freezing process.

Completely reversible reactions, such as we envision for the quantum world, are typically linear. This being the case, what do engineers and physicists do when their systems are not perfectly linear? One obvious strategy involves finding non-linear solutions for non-linear systems. Unfortunately, non-linear systems often defy mathematical analysis. Our best mathematicians just can’t cope with the complexities.[v] Another strategy involves pretending that the system is perfectly linear, and compensating for any resulting inaccuracies with well-chosen fudges. A closely related approach involves finding the conditions in which a non-linear system behaves linearly, and developing solutions for those instances (i. e., making the non-linear system resemble a linear system) [223]. We will not delve into the specifics.

In Chapter 7, “Fundamental Constants,” we will suggest that the fundamental constants are persistent resonances of impulses of fundamental forces occurring at the first moments of the Big Bang. Heady stuff indeed! To reach that hypothesis, we used our knowledge of impulse waves. For now, let’s take a look at how impulse waves may apply to both DSP and quantum physics. Let’s consider the significance of an instantanteous clap (i. e., a sound impulse of zero time duration such as a thunderclap or a sound clap). A clap has no assignable frequency because its brevity is shorter than any wave cycle. If there is no wave cycle, there can be no assigned frequency. Yet, we obviously can hear a clap. How can a sound have no frequency?

---

iv Linear functions obey an additive principle: $f(x+y) = f(x)+f(y)$, or, more generally, for any operation, $*$, on a function, $f(x * y) = f(x) * f(y)$. When functions are non-linear, there is no expectation that the combination of functions will produce an output in direct proportion to the inputs. Non-linear functions do not, as a group, yield generalizable rules for their behavior. Each non-linear function must be analyzed in accordance with its particular properties.

v It can be difficult, at times, to distinguish a linear system from a non-linear system. For systems designers, failing to make the distinction can be catastrophic, as non-linear systems tend to become unstable and chaotic, over time. Aleksandr Lyapunov (1857–1918) was a mathematician who developed methods for distinguishing linear from non-linear systems and for predicting the stability of systems, over time [270].

As it happens, the reason that a truly instantaneous clap has no assignable frequency is that a clap contains all frequencies, at once. Similarly, any wave pulse (such as a flash of light), carries all wave frequencies. When we perform a mathematical transformation that yields the frequencies of an impulse wave, we produce a straight line (in the frequency domain), indicating that all frequencies are present, in equal amount (discussed under the topic of the Dirac delta impulse function in Section 7.6, "The origin problem").

We are now prepared to introduce a central dictum of digital signal processing, namely:

- When the time domain is compressed, the frequency domain is expanded [223]. And
- When the time domain is compressed until it becomes an impulse (of zero duration), the frequency domain is expanded until it becomes a constant value, representing an equal distribution of all possible frequencies.

The practical significance of the Dirac delta function (i. e., the impulse wave) and its response (i. e., the impulse response) is the following. When we know the impulse response (i. e., what happens to the impulse wave when it is passed through a linear system), we never need to pass any signal through the system to know that signal's system output. We can accurately predict the output knowing only the input signal and the impulse response. Why is this so? Because the output of an impulse wave contains the output of every possible frequency passing through the system, then the output of the impulse signal moving through the system will contain within it, a subset of outputs that represent the output of *any* particular signal passing through the system. The method by which a signal's output is teased from the output of the impulse wave is through the use of the convolution operator. The convolution operator essentially computes the dot product of the impulse output as it slides over the values of any given input signal, in the frequency domain. The result is the output of the input signal (as though it had passed through the system), a result obtained knowing only the input signal and the impulse response.

The mathematical expression of the convolution operation ($*$) on two signals (or two wave equations) is shown here:

$$(f * g) = \int_{-\infty}^{\infty} f(x - \xi)g(\xi)d\xi.$$

The convolution operation can be expressed in terms of the Fourier transform.[vi] Specifically, the Fourier transform of the convolution operation on two signals is equal

vi The forward and inverse Fourier transforms are:

$$\hat{f}(\omega) = \mathcal{F}(f(x)) = \int_{-\infty}^{\infty} f(x)e^{-i\omega t}dx,$$

$$f(x) = \mathcal{F}^{-1}(\hat{f}(\omega)) = \frac{1}{2\pi}\int_{-\infty}^{\infty} \hat{f}(\omega)\, e^{i\omega x}dx.$$

to the Fourier transform of each of the signals multiplied together, as in $\mathcal{F}(f * g) = \mathcal{F}(f) \times \mathcal{F}(g)$. A proof is provided as a note in Chapter 11.[55]

Because these operations are reversible, signals passing through linear systems can be composed and decomposed at will. This is what digital signal processing is all about. For serious students of signal processing, we have included a proof of the impulse response convolution theorem, as a note in Chapter 11.[56] More to the point of this book, the techniques of signal analysis apply to waves transformed in linear systems. Most importantly, because everything in the universe seems to be characterized by waves interacting in linear systems (particularly, systems in which reactions are reversible and forces are conserved), the mathematical rules of digital signal processing apply to quantum physics.

3.5 Watching waves

Everyone knows that an increase in speed builds over time, and that the measure of this increase in the speed in known as acceleration. Imagine that your car stops at a red light (speed 0 mph). The light turns green, and you press on the gas peddle (also known as the accelerator), and the car moves forward, gaining speed until it reaches its cruising velocity (40 mph). In the intervening moments, your car traverses every intermediate speed. You can choose any speed between 0 and 40, and you can be absolutely certain that your car matched that exact speed, at some moment in time. In other words, your car cannot instantly reach its cruising speed, a feat that would require infinite acceleration.[vii]

Now, turn on your flashlight. A beam of photons races outwards, at the speed of light. Can we expect that photons, like automobiles, will gradually accelerate from 0 mph up to 670,616,629 miles per hour. Can we detect the photons traveling at intermediate speeds? Of course not. The speed of light is a universal constant, and its measured speed is invariant in every frame of reference. Light cannot be decelerated or accelerated. Photons (and all bosonic particles) travel at their customary cruise speed from the moment they are created. But why? Photons exist in the same world as my automobile. Shouldn't photons and cars obey the same laws of motion? Specifically, what is the energy source that accelerates the photon from a speed of zero to the speed of light, in zero time? Maybe the most constructive way of answering this question is in terms of frames of reference. We see the movement of photons in our frame of reference, but the photon exists in the special frame of reference belonging to electromagnetic waves. The photon, within its own frame of reference, is eternally motionless, living in the so-called "Eternal Now." When we turn on our flashlights, we are simply initiating an oscillation of an electromagnetic field that is stationary inside its frame of reference.

Let us move away from light for a moment and consider a physical wave in water. When we drop a rock onto a still pond, a wave is created, expanding outwards. At the

vii The theoretical limit to the maximal acceleration of mass is thought to be $2mc^3/\hbar$ [194].

point that the rock hits the water, the waves move out at a particular speed that is attained instantaneously, without acceleration. The wave, seemingly composed of garden-variety matter (water, in this case), begins at its top speed, without a build-up of acceleration. How is this possible?

What is really happening when the ripples of waves circle out from the fallen rock? Does the water travel outwards? Not really. The water is the medium through which the wave moves. The falling rock imparts energy, causing the water to move up and down, but there really is no lateral movement of the water (only up and down movement). The wave travels at its customary speed, from the moment of its creation, without acceleration, because there really is no matter that is carried laterally by the wave.[viii]

What we interpret as instantaneous acceleration is actually a general property of waves. We use energy to create the up-down movement (the oscillation), but wave properties do the rest. Extrapolating to the question of how our universe began, we can imagine that our physical reality may have begun with an oscillation. Because light and matter and energy all have demonstrable wave properties, our physical universe may have have begun with a single instantaneous impulse, consisting of a superposition of oscillations (i. e., a blending of different waves). let us examine what wave superposition means.

Given that particles are dimensionless points, made from nothing, then we may think that we should be able to collect all of the particles in the universe at one point. Placing more than one object into one location and time is referred to as superpositioning. Superpositioning is a property of waves, and it turns out that we can understand the properties of particles by imagining them as wave packets (basically, grouped oscillations in a field). In fact, in a universe composed entirely of particle waves, with no contentious qualifiers such as the Heisenberg's uncertainty principle, the Pauli Exclusion Principle, or the existence of attractive and repulsive charges, then all the waves of the universe could be superpositioned, one atop the other, to form one super-wave.[ix]

For example, when we look at white light that has been refracted by a prism, we see a continuous spectrum of colors. The wave is not what we think it is (i. e., not white light). Whiteness is simply our way of interpreting what we see, by blending the constituent colors together. The wave that we see is a composite of many sinusoidal waves, all existing at the same time and the same place. The prism reveals to us that white light is a collection of electromagnetic waves, in superposition. If we were to be truthful when asked to identify the color of the light, we would honestly say that it has all colors at once.

---

viii The tsunami that arrives thousands of miles from its earthquake source is not composed of water that traveled from the earthquake. It is the traveling up-down oscillation that has propagated as a wave, lifting the water at the shore.

ix It happens that our universe is constrained by the Pauli Exclusion Principle, and that fermions (the stuff that composes solid matter) cannot be superpositioned. We discuss these issues in Section 6.10, "Identicality and the Pauli Exclusion Principle". For now, we concentrate on understanding the general properties of waves, including those of bosons (i. e., waves traveling at the speed of light).

Sound, like light, travels as waves. What we detect as a musical sound is a summed collection of pure tone sound waves. If we were to look at a sound waveform, we would see a highly complex signal, but we know that the complex signal is just the value of a many simple periodic sinusoids, superpositioned at a single moment of time. Our ears and our brains have evolved to sense the complex sum of the superpositioned component waves. Evolution could have taken a different path. We can imagine organisms that evolved to sample complex sound waves, hearing only one representative sinusoidal wave, out of many. Or, an organism may have evolved that receives a complex wave and reduces it to a frequency distribution, wherein only the waves with the most common frequencies are "heard" by the organism.[x] Evolution notwithstanding, we have mathematical methods by which complex signals composed of superpositioned waves can be decomposed to their individual components (i. e., Fourier analysis and convolution). The field of digital signal processing consists largely of decomposing, selecting, and re-assembling waves.

Likewise, in the quantum wave equation, quantum particles occupy all of the possible wave solutions, at once. When we place a photon detector in the path of a wave, each of the detected photons represents a specific solution to the wave equation. All of the alternate solutions, previously available to the photon, do not apply. In later chapters, we ill discuss the collapse of the wave equation earlier, wherein a detector placed in the path of one of two waves entering two different slits instantly transforms our observation of a photonic wave into an observation of a photonic particle beam. In way of explanation, we can imagine that the act of detecting a photon in a wave restricts our observation to one particular solution of the wave equation, effectively imposing particle behavior on the detected photon.[xi]

In this chapter, we have been exposed to two truths that seem to contradict one another (another paradox):

1. Waves do not collide. They pass through one another, emerging unscathed by the process,
   And,
2. Particles are waves,
   But,
3. We know that particles do collide. Much of high energy physics involves colliding particles and watching the results as showers of new particles stream from the point of collision,
   Therefore,

---

x Such organisms would, in essence, listening to the Fourier transformations of complex sounds, continuously converting time domain signals to frequency domain signals.

xi Much more difficult to understand is how the act of detection of a photon would instantaneously collapse the wave function of photons that have not been sampled (i. e., the photons of waves passing through the alternate slit). Einstein referred to this observable phenomenon as spooky action at a distance. To this day, no satisfactory explanation has surfaced.

4. Either waves do collide or particles are not waves. Both options contradict our empiric observations.

The term "collide", which calls to mind two billiard balls or an interstate automobile pileup, is not up to the task of describing what happens in the Large Hadron Collider. Keep in mind that the "Hadron" in Large Hadron Collider refers to the types of particle being accelerated. Hadrons, which include our familiar protons and neutrons, are composite particles composed of quarks and gluons. As two hadrons come close to one another, they interact as the strong force builds and the Pauli Exclusion Principle rises to the occasion (i. e., two identical quarks of the same quantum state cannot occupy the same time and place). The energy carried by the accelerated hadron is suddenly expended, creating a shower of fundamental particles. In the process, an army of physicists and engineers earn their salaries. In summary, fundamental particles never truly collide; they **interact** energetically when they come extremely close to one another.

When we stop and think about it, new particles cannot spill out of old particles, since fundamental particles have no constituent parts from which a subparticle can detach. New particles arise from the forces that produce excitations in fields.

3.6 An uncertain universe

The Heisenberg's uncertainty principle is one of the most startling discoveries to come from quantum physics. It places a strict limit on what we may know about the quantum states of objects in our universe. It tells us that we can never determine the exact location and momentum of a quantum object.[xii] Multiplied together, our uncertainty of position and momentum must exceed $\frac{h}{4\pi}$:

$$\triangle x \triangle p \geq \frac{h}{4\pi} \quad \text{Heisenberg's uncertainty for position and momentum.}$$

The uncertainty principle is often expressed in terms of position and momentum. Less famous, but equally important, is the uncertainty principle, expressed in terms of energy and time.

$$\triangle E \triangle t \geq \frac{h}{4\pi} \quad \text{Heisenberg's uncertainty for energy and time.}$$

xii A variation of the statement is that if we know the precise location of a quantum object, we cannot know the exact momentum of the object. Conversely, if we know the precise momentum of a quantum object, we cannot know the exact location of the object. As our certainty of the position of a quantum object approaches zero, the uncertainty of the momentum approaches infinity. The converse also applies. Aside from that, there is no physical way of achieving exactitude for either quantity insofar as our measurements are never completely accurate.

Looking at these two equations, it may seem that either relationship could be extracted directly from the other. This is not the case. As it happens, both equations are derived independently from one another, and both equations derive ultimately from a fundamental quality of Fourier conjugates. A derivation of the generalized Heisenberg's uncertainty principle is provided as a note in Chapter 11, along with supporting material related to the canonical commutation relation.[57] [58] Heisenberg's uncertainty principle tells us that our vision of the universe is blurred. All of our measurements are approximations. To the best of anyone's knowledge, the universe will never come into sharp focus; and for that, we must be grateful. let us imagine, for a moment, the kind of havoc that a single electron might inflict upon our universe, if there were no Heisenberg's uncertainty principle. We know that an electron is the carrier of charge, one of the four known forces in our universe. We also know that an electron is a point particle. When we move away from the point source, the field strength diminishes according to an inverse square law.[59] This means that as we double our distance from the point source, we diminish the field strength by one quarter. Conversely, when we cut in half our distance to the point source, we multiply the effect by a factor of four. As we move closer and closer to the electron, the field strength around the electron increases greatly until we are just at the electron, at which point, the field strength approaches infinity, creating a singularity that destroys our universe in a new Big Bang (or not). Luckily, before we reach the electron, we reach a limit established by the Heisenberg's uncertainty principle, which prohibits us from directly encountering the electron. The charge of the electron is smeared out by fluctuations around the electron (which have no precise location) [263].[xiii]

3.7 Nothing, the key to everything

> The vacuum holds the key to a full understanding of nature.
>
> – Paul Davies in his 1985 book,
> *Superforce: The Search for a Grand Unified Theory of Nature* [70]

Heisenberg's uncertainty principle tells us something about the nature of observations (fuzzy, at best), but does it tell us anything about the origin of our reality? Yes, it does. To understand why this is so, let us look again at the uncertainty associated with energy and time.

$$\triangle E \triangle t \geq \frac{h}{4\pi}.$$

xiii By the same line of reasoning, we can infer that no two fundamental particles ever "touch" one another. During high-speed collisions, virtual particles mediate a transfer of forces at very small distances, avoiding any violation of Heisenberg's uncertainty principle.

It would seem that we can never be certain that the energy, at any location, is zero. If we had a certain level of zero, then the product of energy and time would also be zero, but the product of energy and time must be at least equal to $\frac{h}{4\pi}$. Therefore, we can never be certain that any location in our universe, at any time, has a value of zero!

This being the case, we might expect to see random fluctuations in the energy level, in every volume of space. Put another way, energy must arise from the vacuum of space, and the length of time during which this energy exists must not be zero (a zero time interval would pose yet another violation of the uncertainty principle).

Furthermore, because there is a direct relationship between energy and mass, and because fluctuations of energy must occur (via the Casimir effect, *vida infra*), we can infer that particles having mass will arise spontaneously from the vacuum of space and that these particles must exist for a finite length of time.[xiv] Assuming this to be true, we do not need to ask when particles first came into existence in the universe, as their existence is an intrinsic condition of spacetime.

Finally, we can speculate that in a seemingly infinite universe, we might expect to encounter random energy fluctuations that are extremely large, and when a spontaneous energy burst is sufficiently large, a new universe may spontaneously appear, from nothing!

Before we are completely carried away with the cosmic consequences of the humble Heisenberg's uncertainty, let us touch base with terra firma and review the experimental evidence that provides some credibility to our wild hypotheses. As it happens, there are two findings that confirm that energy arises from a vacuum. These are the Casimir's effect and Lamb's shift.

In 1948, Hendric Casimir predicted that a vacuum, devoid of matter and energy and shielded from external energy by two metal plates, will produce energy fluctuations of its own. In 1997, Steve Lamoreaux conducted the experiment and he showed that the vacuum between the shielding plates created a net force (attractive or repulsive) that measured within 5 % of Casimir's theoretical calculations. Thus, something can arise from nothing.

The Lamb shift is a quantum phenomenon of atoms that was named after Willis Lamb who, along with Robert Retherford, conducted an experiment that demonstrated vacuum fluctuations occurring within atoms. According to predictions based on the Dirac equation, the 2s1/2 and the 2p1/2 (the spin 1/2 electron in the s-level of the second orbital shell and the spin 1/2 electron in the p-level of the second orbital shell) energy levels of the hydrogen atom should be equal. They are not equal, and it turns out that vacuum fluctuations (i. e., the Casimir effect) nicely account for the differences in en-

---

xiv Particles resulting from fluctuating vacuum energy levels can only produce particle-antiparticle pairs, to conserve mass and charge. The resulting short-lived particles are known as virtual particles, and they come fully loaded with the properties corresponding to their field (i. e., a virtual quark has the properties endowed by the quark field).

ergy [149]. The Casimir effect and the Lamb shift remind us that even"nothingness" has properties. [xv]

Let us take a moment to review some of the many physical properties of nothingness.

- The principle of locality.
  The principle of locality asserts that every quantum event, that is to say everything that happens, occurs because of local conditions, usually forces, that cause the event to occur. Hence, space must have a physical nature that accommodates these exchanges, which we call "the ether."[xvi]
- Permittivity and permeability.
  As previously described, the speed of light in a vacuum is determined by the permittivity (ε_0) and the permeability (μ_0) of space,

$$c = \frac{1}{\sqrt{\varepsilon_0 \mu_0}}.$$

  Light travels through the nothingness of space. If nothingness had no properties, then there would be no resistance to the propagation of light, and its speed would be without limit. Because empty space has the properties of permittivity and permeability, both of which being fundamental constants, then the speed of light is finite and constant, as measured in all frames of reference.
- The physicality of particles made of nothing.
  Fundamental particles are composed of nothing (i. e., there is no "stuff" of which fundamental particles are composed; only properties). Therefore, the nothingness of particles has observable properties.
- Geometry.
  Geometric properties of empty space exhibit symmetries, and symmetries determine the laws of physics.
- Nothingness prohibits action at a distance.
  Causes must act at the location where effects take place. For example, a rock slides into a second rock, leading to a landslide. The action of the first rock must occur at the location of the second rock. Likewise, if you are startled by a flash of light,

---

xv If nothingness had no properties, then it would have the property of having no properties!

xvi Like all simple theories, the principle of locality has its problems. There is experimental evidence that certain types of events are non-local, happening over distances, with no lapse of time. Two examples, both closely related to one another, are entanglement and collapse of the wave function. If the space between two points can be ignored, then how can we assert that the ether must exist? Like so much in science, the confusion comes from the different ways we choose to interpret a particular phenomena. We can interpret entanglement as a physical feature that occurs locally, at the time when two particles are entangled. We can interpret wave collapse as a phenomenon that occurs in a wave front, locally, at one moment in time. We view both phenomenon over distances, at later times, and we mistakenly confuse local events with non-local phenomena. Everything we observe is subject to interpretation.

the photons from the flashing light must impinge directly onto your retina. Every reaction must arise at the locality its cause. The universe does not seem to permit action at a distance.[xvii] The empty space between an action and its intended reaction effectively blocks the occurrence of any subsequent event. Therefore, the empty space (i. e., nothingness) has a physical role.

- Gravity.
  Gravity is a deformation of space, a phenomenon that would not be possible if empty space lacked the property of deformability.
- The expansion of the universe.
  The notion that the universe is expanding carries the implication that there is something into which expansion occurs. The deformability of empty space (by gravity), and the expansion of the universe (accommodating our receding galaxies) both suggest that nothingness is a type of fabric, with geometric properties.
- Energy fluctuations.
  Casimir's effect (described earlier in this section) proves that energy fluctuations must exist, even in "nothingness." This implies that nothingness has observable physical properties.[xviii]
- Unruh's effect and the relativity of nothingness.
  For quantum physicists, a vacuum represents the lowest possible quantum state within a locality. Because there are energy fluctuations and virtual particles that appear, even in a vacuum (as per the aforementioned Casimir's effect), a vacuum has observable physical properties. Relativity tells us that observability is something that is dependent on the observer's frame of reference. Therefore, a vacuum that is observed in one frame of reference may appear different from the same vacuum observed from another frame of reference. Hence, the vacuum, a form of nothingness, will be measurably different for different observers. This is the Unruh's effect, and it implies that the very notion of nothingness is relative, much like any other physical property.
- The reification of the magnetic potential and the electric potential in the absence of fields.
  Even when we remove fields, we have measurable potentials (electric and magnetic), both of which are basically mathematical abstractions. This would suggest that fields are not the fundamental "something" of the universe, and that "nothingness" can spawn its own form of existence.

---

xvii There may be exceptions, such as seen with spin-entangled particles. But we can try to argue that these apparent exceptions reflect improper interpretation of our observations.

xviii The energy fluctuations that arise from nothingness are observed in the aforementioned Casimir's effect, Lamb's shift, and symmetry breaking, the latter being responsible for quantum tunneling, spontaneous photon emissions (as occur in lasers), and predicted cosmological phenomena such as black hole evaporation.

Because the nothingness of space has many properties, both physical and spatial, the long-abandoned notion of the "ether" can be resurrected, as we can now equate the ether with the fabric of reality.

The demonstration of the Casimir effect does not immediately explain how the Big Bang occurred. Still, if asked whether it is theoretically possible that reality sprang spontaneously from nothing, readers may feel free to answer in the affirmative.[xix]

xix Let us end this chapter on a humorous note. A physician, a civil engineer, and a physicist were engaged in an animated argument. Each claimed to represent the world's oldest profession. The physician referred to the book of Genesis in the Old Testament, remarking that God created Eve from a rib taken from Adam. Clearly, this act required the skills of a surgeon, hence the medical profession must date back to the beginning of everything. Not to be outdone, the engineer said that even earlier in the book of Genesis, it states that God created the order of the heavens and the Earth from out of the chaos. Creating order from chaos is an engineering task. Therefore, civil engineering is the oldest profession. The physicist yawned and said, "All well and true, but who do you think created the chaos? (*Anecdote rephrased from remarks by Grady Booch* [41].)

4 What can begin?

> All the greatest and most important problems of life are fundamentally insolvable. They can never be solved, but only outgrown.
>
> – Carl Jung

4.1 Limits set by initial conditions

We do not live in a world where everything is possible. Many outcomes that seem to be attainable are rendered impossible by strict physical laws, laid down at the dawn of time. This puts a damper on the variety of objects we might hope to find in our universe. When we peer at galaxies, we see a collection of spinning flat disks, varying somewhat in size, offering little in the way of visual variety (Figure 4.1). Our current collection of galaxies has evolved through a type of natural selection, from the same hot soup of energy and particles. Unstable configurations of matter have perished (because that is what it means to be unstable), leaving us with the limited number of galactic forms that have endured.

For cosmologists, distance equals time. An object 14 billion light years from Earth is equivalent to an object that existed soon after the Big Bang. When we observe an object that is 7 billion light from Earth, we are actually looking at the object as it existed 7 billion years ago. When cosmologists marvel at the many different astronomical bodies found in space, they are, in fact, observing a very limited number of available objects appearing at a limited set of moments in history.

When we switch our attention from the macroscopic to the microscopic, we see another limited cast of players. These are the naturally occurring atoms (fewer than 100), and a few dozen atomic or subatomic particles. The Earth is composed largely of several dozen elements, the others being exceedingly rare. The naturally occurring chemical compounds (i. e., chemicals not synthesized by living organisms) are also limited, as only a few thousand have been cataloged. There are certainly additional compounds that have yet to be recorded, but we must resign ourselves to a paltry menu of assembled molecules.

The reason that we have limits to what we can expect to find in our universe relates back to initial conditions. All that we can observe came into existence at precisely the same moment, under the same set of physical laws. Hence, the universe today is more or less uniform. As far as we can tell, at any chosen moment in the past 14 billion years, the universe was roughly equivalent at all points.

The field of physics has several definitions. One definition suggests that physics is the study of how reality operates, as described by relationships that we know as our universal physical laws. A more inclusive definition, that takes us back to the beginning of reality, is that physics is the study of how initial conditions produced the universe that we observe [236]. Initial conditions account for reality, and if we understand initial

https://doi.org/10.1515/9783111498690-004

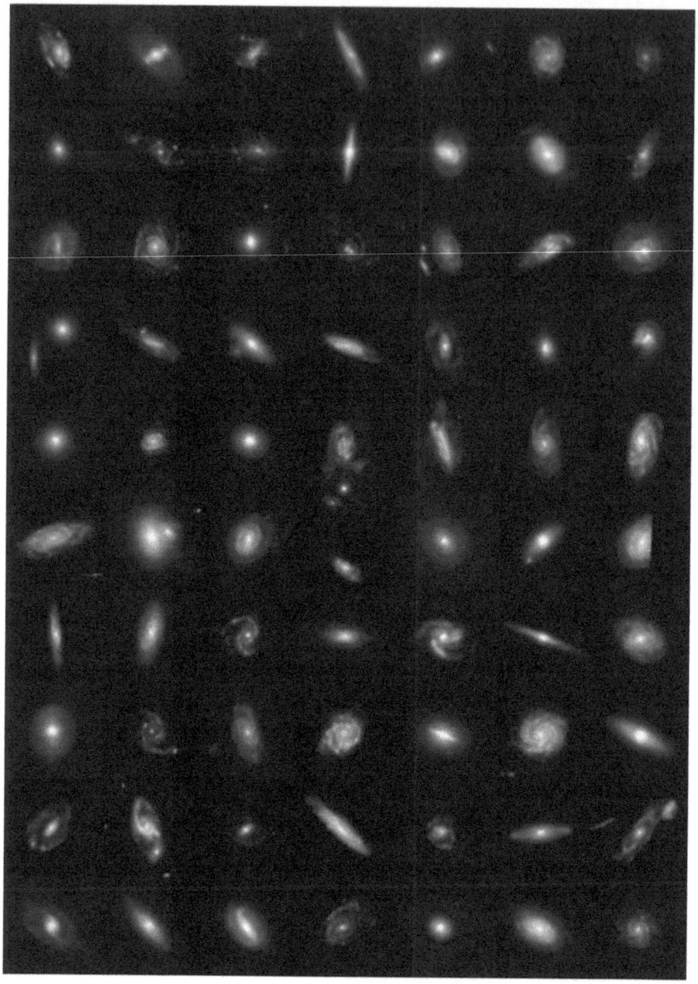

Figure 4.1: Galaxies come in a limited variety of shapes and sizes. Why do they all look so much alike, and why is there not something in space other than galaxies and their strictly limited variety of component parts? Source, public document from the U.S. National Aeronautics and Space Administration.

conditions, then we can understand what is possible and what is impossible (i. e., the limits of reality).

4.2 Limits due to special relativity

Albert Michelson and Edward Morley demonstrated, in 1887, that the speed of light is constant, in every frame of reference, setting the stage for Albert Einstein to introduce special relativity.[60] The theory of special relativity sets limits for what can happen in our universe, as listed here:

1. Electromagnetic waves, and all massless particles, travel at the same speed, this being the speed of light in a vacuum.[i] [61]
2. There are certain invariant quantities (e. g., the speed of light) that are observed to be the same in all frames of reference.
3. Matter, the stuff composed of fermions such as electrons, protons, and neutrons, cannot be accelerated to match the speed of light.[ii]

The laws of special relativity can be derived with high school level mathematics. A derivation of the famous time dilation formula of special relativity is provided as a note in Chapter 11.[62]

4.3 Limits in elements

There are only 94 naturally occurring elements. Heavier elements have been formed in laboratories, but the man-made elements are unstable and pass out of existence in mere fractions of seconds. Is there an upper limit to the number of elements, and can we assign a specific number to highest allowable atomic number?

It happens that as the mass of an element increases, the velocity of the orbiting electrons must also increase. The speed of the innermost orbital (i. e., the ground state) electrons for any element is determined by the element's atomic number.[iii] The equation describing the relation between electron speed and an element's atomic number is:

$$v(\text{relative to c}) = \frac{Z}{137}.$$

Here, v is the velocity of the inner orbital electron relative to the speed of light, Z is the atomic number of the element, and $\frac{1}{137}$ is a constant of the universe known as the inverse of the Fine Structure Constant.[iv] We can see that when the atomic number of the element is 137, the electrons in the inner orbit would travel at the speed of light, a relativistic impossibility. Hence, it would seem that there can be no elements having atomic numbers of 137 or higher. Consequently, the theoretical limit for the highest

i The speed of light can be reduced when traveling through various transparent media (e. g., water).

ii We need not waste time proving that massive particles cannot attain the speed of light The energy required to move a mass varies as the inverse of $\sqrt{1 - v2/c2}$. As $v \rightarrow c$, the energy approaches infinity. Therefore, no finite amount of energy can move a mass at the speed of light or greater.

iii Electrons do not actually orbit the nucleus. Electrons inhabit a probabilistic cloud in a area of space that envelops the nucleus. The term "orbit" is a hangover from Niels Bohr's atomic theory (circa 1913) likening an atom to a solar system with orbiting planets. Bohr's description of the atom remains a simple model that students appreciate. Incidentally, don't be fooled by the sound of his name. Bohr was quite exciting, and his contributions to science will pop up throughout this book.

iv More precisely, $\frac{1}{\alpha} = 137.035999174$.

atomic number attainable, 137, is a fair approximation of the inverse of the Fine Structure Constant, α.Various scientists, using different theoretical models, have arrived at alternate upper limits for the Periodic Table. Still, physicists agree that some maximum must exist, insofar as it becomes increasingly difficult to create new elements, even under the most extreme laboratory conditions.[63]

A ground state electron in gold ($Z = 79$) will travel at 58 % the speed of light. As an electron travels at relativistic speeds, its mass increases, according to the following formula.

$$m_{\text{rel}} = \frac{m}{\sqrt{1 - \frac{v^2}{c^2}}}.$$

The relativistic mass (left side of equation) can be calculated from the mass measured in an invariant frame of reference (also known as the rest mass) and the velocity of the mass relative to the speed of light. Notice that as the velocity approaches the speed of light, the relativistic mass becomes infinitely large.

Because the speed of an atom's orbiting electrons must continually increase as the mass of the nucleus increases, the upper limit to the mass of an atomic species is set by the speed of light. The practical limit for the size of a naturally occurring element is attained by Uranium. Elements with an atomic weight above Uranium are unstable and are not encountered in nature.

Readers who enjoy science fiction will certainly recall a common plot ingredient wherein the metal hull of an alien spaceship is examined by military scientists, and several new elements, never encountered on Earth, are found. Afficionados will tell you that the presence of unearthly elements in an unidentified spacecraft is a reliable sign of alien invasion. So wrong! The Periodic Table of the Elements is complete, in that there is an identified element for every atomic number up to the limit for naturally occurring atoms.[v] There are no vacancies in the periodic table of naturally occurring elements.[vi] When our spectroscopes examine the emission spectra of elements in distant stars, we find evidence of the same elements that we find on Earth, and no others. Were an alien fleet to visit earth, their ships would certainly be constructed from the same elemental atoms that we have on our own planet.

The physical attributes of the common elements are strongly influenced by limits set by the principles of relativity. In our high school physics courses, we are taught Newtonian physics, and we are advised that Newtonian physics accounts for the observable physical phenomena encountered under normal conditions. Special relativity, we are

---

v We are still adding unnaturally occurring elements created in the laboratory, but all of these elements are unstable, and most decay in a fraction of a second.

vi Because every cell in the periodic table is occupied (i. e., there is an atom for every atomic number in the table), we can infer that nucleosynthesis (the process by which stars fuse protons and neutrons into the nucleus of newly synthesized atoms) proceeds incrementally, proton-by-proton.

often told, only applies to quantum objects. We now know that special relativity is actually quite ordinary. We see relativistic phenomena every day, when we care to look for them [6, 205, 200]. It is Newtonian physics that is special, insofar as it applies only to a narrow range of objects moving at slow speed. So-called "special relativity" is not at all special; it is the physics that is observed in all possible frames of reference, and it leads us to the most general ways of understanding the visible world.

4.4 Limits due to quantization

Quantum, from the Latin meaning "amount", usually refers to a small, discrete (non-continuous) unit. Quantum physics refers to the various features of matter and energy that always have discrete values. It seems counter-intuitive that the relationships of quantum physics occur in integer increments. To provide some perspective, a quantum automobile might drive at 0, 10, 20, 30, 40, 50, or 60 miles per hour (e. g., $10 \times n$, where $n = 0, 1, 2, \ldots$), but at no speeds between these discrete multiples. Such cars do *not* exist in reality, so why is much of physics, which is intended to describe reality, quantized?

Let us look at some familiar phenomena that are constrained to occur in quantized portions.

1. Elemental atoms.
   Each neutral atom has an equal integer number of electrons and protons. The charges of electrons and protons are equal and opposite, and each particle has exactly one charge.[vii]
2. Molecules.
   When we say that we have 0.221572 gallons of water, we are really expressing a liquid volume composed of some integer number of identical water molecules (discounting isotopic variants of constituent atoms, characterized by differing numbers of neutrons). There are no fractional water molecules.
3. Standing waves.
   Standing waves are waves caused when a fixed end of the string (or any vibrating material) causes the wave to reflect back upon itself, producing something akin to mirror images of the wave moving in opposite directions. The observed standing wave is a superposition of the forward and backward waves, producing a stationary wave pattern (we say that the wave oscillates but does not travel). The amplitude of the standing wave at its boundaries must equal zero, and this zero-value boundary condition enforces an integer (or quantal) number of nodes appearing at fixed

vii Experimentally, the charge of the electron and the proton are identical to within a factor of 10^{-21} [196].

intervals. Because particles (e. g., electrons) have properties of standing waves, we find that particles have quantized states.[viii]

4. Spatial transformations.
   Geometric transformations such as rotation and reflection, that restore an object to its original location (i. e., the preserve the object's symmetry) are also quantized (i. e., occur as transitions with integer descriptors, such as 3 flips, 5 rotations, 1 reflection).[64]
   Crystals are elaborate cages for atoms. The structure of crystals is dictated by strict mathematical rules, and the rules originate in geometric symmetries of our universe. By empiric observation of crystals, 19th century scientists knew that there was a limited set of interfacial angles found in crystal units.[ix] Every crystal, regardless of its chemical composition, must have interfacial angles that are restricted to a relatively small group of observable configurations. Applying their knowledge of group theory, mathematicians predicted that there are exactly 32 point groups representing all of the crystal classes [47].[x] Because this collection of 32 groups is complete, and because each group is a class that represents symmetry operations on unit shapes, we can say that the 32 mathematical groups comprise a the full classification of crystals.[65 66 67]
   When we create a mathematical classification with 32 point groups, then we can expect to find no more than 32 point groups anywhere in naturally occurring crystal shapes, if our classification is correct. As it happens, back in 1830, a crystalographer named Johann Friedrich Christian Hessel painstakingly collected all the naturally occurring forms of crystals. At the end of his labors, he reported exactly 32 types of morphological symmetries. Had Johann found 33 crystal symmetries, then our mathematical representation of reality would have been in a great deal of trouble insofar as our classification would have provided an incomplete model of nature. Had Johann found only 31 crystal symmetries, then crystalographers would have been saddled with the job of finding, somewhere in nature, a crystal having the thirty-second point group. As it happened, the mathematical classification of crystallographic point groups agreed fully with empirical observations of known crystals. Hence, the complete mathematical classification of crystals, based on point groups, accounts for the complete assortment of naturally occurring crystals. The symmetry group of crystals strictly limits the kinds of crystals observable in reality.
   What about snowflakes? We have all grown up hearing that no two snowflakes are identical, and this would imply that the number of forms are limitless. We now know

---

viii We discussed standing waves in Section 3.2, "Nothing but waves", when we derived the Schrödinger's wave equation.[52]

ix All crystals have a repeating unit that composes the crystal and that can be observed at the macroscopic or microscopic level. The interfacial angle of a crystal is the acute angle measured perpendicular to the line of intersection of two unit faces.

x We will be discussing the relevance of group theory in Section 9.2, "The "Symmetry First" Universe".

Figure 4.2: The major classes of snowflakes. All snowflakes fall into one of a limited number of snowflake classes. Source, Wikipedia, from Plate XIX of "Studies among the Snow Crystals" by Wilson Bentley, published in "Monthly Weather Review", 1902.

that this is simply not true. Because snowflakes obey the same set of rules as any other crystals, we can assign snowflakes to a finite number of symmetry classes. With sufficient time and patience, we can always find two snowflakes whose geometries are identical (Figure 4.2).

5. Quantized atomic emissions.
   In 1860, the German physicist Gustav Robert Kirchhoff demonstrated by black bodies (heated, non-reflective objects) emitted a continuous spectrum of radiation, with a peak intensity determined by the temperature of the black (i. e., non-reflective) body; not dependent upon any of the physical or chemical features of the body.
   Just one year earlier, Kirchoff had shown that individual elements, when heated, emitted discrete wavelength of radiation that could be seen as spectral lines. Each element had its own unique pattern of spectrographic lines, by which the element

could be identified. Kirchoff's observation accelerated the discovery of elements and soon led cosmologists to find that the spectra of at least a dozen of the elements found on Earth were also present in the light emitted by the sun.[xi]

Why do we find that black body radiation is random and continuous, while the emission spectra of individual elements is discrete? Black body radiation is thermal energy, and thermal energy (also known as heat) is nothing more than kinetic energy, produced by the movement of mass, done a molecular scale. Molecules contain charged particles (electrons and protons), and the movements of charges (in the form of charge acceleration and dipole oscillation) generates electromagnetic waves. The movements are, as far as we know, random and occurring over a range of energies, in innumerable atoms, and the resulting spectrum of electromagnetic waves is broad and continuous (Figure 4.3). In the case of the discrete emission spec-

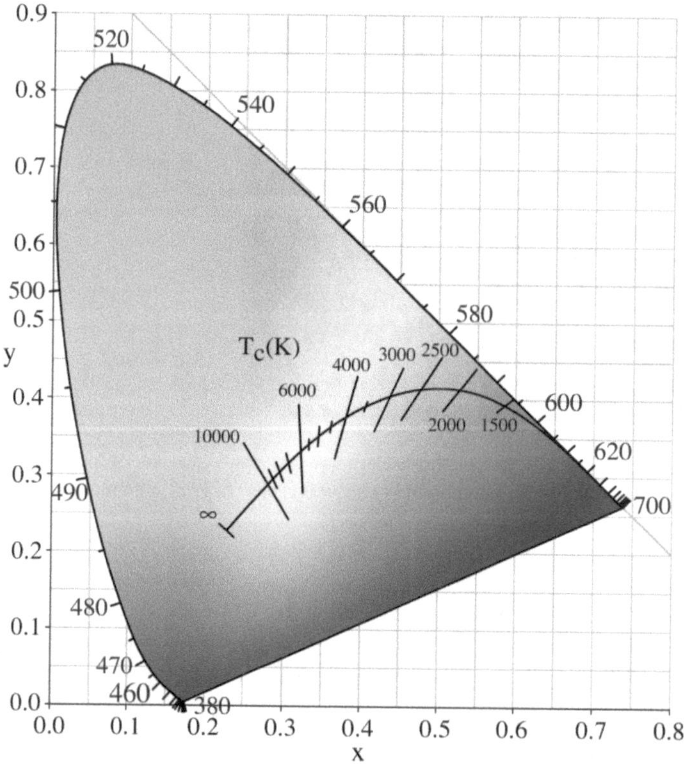

Figure 4.3: The emission spectrum of a radiating black body is continuous, with a peak intensity determined by the temperature of the black body. Source, Wikipedia, from an image entered into the public domain by en:User:PAR.

xi In 1860, the elements were a hot topic for research. More than half of the naturally occurring elements were discovered during the nineteenth century, and the first popular table of the elements was proposed by Dmitri Mendeleev in 1864.

Figure 4.4: The emission spectrum of hydrogen is discontinuous. The spectrum features a few discrete lines, corresponding to the emission frequencies attainable by the single hydrogen electron. It is the emission spectrum of hydrogen that was the subject of Balmer's formula. Source, Wikipedia, from an image entered into the public domain by Adrignola Merikanto.

Figure 4.5: The emission spectrum of iron. Note that iron emits discrete wavelengths, observed as emission lines, and the pattern for iron is different than the pattern for hydrogen. Each element has its own characteristic emission spectrum consisting of discrete lines. Iron emits more lines than hydrogen because it contains more electrons than hydrogen, all emitting at characteristic frequencies. Source, Wikipedia, from an image entered into the public domain by Nilda.

trum of individual elements, the emitted electromagnetic waves are produced when excited orbiting electrons move to a higher shell (more precisely, a higher quantum state) (Figures 4.4 and 4.5). For each orbital transition of an electron, there is a precise emission frequency.

To begin to understand how these precise quantum jumps occur, we must focus our attention on a improbably influential schoolteacher named Johann Jakob Balmer (1825–1898).[xii] Balmer, like many others of his time, was fascinated by the discrete set of emission lines characterizing each element. In about 1880, Vogel and Huggins had published the emission frequencies of Hydrogen (i. e., the hydrogen spectroscopic emission lines) [125].[xiii] In 1885, Johann Balmer, obtaining a copy of those frequencies, developed a formula that precisely expressed frequency in terms of the numeric order of its emission line (i. e., $n = 1, 2, 3, 4$, and so on).

$$\frac{1}{\lambda} = R_{\mathrm{H}}\left(\frac{1}{2^2} - \frac{1}{n^2}\right) \quad \text{for } n = 3, 4, 5, \ldots$$

Balmer's attempt at data modeling produced one of the strangest equations in the history of science. At the time, there was simply no precedent for expressing the frequency of an electromagnetic wave in terms of an integer emission rank. Balmer himself admitted that he was just playing around with numbers, and his formula

xii Johann Jakob Balmer earned his living as a teacher in a school for girls, working in obscurity, without benefit of laboratory or influential colleagues.

xiii See H. W. Vogel's manuscript, "Monatsbericht der Konigl," published in the Academie der Wissenschaften zu Berlin, July 10, 1879.

was introduced to the world without any theoretical foundation. Nonetheless, he had hit upon a formula that precisely described multiple emission lines, in terms of ascending integers.[xiv] Twenty-eight years later, Niels Bohr, in 1913, chanced upon Balmer's formula and used it to explain spectral lines in terms of energy emissions resulting from quantal transitions between discrete electron orbits. Balmer's amateurish venture into data repurposing led, somewhat inadvertently, to the birth of modern quantum physics.[xv]

At this point, we have observed that emissions of energy (in the form of photon emissions) that occur when electrons change orbital position are quantal and can be described in an equation that employs integers to describe the discrete emissions of photons that occur during these transitions.

Why is this so? What are the physical reasons for the quantization of the emissions that occur when electrons change their atomic orbits?

To answer this questions, we must introduce the concept of a standing wave. A standing wave (also called stationary wave) is a wave that oscillates but does not move in the sense there is no succession of waves that pass a point. Examples of stationary waves are children playing with a jump rope, or a musician plucking a guitar string. Both ends of the rope and both ends of the guitar string are fixed, and the amplitude of the wave at either end is zero (because the ends are fixed and cannot raise their amplitudes). The wave oscillates up and down around an integer number of nodes having zero amplitude, with the two ends of the string being the first and the last of the zero amplitude nodes. Standing waves are always quantal insofar as there are an integer number of nodes holding an integer number of identical wavelengths (Figure 4.6).

In our example of the stationary wave, we employ a system bounded by two points (the ends of the string), with the two points at either end of the string having zero

---

xiv Bear in mind that Balmer was not the first scientist to advance a formula that fit the available data, but lacked an underlying theoretical justification. Johann Titius (1729–1796) and Johann Bode (1747–1826) developed a mathematical law (the Titius–Bode law) that predicted the existence and location of all solar planets yet to be discovered, based upon orbital data then available on the known planets. In 1781, Uranus was discovered. Its position fit almost perfectly into the Titius–Bode series, thus vindicating the predictive power of their formula. The law predicted a fifth planet, between Mars and Jupiter. Though no fifth planet was found, astronomers found a very large solar-orbiting asteroid, Ceres, at the location predicted by Titius and Bode. By this time, the Titius–Bode law was generally accepted as a scientific fact. Then came the discoveries of Neptune and Pluto, neither of which remotely conformed to the law. The new data fully discredited prevailing dogma. The Titius–Bode law was purely descriptive; not based on physical principles. Balmer was fortunate that his arithmetical description of reality has been confirmed by experiment and theory. Titius and Bode were not so lucky.

xv Bohr went so far as to derive the Rydberg constant, R (empirically determined for Balmer's formula) from theory alone, showing that $R = 2\pi mq^4 \div h^3$, where m is the mass of an electron, q is the charge of an electron, and h is Planck's constant. Doing so removed the Rydberg constant from the list of fundamental constants of the universe, and placed the Rydberg constant among the long list of derived constants.

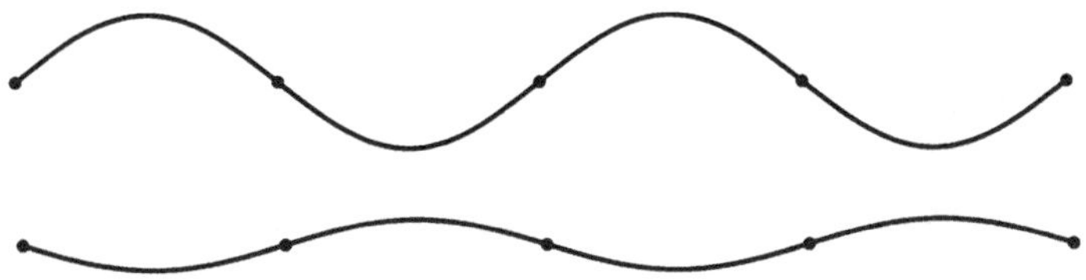

Figure 4.6: A vibrating string stretched between two fixed points will produce a standing wave. The locations of the nodes of zero amplitude stay fixed as the wave oscillates. Above, we see the wave at its peak. Below, we see the wave a moment later, when the amplitudes have fallen. Notice that the nodes of the wave do not move and that the fixed nodes at either end always have zero amplitude. There are five zero-amplitude nodes along the length of the oscillating string. The nodal configuration of a standing wave is a quantal system. The number of nodes is always an integer. Source: author.

amplitudes (i. e., the two ends of the string are fixed and bounded). We could have imagined a box, with waves bouncing inside. In this case, there would be imposed a zero wave amplitude at the borders of the box. Consequently, inside the box there would be an integer number of zero amplitude nodes. The higher the frequency of the waves inside the box, the greater the integer number of nodes. We can never have a non-integer number of nodes because the nodes are evenly spaced (a property of waves), and a node must occur at each of the two fixed boundaries.

The sound waves produced by a piano are standing waves having an integer number of evenly spaced nodes between the ends of the set of taut piano wires that constitute the piano. We get quantized notes because quantization is a feature of standing waves.

Suppose we choose to loop our string (e. g., jump rope or guitare string) into a circle, such that the two ends of the string are now touching one another, and the waves were confined to the circle. The same constraint applies. Waves traveling in a closed loop are standing waves, and the closed loop must have an integer number of nodes. If the situation were otherwise, and there were not an integer number of nodes, then with each rotation around the circle, the waves would move further out of sync, interfering with one another destructively, and destabilizing the vibrations in the string. We can imagine that orbits of electrons are standing waves akin to the plucked wires of a piano (but wrapped into a loop). All the electrons in an orbit are tuned to the same frequency; otherwise, they would interfere with one another, as their repeated orbits put them further out of sync.

When an electron jumps from one orbit to another, it must have a frequency tuned to that of the other orbit. To do so, it must gain or lose energy that is itself quantized (i. e., constrained to a specific amount suitable for the orbital transition). The quantum of energy is exactly equal to the difference in the energies of the respective orbits. The energy that is gained or released is carried by photons. Therefore, each photon absorbed or released in an orbital transition is characterized by electromagnetic radiation of a specific frequency. For this reason, the emission spectra of heated atomic element always has the same set of thin emission bands, where each band represents quantal electromagnetic radiation released by one of many possible orbital transitions occurring in the heated atoms.

Summarizing, the orbital transitions of electrons in atoms are quantal because:

1. Electrons, like all fundamental particles, are waves.
2. The waves of electrons orbiting the nucleus travel are constrained to be standing waves (the so-called boundary condition).
3. Quantization is a feature of standing waves.

Because orbital transitions have quantal energy emissions, there is a tendency to think that energy is quantized. This is not the case. The quantal energy emissions are a consequence of the quantal properties of electrons and photons, which are ultimately determined by their respective fields. Energy itself is not quantized, and although there is a lower limit to the observability of energy and an upper limit to concentration of energy (beyond which the laws of physics do not apply), energy itself is not quantized.

Now that we have addressed the quantization of orbital transitions in atoms, we can nibble at the question of why all fundamental particles are quantal. Why can we measure one electron, but never 1.3 electrons. Why can there be one quark, but never 5.2 quarks. The answer again relates to boundary constraints. All fundamental particles can be described as wave packets, and all wave packets travel under a variety of boundary restraints. Essentially, fundamental particles are quantum objects because:

1. A fundamental particle is a wave packet.
2. And every wave packet is a bounded standing wave.
3. And standing waves are always quantized.

6. Quantized voltage and resistance

Due to the quantization of charge, we would expect to observe that properties that are directly dependent upon the charge will also be quantized. As it happens, the voltage associated with a point charge is described by a simple formula:

$$V = k\frac{q}{r}.$$

Here, V is the voltage, k is a constant, q is the charge, and r is the distance. Because the charge, q, is quantized, we can infer that at any specific distance, the allowable voltages will also be quantized (i. e., restricted to a particular set of numbers).[xvi] Also, we observe that resistance is directly proportional to two constants, the charge

xvi The demonstration of a non-continuous jumps in voltage observed with changes in an applied magnetic field is known as the quantum Hall effect, credited to Edwin Hall in 1879.

of the electron and Planck's constant. Because charge is quantized, resistance too must be quantized when measured with high precision [247].[xvii]

4.5 Minimal limits on observable energy

In the early Bohr model of the atom, charges moved around the nucleus in circular orbits. The orbits were divided into shells, that were numbered 1, 2, 3, . . ., with 1 being the shell closest to the nucleus. Most children are taught the early Bohr model of the atom because it can be easily compared to the orbits of planets around the sun.

Among the many drawbacks of the early Bohr model of the atom is that, according to classical electrodynamic theory, electrons in movement about the nucleus must emit electromagnetic radiation, causing them to lose energy. Orbiting electrons, attracted to the nucleus by charge, must therefore spiral into the nucleus as they lose energy. If the early Bohr model were true, then electrons would be absorbed into protons, and atoms would not exist.

A new model of the atom was needed, and in the first three decades of the twentieth century, physicists feverishly developed the quantum model of the atom, and this model has stood the test of time. Let us review those features of the canon that are most relevant to the discussion in this section:

- The ground state orbital of the atom (corresponding to the first shell in the Bohr atom) is stable and represents the absolute minimum energy level that an electron may possess and still exist.
- One quantum is just another way of indicating the minimum energy requirement for the existence of a field particle. For example, one electron is one quantum of charge. One photon is one quantum of the electromagnetic field. One quark is one quantum of the quark field.
- Matter particles are wave packets, sometimes referred to as matter waves. We explain the equivalence of particles and wave packets as a note in Chapter 11.[68]
- The location of matter particles is probabilistic, and is described by the Schrödinger's wave equation.
- For atomic electrons in the ground state (i. e., the lowest possible energy state), the position of the electron is anywhere and everywhere in a probabilistic sphere that includes the nucleus.
- The ground state orbit of the electron is stable, and electrons in the ground state never drop into the nucleus. We can actually prove this point.
  1. We have shown that the energy released when an electron shifts from a higher orbital to a lower orbital is quantized (see Section 4.4, "Limits due to quantization".).

---

xvii Demonstration of the quantal nature of resistance earned Klaus von Klitzing the Nobel Prize in Physics, in 1985.

   2. If an electron were to drop from the ground state into the nucleus, then it must release a quantized amount of energy in the process.
   3. Doing so would drop its energy below the quantum level required for its existence.
   4. Therefore, the electron would cease to exist.
   5. But an electron cannot cease to exist without violating the of charge conservation.
   6. Therefore, an electron cannot drop into the nucleus.
- The quantum of light is the photon. The energy of a photon is directly related to the frequency, f, of the photon. Specifically, $E = hf$. Here, "h" is Planck's constant (named for Max Planck). There is a minimal frequency below which no energy is observable and for which no photon can exist.

We note once more that although there is a minimal energy for a field quantum, energy itself is not quantized. Energy is continuous. However, we cannot measure zero energy at any point in space, for two separate reasons:
1. Measurement itself requires energy.
2. Due to the Heisenberg Uncertainty Principle, we can never be certain that any particular point in space has a zero energy.

In any case, zero energy is unobservable.

4.6 Limits on particle collisions

Let's revisit the topic of particle collisions (briefly discussed in Section 3.5, "Watching waves").
1. As the distance between two massive particles diminishes, gravitational attraction increases.
   The force of gravity increases according to an inverse square formula. This means that the force of gravity on the particles increases four-fold as the distance between the masses is reduced by half. As the distance between the two point particles decreases to zero, the force of gravity increases to infinity.
   If this were so, then why doesn't the force of gravity increase to infinity every time two massive objects (like a tennis ball and a soccer ball) touch one another? When ordinary objects collide, their centers of gravity never touch. Their exteriors touch well before their centers of gravity come close enough to produce any significant gravitational effect. In the case of point particles having no size (i. e., no diameter), a collision would place the centers of gravity of the two particles at one location, with no intervening space.
2. If the two electrons actually collided, an untenable situation would arise, wherein an infinite force would be exerted at one point in space.

Such a scenario would likely result in a new Big Bang, replacing our universe with a brand new universe. The fact that our universe is more or less intact is strong evidence that fundamental particles do not collide. What does happen when two massive fundamental particles closely approach one another? During "near misses", short-lived low-energy and unobservable particles (so-called virtual particles) enter existence and mediate a transfer of forces over very small distances. We will not be discussing this topic further, though it is of enduring interest to particle physicists.
3. Therefore, fundamental particles having mass cannot collide with one another.

We have just explained why a massive fundamental particle, such as an electron, cannot collide with another massive fundamental particle. What about massless fundamental particles, such as photons? Can they collide with one another? When we come to the discussion of the Pauli Exclusion Principle, we will find that the massless fundamental particles have the property of superpositioning. Two massless particles can exist at the same location at the same time, without colliding. They simply pass through one another in the manner of waves.

What about non-fundamental massive particles; can they collide with one another? Yes, protons and neutrons are composite particles composed of quarks and gluons. As such, they have dimensionality (size) and mass and can approach one another without tearing apart the fabric of spacetime. Collisions between non-fundamental particles are commonplace.

4.7 Limits at the root of life

In the biological sciences we see numerous examples in which early events, at the root of life, severely limit late outcomes. The living world of today is bounded by the properties that emerged from events occurring at the beginning of life, 3 to 4 billion years ago. Let us look at some of the universal properties of Earthly organisms, past and present.[69]

Earthly life began in a water environment, and the internal milieu of all cells is water. Furthermore, the early seas were salinated, and all cells today contain salt (sodium chloride) as a water solute. There is no physical law that precludes the existence of organisms that contain no water or salt, but no such organisms have been found. It would seem that the earliest evolved forms of cellular life set the standard for all subsequent progeny. To appreciate why life on earth developed in the manner that it has, we must understand something about chemical chirality.

A molecule has chirality if it cannot be exactly positioned upon its mirror image. The chiral forms are "D" (dexter or right) and L (levo or left) isomers, with the "D" and "L" referring to the respective directions in which light is reflected off the molecule (Figure 4.7). In a non-enzymatic molecular synthesis, such as might occur in a test tube, "D" and "L" forms are produced in near-equal quantities. To the surprise of chemists,

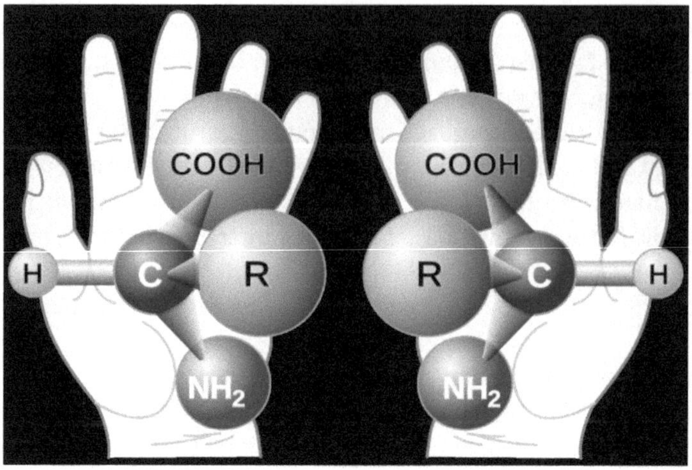

Figure 4.7: Amino acid chirality. Source, U.S. National Aeronautics and Space Administration, from a retouched public domain image.

the same molecules, when found in living organisms, occur almost exclusively with a particular chirality; and the same chirality extends throughout all the classes of living organisms. For example:
- D-isomers of saccharides prevail over L-isomers in living organisms;
- L-isomers of phopholipids prevail over D-isomers in living organisms;
- All amino acids of proteins have the S (left) configuration [79];
- All ribose units in nucleic acids have the *R* (right) configuration [79].

Let us stop to consider why an organism might evolve to require one isomeric form of any metabolic molecule. In the case of enzymatic metabolism of "D" and "L" forms of substrate molecules, chirality is an issue. The catalytic units of enzymes operate by fitting the substrate molecule (e. g., a saccharide) into a matching site. A D-isomer of a molecule will not fit into a site that matches the L-isomer of the same molecule. Hence, only one form of the isomer can be metabolized by a particular enzyme.

We see that enzyme monochirality (i. e., uniform isomeric handedness) is required for metabolic functions. Enzymes synthesized with amino acids having both right and left chirality could not provide a geometrically suitable reactive site that could accommodate the available (monochiral) substrate. Likewise, a nucleic acid built from a mixture of ribose molecules having random left or right nucleotide chirality could not achieve the specific helical structure observed in our genomes.

The prevalance of a uniform handedness extending to virtually every living organism would suggest that chemical substrates of particular chirality were selected at remote moments in time (i. e., near the dawn of terrestrial life). Furthermore, the genes that enforced the early chiral preferences in enzymes and substrates were promulgated throughout all of the living organisms that subsequently evolved [79]. Chiral-specific sites in enzymes and synthetic machinery also tells us that gene evolution is highly constrained to code for just those particular forms of proteins that recognize a specific molecular chirality in substrate molecules [101]. The observation that all living organisms have the same chirality would suggest that all life on the planet arose from a common ancient ancestor.

We can draw one more valuable inference from the chiral specificity of molecules in all known living organisms. The absence of any organisms that defy the general features of chirality would indicate that the classification of life is complete. Otherwise, we would have encountered exceptions to the classification that would necessitate a reconstruction of the class hierarchy and the addition of new branches to accommodate the organisms with alternate chiralities.

When we think about the many possibilities for cellular and non-cellular life forms (possibly existing on other planets), we can begin to see that we Earthlings greatly limited our options when our first organisms emerged from a chemical soup. On a grander scale, we might imagine that a single early event, possibly occurring by sheer accident, at the dawn of the universe, may account for much of what we perceive as our reality.

4.8 Limits set by evolution

Evolution, like all biological processes, follows a set of natural laws that cannot be violated. Consequently, there are limits placed upon what can happen in the natural world. Let us look at a few examples.

Evolution proceeds through the accumulation of many small genetic alterations that contribute to a new phenotype (set of physical features) that is heritable. There is no single gene mutation that accounts for the set of features that characterize species. Because evolution is a polygenic process, we do not see "throwback" mutants among the individuals of a species. To reverse the evolutionary process, we would require an orchestrated loss of multiple genes, with replacement by the forms of the genes, and their regulators, that were present in the ancestral species. For this reason, the general form of the parent class persists in all of its descendant subclasses. For example, all eutherians (placental mammals) have the same skeletal structure, allowing for modifications in size, shape, and function. Mice, horses, and humans all have about the same skeletal structure. The whale has vestigial arms and legs, and its nose has migrated backwards to become a blowhole, but the whale retains the basic structure of any proud eutherian. On a microscopic level, there is little difference between liver cells (or any other cell types) of mice, horses, or any other eutherian animal.

In many cases, evolution's stubborn irreversibility has worked to the detriment of extant species. Here are just of few examples:

– The right and left recurrent nerves are branches of the vagus nerve that supply most of the laryngeal musculature. Without these nerves, we could not talk. Most small nerves branch off their major nerve trunk and proceed directly for their target tissue. The recurrent nerves, due to the circumstances of their embryologic development, must swing below the major branches of the aortic vasculature. The right recurrent laryngeal nerve dives down the neck and loops below the right subclavian artery. The left recurrent laryngeal nerve dives down the neck, into the thorax, and under the aortic arch, near the heart. This nerve, due to its long and treacherous path, is vulnerable to cuts, torsions, compressions, and all manner of mayhem. Inadvertant bisection of the right recurrent laryngeal nerve is a leading cause of legal action taken against surgeons who specialize in neck and chest surgery. The same hazardous anatomic route, but many feet longer, applies to the giraffe, whose recurrent nerve must travel the full length of the neck, and down into the chest. The left recurrent laryngeal nerve is one, among many, reminders that animals are the product of evolution, not design.[xviii]

- Whales have vestigial, and useless arms and legs, a consequence of their membership in, and descent from, Class Tetrapoda. Likewise, the vestigial femur and pelvis serves no useful purpose for the whale. They do, however, serve to inform us that whales evolved as a species of Class Mammalia, that lived on land.
- The enzyme nitrogenase actually preferentially binds with acetylene moreso than with dinitrogen, its intended substrate, despite its being the key enzyme used in nitrogen fixation in many bacteria and archaea.
- Birds evolved light, hollow bones, for flight. These bones are of little value for flightless birds, such as penguins and ostriches, but a bird is a bird, whether it flies or runs, or waddles. Contrariwise, bats are flying mammals, descendants of Class Laurasiatheria, and are much more closely related to whales and giraffes than to birds. Consequently, flying bats have heavy bones, when they would certainly prefer a skeleton of lighter weight.
- All birds have wings, even the flightless ones. Given a design option, would not ostriches prefer arms to wings?
- Humans, unlike our prognathous ancestors, have rather flat faces. As a consequence, our evolved teeth are crowded, and we have poor sinus drainage. If this were not bad enough, we inherited wisdom teeth, which we do not need and have no room for. To counteract our ancestral history, evolved humans designed dentists, who deal with our evolutionary shortcomings, on a daily basis.
- The colon's appendix is an evolutionary relic from ancestors that fermented their intestinal contents in alimentary holding sacs, serves no important purpose in humans. The appendix contains some lymphoid tissue, but so does the every other part of the alimentary tract. In 2013, appendicitis accounted for 72,000 deaths worldwide.

xviii See "Useless Body Parts: What do we need sinuses for, anyway?" by Jocelyn Selim and Max Aguilera-Hellweg, in the June 26, 2004 issue of Discovery Magazine.

Despite claims that the appendix is a useful part of the human anatomy, it is notable that people who have had their appendix removed do not seem to miss it.

- RuBisCo (Ribulose 1,5 bisphosphate carboxylase/oxygenase) is the most abundant protein on Earth. RubisCO catalyzes the carboxylation and cleavage of ribulose 1,5 bisphosphate into two molecules of 3 phosphoglycerate. RubisCO is ancient, predating the first synthesis of oxygen by cyanobacteria, and found in every major class of life, including archaea, bacteria, and eukaryotes. Nearly half of the soluble proteins in plant leaves are RuBisCo. Why is RubisCo present in such great quantities?
  There are many enzymes involved in oxygenic photosynthesis. The reason for RuBisco's superabundance has less to do with its function in carbon fixation and oxygenic photsynthesis than with its remarkable inefficiency. RuBisCo is strongly inhibited by oxygen (thus limiting its activity); it saturates poorly; and it catalyzes slowly. Its inhibition by oxygen may relate to the fact that it evolved prior to the appearance of any appreciable oxygen, at a time when every organism was an anaerobe. Despite its many shortcomings, RuBisCo had, at a very early epoch, secured its role in terrestrial organisms [83]. Plants apparently had no choice but to synthesize RuBisCo in great abundance. In addition, ancillary enzymes evolved to enhance RuBisCo's catalytic output, forming a complex known as the RuBisCOsome.
  Will Earth's organisms evolve an efficient enzyme that can replace RuBisCo? If more than a billion years of history is any indication, the outlook for the future looks dim, but not absolutely dismal.

We have seen examples wherein alternate enzymes have evolved, to substitute for ancient enzymes, when absolutely necessary. The proof for this assertion is somewhat convoluted, but well worth the effort to discuss here. It seems that the minimum number of genes required to support the least complex forms of bacterial life is about 250. This is approximately the number of genes in Bacillus subtilis, Mycoplasma genitalium, Mycoplasma pneumoniae, and Haemophilus influenzae. About 80 of the genes (from the 250 gene set) seem to be present, as homologs, in all nonviral life on Earth. To see just how minimal the minimal gene set may go, researchers went about knocking out individual genes to see which specific genes were absolutely necessary for viability. Much to their surprise, they could knock out individual genes that belonged to the universal gene set (i. e., were present as homologs in every nonviral organism) without destroying viability. This means that there are conserved genes that are present in every organism but not absolutely necessary for cell survival. How can this be? The conclusion drawn is that alternative genes must have evolved that can substitute for the universal genes, in a pinch (e. g., when the universal gene is experimentally knocked out) [140]. This conclusion can be challenged, but it serves to remind us that evolution produces diversity and it is diversity that provides biological options that may come in handy in unexpected situations. When we examine the imperfections in our bodies, on either the cellular or the anatomic level, we can imagine that flaws in an existing species may have conferred a survival advantage for an ancestral species. For example, a useless appendix may have served as an accessory gut in an ancestral animal with a particular set of digestive requirements. The vestigial limbs of the whale may have helped its land-based ancestor in its search for food. The ridiculously long recurrent nerve in the giraffe may have been totally suitable for its short-necked ancestors. The presence of inherited flaws in animal species is evidence that animals were not designed in a single creative moment, but evolved over time from ancestral species.

About 550 million years ago, planet Earth enjoyed a temperate climate. The supercontinent Rodinia had just broken into fragments, but most animals lived in water. The seas were a soup of living organisms, and the organisms were small, many were microscopic. The waters of the relatively shallow oceans had, for the prior 1.5 billion years or so, been absorbing increasing amounts of oxygen produced largely by cyanobacteria. Beginning about 300 million years before the Cambrian explosion, and after the land and seas were saturated, the oxygen levels in the atmosphere had been steadily increasing.

The earliest metazoans (organisms developing from an embryo with a blastocyst) had evolved much earlier (about 760 million years ago), as had the eumatazoans (animals developing from two-layered embryos), and the first bilaterians (animals developing from a three-layered embryo) had appeared very recently (about 555 million years ago).

The availability of oxygen made possible the evolution of muscular animals (i. e., animals with cells having a high respiratory demand), and muscles permitted the evolution of predation (e. g., pursuing, grasping, tearing, and eating other animals) [53, 226, 158]. In turn, predation rather than passive consumption of the decomposing debris of single celled eukaryotes, provided an enhanced supply of metabolic energy that hastened the natural selection of bigger and more muscular carnivores.

The theory that increasing oxygen levels fueled the Cambrian explosion is strengthened by the recent observation that vampire squids, a cephalopod, lost their appetite for killing, when they adapted to a low-oxygen existence in pelagic waters. The vampire squid, despite its gruesome name, is a detrivore, living on cellular debris dropping to the seabed, in the fashion of organisms that lives in the low oxygen environment of pre-Cambrian oceans [226]. We presume that carnivorous diet, and a consequent competitive predator/prey lifestyle, explain the fast tempo of evolution achieved in the Cambrian period.[70]

The Cambrian marked the birth of our current predatory world, wherein animals typically armed with five senses (sight, hearing, smell, taste, and touch) use their sensitive nervous systems and their high-energy muscles to pursue, grasp, kill, and devour weaker organisms [226]. Not surprisingly, post-Cambrian metazoan evolution is highlighted by the acquisition of offensive structures (e. g., teeth), defensive structures (e. g., elaborate body armor) and mixed function structures (e. g., brains).

In summary, the predator/prey lifestyle provided impetus for the rapid evolution of new, mobile body plans. Another question remains. How was it possible that all of the animal body plans in existence today could have evolved in a relatively short evolutionary burst (no more than 50 million years), occurring half a billion years ago?[xix] We observe that the Cambrian period followed the evolution of genes that regulated the development of tissues and diverse cell types. In particular, the HOX family of regulatory genes directed the development of body plans. HOX genes are conserved in all animals, a strong indicator of their evolutionary importance.[71] Because HOX genes control the implementation of body plans, it would seem that variations in HOX genes may have accounted for the profusion of body plans occurring in the Cambrian period [52, 244].[72] [xx] [73]

We can begin to imagine that there are good reasons why a profusion of body plans evolved in the Cambrian period. One more question remains. Why have no additional animal body plans evolved in the nearly half-billion years that followed? A facile explanation points us to the oppressive effects of competition [85]. In the beginning of the Cambrian period, there was a free market in body plans. The field of competitors was sparse at best. After the Cambrian period, the many freshly minted organisms ensured that no new competitors would establish new ancestral lineages. This was accomplished by eating the new species or by consuming their supply of food and other necessary raw materials. This hypothesis is supported by the frequent observation that following a mass extinction of dominant predators (e. g., the dinosaur extinction occurring 66 million years ago), new subclasses of previously non-dominant animals will emerge. When the dinosaurs died, hundreds of subclasses of small mammals, including many new carnivores, grew in number, size, and ecologic influence, over the next 16 million years (when evolution settled down to the new normal).

4.9 If they could have they would have

Anything that is not prohibited from happening may happen. The results of the "happening" will cease to persist if the result is unstable. If we incubate all the natural elements in a box, we might expect a very large number of compounds to form, but most of these compounds will vanish, almost immediately, because they are not stable. A stable molecule is simply a molecule that requires more energy to break its bonds than is currently available. If we pump a large amount of energy into a soup consisting of

xix Scholars quibble, suggesting that some new animal phyla may have appeared prior to the Cambrian period, and some may have occurred soon thereafter the Cambrian. Regardless, there is general agreement that the Cambrian period marked the emergence of most of the body plans for animal species extinct and extant.

xx A malformation is a growth disorder occurring during the development of an organism. Technically, the term "malformation" includes growth abnormalities that arise at any time of life (e. g., angiomas appearing in adults). In common parlance, "malformation" is reserved for congenital conditions (i. e., present at the time of birth).

compounds that are normally stable (e. g., by raising the temperature of the soup or by bombarding the soup with high-energy cosmic rays), we will find that every compound will break into its constituent atoms.

The inevitability of chemical reactions is something that we are not taught in introductory chemistry courses. Traditionally, we are informed that there are two allowable types of chemical reactions: those that draw energy from the environment (i. e., endothermic reactions) and those that yield energy to the environment (exothermic reactions). We are also taught that reactions require an input of energy before they can begin (i. e., the activation energy). Such reactions can be represented by equations that balance substrate molecules with reactant molecules, and pre-reaction energies with post-reaction energies, taking into account the stored energy contained in molecular bonds.[xxi] Teaching chemistry in terms of energy balances is much like teaching economics in terms of account balances. Yes, balancing the books is important, but when currency is not available at any given moment, cash can be borrowed today against future gains. The future gains need to be predicted based on a deep understanding of how the flux of business changes from moment to moment. Understanding balance sheets is not the best way to learn how business is conducted. Likewise, chemical reactions are not driven by an energy balance sheet. We can simply stipulate that in the end, energies balance out, in accordance with the universal law of energy conservation.

High school chemistry courses often omit the most obvious, and the most enlightening, aspects of chemistry. We can simplify chemistry by focusing on a few fundamentals:

1. If an inter-atomic reaction may possibly happen, without violating a universal law, it will happen.
2. After an inter-atomic bond occurs, the newly formed bonds will break if (and only if) the energy required to break the bond is available. A stable molecule is a molecule that needs more energy to break its bonds than is available. As the environmental energy increases (e. g., in the form of heat or ionizing radiation), more and more bonds will break.
3. The energy for reactions come from heat, and heat is created by the vibrations of molecules. Because the universe is heated everywhere (i. e., there is no location in the universe whose temperature is absolute zero), molecules can derive energy from other molecules. The higher the temperature of the environment, the more vibrational energy is available.
4. Chemical reactions occur exclusively between atoms and molecules that come into contact with one another (obviously).

---

xxi The concept of "stored" energy, much like the concept of "potential energy" is not wrong, but it can mislead. Energy is that which causes movement, and if movement is not being produced, then energy is not at play. The "stored" energy of a formed molecule represents the difference between the energy that would be released when the molecular bond is broken and the energy required to break the molecular bond. Energy is not "stored" within the molecule.

5. Fractions of atoms do not occur (e. g., there is no such thing as half of a lead atom). Therefore, chemical reactions occur between integer quantities of atoms and molecules (i. e., chemical reactions are stoichiometric).
6. The most influential determinant of an molecule's stability is the outermost shell of the participating atoms (the so-called valence shell). This is somewhat surprising considering that nearly all of the mass of an atom is in the nucleus, and the net positive charge of the nucleus exceeds the net negative charge of the valence shell of the atom.[xxii]

The last point, that the outermost or valence shell of an atom is the most important determinant of chemical reactions, is something that we are all taught. It is worth a moment to accept the irony of the situation. Virtually all of the mass and energy of an atom is contained in the nucleus.[xxiii] The total electronic charge of the nucleus of an atom is equal to the sum of all of the (negative) electronic charges of all of the electrons of an atom, and is therefore greater than the total electronic charge of the outermost shell of an atom. This being the case, it is surprising that the outer shell of an atom determines the chemical reactivity of atoms. We find here a classic example of the tail wagging the dog.[74]

As in many situations, what we observe on Earth as a commonplace occurrence is actually a rarity when we broaden our perspective. We look around us and we see all manner of molecules, many of which are complex compounds composed of several bound atomic species. If we can tear ourselves away from our geocentric view of reality, we will observe that the majority of matter in the universe is accounted for by individual atoms (not complex molecules). In point of fact, about 75 % of observable matter in space is free-floating hydrogen atoms, and much of the remainder is helium. It is estimated that, on average, a quart of outer space contains on the order of 100 atoms [201], indicating that space is not a vacuum. Back on planet earth, we see all manner of compound molecules, but earth is the exception.[xxiv] On a cosmologic scale, atomic reactions involve the strong forces of the nucleus (that bind the nuclear subatomic constituents) or the weak forces (that account for nuclear decay processes). These forces account for fusion in stars, and for the creation of heavy elements by neutron capture (in merging

xxii Except in the case of neutral hydrogen atoms, in which the net valence charge equals the proton's charge.

xxiii A neutron has the same mass as a proton, this being more than 1,800 times the mass of an electron. The binding energy of a proton or neutron (the so-called strong energy that binds quarks) is more than seven orders of magnitude larger than the electromagnetic energy of electrons.

xxiv Molecules are chemicals that are composed of more than one atom. Compound molecules are chemicals composed of two or more different atoms. For example, O_2 is a molecule, but not a compound molecule. H_2O is a compound molecule.

neutron stars).[xxv] The chemical reactions conducted in high school are all low-energy processes that yield makeshift and impermanent associations between atoms, occurring in their outer shells.[xxvi] For better or worse, we must accept that we live in a low-energy environment featuring low-energy reactions determined by low-energy valence shells.

It is useful to have a grip on what is possible and what is impossible in the chemical world. Compounds are torn apart when they are bombarded by energy that exceeds the energy of their bonds. X-rays aimed at tissue can disrupt the bonds of organic chemical, including DNA. Biologists refer to X-rays as ionizing radiation, as they shred the valence shells of molecules to yield ionized molecules (molecules whose valence shells are have unbalanced charge). At very high energies (such as we see in cosmic rays), all manner of interactions occur at the level of subatomic particles. High energy colliders allow particle physicists to learn what happens to subatomic particles encountering cosmic rays. Here is a short list of some of the energies associated with molecular bonds, electromagnetic rays, and fast-moving protons (measured in electron volts).

- Rotational energies of molecules 10^{-5} eV.
- Vibrational energies of molecules 10^{-1} eV.
- Energy between outer electron shells in atoms 1 eV.
- Binding energy of a weakly bound molecule 1 eV.
- Energy of red light 2 eV.
- Energy of ultraviolet light 3 to 30 eV.
- Binding energy of a tightly bound molecule 10 eV.
- Energy to ionize atom or molecule 10 to 1000 eV.
- Energy of soft x-rays (absorbed in air) 100 to 1000 eV.
- Energy of hard x-rays (for medical radiology) 5000 to 10000 eV.
- Energy of cosmic rays (mostly fast protons) 10^7 up to 10^{20} eV.

We can now appreciate that the chemical reactions that occur in living organisms are nothing like the chemical reactions occurring in stars and black holes. Terrestrial animals operate on a very small energy margin. Minor alterations in our low-energy environment can have profound consequences. A rise of just a few degrees in our body temperature is a potentially life-threatening medical condition? At 104° Fahrenheit, our proteins begin to denature. Unrelieved hyperthermia leads to multi-organ dysfunction and death, as our enzymes operate optimally within a very narrow temperature range (i. e., near 98.6° Fahrenheit). Above 104°, cells cannot function properly. Warm-blooded

---

xxv The bonding energy required to hold quarks together in protons and neutrons is about 10^7 times greater than the energy bonding electrons to an atom.

xxvi We call these associations "bonds", but there is no physical structure that corresponds to a chemical bond. The eminent chemist Charles Coulson, wrote, "Sometimes it appears to me that a bond between two atoms has become so real, so tangible, so familiar that I can almost see it. But then I awake with a little shock: for the chemical bond is not a real thing, it does not exist, and no one has ever seen or will ever see it. It is a figment of our imagination [67]".

animals (homeotherms) employ complex metabolic pathways to maintain a nearly-constant body temperature despite seasonal and diurnal variations in the temperature of our atmosphere.

Back in Section 2.16, "First contact with others", we discussed the enormous diversity of molecular species synthesized by living organisms (many trillions), compared with the paltry number of molecular species that occur from inorganically mixing atoms. Without the benefit of enormous pressures (as found deep in the Earth's interior), or heat (as produced by vulcanism), or varying concentrations of substrate (as found in geologic strata), or time (as occurred during the evolution of our solar system), the living species in Earth's razor-thin biosphere have yielded billions of unique proteins, and trillions of individual organisms, each with its own unique chemical genome. It all happened with very little energy, in a narrow range of temperature.

We date the first atoms to about 380,000 years after the Big Bang, the time when the early universe had cooled to the point that single protons could capture single electrons to produce hydrogen atoms. At about the same time that atoms of hydrogen emerged from the primordial plasma, so did helium atoms. The most abundant form of helium is neutral helium-4, a nucleus composed of 2 protons and 2 neutrons, balanced by two electrons.[xxvii]

Earlier in this section, we asserted anything that is not prohibited from happening (referring to chemical reactions) could happen. Based on this assertion, we would surmise that from the moment that the first atoms appeared, 380,000 years after the Big Bang, we would expect to find all combinations of molecules formed from the only two available atoms — hydrogen and helium. If so, we might expect that Hydrogen atoms would combine with neutral helium atoms to form HeH+. However, we are taught in our chemistry classes that helium atoms are non-reactive insofar as they belong to the inert noble gasses, whose outer valence shell is filled. Therefore, we would expect that hydrogen and helium would never combine with one another to form a compound molecule. Which theory is correct?

In recent years, astronomers observed HeH+ molecules in space (compound molecules formed when hydrogen combines with helium). HeH+ never forms on Earth, because helium is inert and does not react with hydrogen. However, in the void of space, where hydrogen and helium account for nearly all of the atoms that are available for chemical interaction, we find traces of HeH+. Why does HeH+ form in outer space? Because hydrogen and helium are everywhere in space, to the near-exclusion of all other elements, and there is nothing that prohibits their occasional bonding. This finding provides some justification to our assertion that all possible atomic interactions occur when atoms meet. The amount and the stability of the product is a separate issue.

---

xxvii Helium 4, having two protons and 2 neutrons, is an unusual atom because its net spin is 0, indicating that it violates the conditions of a fermionic substance and can be considered a type of boson. We discuss spin and bosons in Section 6.10, "Identicality and the Pauli's exclusion principle".

Figure 4.8: Cyanooctatetrayne, one of many complex molecules detectable in distant space, and thus created early in the history of our observable universe. Source, Wikipedia, from a graphic entered into the public domain by its authors, Ben Mills and Jynto.

When were the first molecules formed? As mentioned earlier, on a cosmologic scale, distance equals time. If light is captured that has traveled 13 billion light years before reaching Earth, then that light must have been emitted soon after the Big Bang, which is estimated to have occurred sometime less than 14 billion years ago. Astronomers have detected light coming from stars more than 13 billion light years from Earth, and spectroscopic analysis of this light reveals the existence of many rare-occurring forms of molecules, such as cyanooctatetrayne, that must have existed soon after atoms first formed (at about 380,000 years after the Big Bang) (Figure 4.8).

The heavy elements (heavier than iron) did not appear in the universe in any significant amount until stars had formed. These first stars eventually depleted their fusion reactions and collapsed, and some of those collapsing stars formed neutron stars. Some of those neutron stars merged with other neutron stars [133].[xxviii] We will not dwell on the specifics. Suffice it to say that most of the heaviest elements of the universe formed from nucleosynthesis occurring within merged neutron stars. In a shameless display of one-upsmanship, the absolutely heaviest elements are formed in laboratories here on Earth, by curious humans. Regardless of their origins, the heavy elements account for a tiny fraction of the atoms in the universe. To this day, the majority of the atoms in our universe are the very same atoms of hydrogen and helium that formed 380,000 years after the Big Bang.

4.10 They could not, so they did not

We are sometime surprised when we find absences of events that we might reasonably expect to occur.

- We never see anything moving faster than the speed of light.
- We never see stars more distant than 14 billion light years from Earth.
- We never see animals that move on wheels.
- We never encounter meteorites that exceed 66 tons.

xxviii See Ethan Siegel's article, "The scientific story of how each element was made," in the June 12, 2017 issue of Forbes magazine.

- We never see planets exceeding 13 times the mass of Jupiter.
- We never encounter naturally occurring elements whose atomic number exceeds 94.
- We never encounter chronic lung diseases caused by dust particles whose size exceed 100 microns.
- We never encounter cancers whose cell nuclei are smaller than 8 microns in diameter.
- We almost never see indigenous species of frogs living on small islands.

In all these cases, our expectations are quite reasonable. Why can't we travel faster than the speed of light? Why can not we build telescopes that can see stars greater than 14 billion light years from Earth. Why can't animals evolve wheels instead of legs? Why can not we find meteorites exceeding 66 tons in weight? And on and on. It just so happens that there is an excellent reason to account for the absences of every item on our list.[xxix] Suffice it to say that an absence may sometimes indicate that a potential beginning was thwarted.

Just for fun, let us look at some events that, to the best of our knowledge, have never happened. Maybe we can learn something in the process.

If time travel were possible, then we would expect that in the future, everyone will have access to their own time machines, and we would expect to see time travelers popping up in our living rooms and shopping malls and restaurants. We have not seen such visitors from the future. In fact, there is no evidence that travelers from the future have visited any sites in Earth's past. Hence, time travel is impossible.[xxx] Of course, this argument makes two assumptions: 1) That time travelers would have chosen to visit Earth's past; and 2) That our visitors would make themselves known to us.

A gedanken argument against time travel has been proposed by Cristopher S. Baird. Suppose we have a time portal that takes objects a modest three seconds into the past. We mount the portal a short distance above a trampoline, activate it, and drop a ball, which appears three seconds before the portal is activated, adjacent to the ball that we

---

xxix Some of these items have been covered in earlier sections, but we can briefly comment that animals cannot evolved wheels because wheels are only advantageous on flat terrain [66]. Meteorites cannot exceed 66 tons because the kinetic energy associated with heavier objects is so large as to disintegrate the object on impact. Planets exceeding 13 times the mass of Jupiter would become stars. Dust particles larger than 100 microns in width could not be ingested by cells and would be cleared from the bronchi by epithelial cilia. Human cells, cancerous or otherwise, all have nuclei exceeding 8 microns, which we presume to be the smallest size required to accommodate an intact human genome. Frogs are not general indigenous species of islands because they are killed by saltwater, and cannot endure the trans-oceanic immigration required to establish residence [33].

xxx A similar argument, known as Fermi's paradox, applies to the existence of intelligent life on other planets. If aliens were zooming around the galaxy, wouldn't they have made themselves known to us? Enrico Fermi summarized the argument in one short question: "Where are they?"

shall drop. Now both balls are about to drop when two more balls appear, but now 4 balls appear just before we are about to make our drop. Once we actually commit to dropping the first ball, an infinite number of balls appear. In no time at all, literally, the universe is filled with balls. What happens next? Perhaps a very Big Bang.[xxxi] [75]

On occasion, a tactical argument-by-absence can be helpful. For example, cosmologists have, through the past several centuries, wondered whether the universe is expanding or contracting. Various individuals, including Mark Twain and Emmanual Kant, have been credited with inventing the following argument, demonstrating that the universe must, in fact, be expanding.

If the universe were of a fixed size, or if the universe were shrinking, then all of the light being constantly generated by the countless stars in the universe would be retained in its limited volume. Over time, the night sky would become full of light; blinding us with its power. But the night sky is not getting brighter and brighter over the ages. All we see in the moonless night sky is darkness, punctuated by scattered points of feeble light, indicating that the universe is expanding. We shall probably never observe a white night.

The observation of night-time darkness does not actually prove anything. Nonetheless, the dark night sky is consistent with the hypothesis that the universe is expanding, and is inconsistent with the argument that the universe is contracting. Aside from the aforementioned "If they could'a they would'a" argument, the observation of absences has served as the basis of some of the greatest discoveries in the physical and the natural sciences. Let us look at a few.

Let us look at the hypothetical "aging" gene. If there were a single gene that causes aging, then we would expect that we would occasionally encounter individuals born with a genetic mutation that alters, suppresses, deletes, or otherwise disables the gene. We can be certain that this is true because disease-causing genetic mutations can occur anywhere throughout the genome. We have at least 7,000 well-studied genetic diseases that would suggest that no gene is exempt from deleterious mutations. This being the case, we would expect to encounter rare instances of individuals who lack the ability to age (because they lack a functioning aging gene). Aside from television and movies, we never ever encounter immortal humans.

If aging were caused by a single gene, you would expect rare occurrences of mutations of the gene, leading to instances of human immortality. Outside of science fiction, immortal humans do not exist. Therefore, there is no "aging gene."

Likewise, if a single gene was responsible for the transformation of the immediate ancestor of humans (let's call them Homo erectus, for convenience) into the modern humans that populate the world today, then we would expect to see rare revertants of the gene, yielding Homo erectus living among us. In reality, we never encounter Homo

---

xxxi Dr. Baird is a physicist and an active blogger. His excellent pearls of wisdom can be enjoyed at: https://wtamu.edu/~cbaird/

sapiens reverting to Homo erectus. Hence, our assumption that a single gene accounted for the emergence of the modern humans is not supported by observations.

Where did we go wrong? Why are there no mutant immortal humans? Where are the Homo erectus revertants? Quantitative attributes, such as the many subtle changes that occur among individuals as they age, and the many subtle differences between closely related species, are not caused by a single gene mutation. Such changes are polygenic, resulting from many small changes in genes and their regulators. Early studies suggesting the polygenic and non-Mendelian origin of inherited quantitative traits were contributed by RA Fisher, in 1919 [90, 255, 245]. This theory has been strengthened by empiric observations.

Regarding Fisher's theory, biologists have had a long learning curve, and much time is wasted searching for single gene mutations that do not account for quantitative variations in populations. For example, scientists in the latter half of the twentieth century devoted much effort to finding the gene responsible for common adult hypertension. Hypertension is a classic quantitative trait. We have a wide range of normal blood pressure values. Above a certain cut-off, we say that an individual is hypertensive. About 25 % of adults, over a billion individuals worldwide, have hypertension [155, 126]. Despite our best efforts, no one has ever discovered a gene responsible for common hypertension. We can assume that the causes of common hypertension are multi-factorial and polygenic, and certainly not caused by the "hypertension gene". Likewise, it would be senseless to search for *the* gene responsible for human aging. The process of aging varies from person to person and is determined by many different quantitative traits, involving many different genes. [76] [77] [78]

How do species acquire their identifying traits? A species is an evolving gene pool and new species develop when a gene pool separates from its parent gene pool (the collection of genes available to a species). The new species, at the first moment of its appearance, is no different from the parent species. The defining feature of the new species is the isolated gene pool from which its descendants select their genes. Over the years, with its own gene pool isolated from the gene pool of the parent species, we would expect to encounter many new mutations and a somewhat different collection of population gene variants. Exposed to particular environmental influences and selective pressures, the new species will favor gene combinations that account for traits that distinguish it from its own parent species. Through this process, Homo sapiens became different from the Homo erectus. There is no Homo sapiens gene.

4.11 Where do all our limitations lead us?

The imposition of natural limits constricts reality to a set of physical values. Our options for amounts, content, sizes, shapes, movements, and durations are less than we might have wished. Here is a small sampling of some limiting attributes of reality, in order of ascending magnitudes.

- 0 meters, the size of all fundamental particles (e. g., electrons, quarks, gluons, photons, neutrinos).
- 0 Joules, the net amount of positive and negative energy in the universe (e. g., electromagnetic energy, vibrational energy of molecules, gravitational energy, nuclear confinement energy, nuclear decay energy, and any yet undiscovered forms of energy).[xxxii]
- 10^{-35} meters, the Planck length, corresponding to the smallest possible observable object.
- 10^{-26} kg/cubic meters. The density of mass in the universe, which is equivalent to a few dozen protons per cubic meter of space. A proton is equivalent to a hydrogen atom without its single electron.
- 10^{-15} meters. Diameter of a proton.
- 10^{-10} meters. The diameter of a molecule of water is about 0.3 nm, which we will round down to 10^{-10} meters.
- 10^{-8} meters. The smallest viral capsid is Porcine circovirus type 1, which has a diameter of 17 nm, which we will round down to 10 nm or 10^{-8} meters.
- 10^{-7} meters. Mycoplasma genitalium has a length of 200 to 300 nm (nanometers). We'll round this down to 100 nm (10^{-7} meters).
- 10^{-5} meters. Width of a human diploid nucleus. Most human cells contain a nucleus that is about 8 microns in. 10 microns is 10^{-5} meters.[xxxiii]
- 10^{-3} meters. The length of the largest known bacteria. Thiomargarita namibiensis, an ocean-dwelling bacteria, can reach a size of 0.75 millimeter, visible to the unaided eye.
- 10^{2} elements. There are 94 naturally occurring elements. Let us round it up to 100.
- 10^{3} nucleotides. Number of nucleotides in the smallest viral genomes. The CRESS DNA viruses and Porcine circovirus type 1 have fewer than 2000 nucleotides in their respective genomes, which we will round down to 1,000 or 10^{3}.

---

xxxii If the total energy of the universe is zero, then what is the negative energy that balances the positive energy that we encounter? Hard to say. One theory holds that the energy required to bend space (i. e., the energy of gravity) has negative value. This topic is discussed further as a note in Chapter 11.[79] A proof that the net energy of the universe is zero is provided as a note in Chapter 11.[80]
Though negative energy may exist, we can prove that it is impossible to have negative mass (see proof provided as a note in Chapter 11.)[81] This being the case, and knowing that the energy equivalent of a rest mass is $E = mc^2$, we can infer that energy derived from mass is always positive energy. Therefore, we must look to sources other than mass to find negative energy. As it happens, much has been written on the subject. Antimatter (as per Paul Dirac), Black holes (as per Stephen Hawking), virtual particles, and gravitational potential energy, are all candidate sources of negative energy. Science fiction buffs will tell you that worm holes and warp drives owe their existence to negative energy. Are such wild speculations valid? We should not judge them too negatively.

xxxiii Almost all diploid eukaryotic cells have about the same diameter as the diploid mitotic cells in humans, one of many indicators that all eukaryotic cells are descended from the same root eukaryote.

- 10^4 cataloged minerals. There are currently about 5,000 cataloged minerals, with a few added each year. Let us round it up to 10,000, for good measure.
- 10^4 protein coding genes. The total number of protein-coding genes in a typical animal genome. The number varies widely. In the human genome, the current estimate is 20,000 to 25,000 protein-coding genes. We will use 10,000 protein-coding genes as a general estimate for metazoans.
- 10^4 genetic diseases. The number of single-gene genetic diseases currently described. There is no reliable number. We will round up to 10,000 single-gene genetic diseases.
- 10^6 red blood cells produced by the human body every second. The number of red blood cells produced by the body is 2.4 million red blood cells per second, and we will round down to one million.
- 10^8 number of species currently living on planet Earth. Estimates range from 10 million to over a trillion extant living species, but 100 million species is a number that most biologists could live with.
- 10^9 years. The age of the Earth is estimated to be about 4.8 billion years, but we will round down to 1 billion.
- 10^{10} years. It is estimated that the observable universe began with the Big Bang, about 13.8 billion years ago. We'll round this down to 10 billion years.
- 10^{10} years. It is estimated that the Earth formed about 5 billion years ago, which we'll round up to 10 billion years. It is fascinating that the Earth is comparable in age to the observable universe, to an order of magnitude
- 10^{13} red blood cells in the human body. It has been estimated that there are 20–30 trillion red blood cells in the human body. If we take the estimate of 37 trillion as the total number of cells in the human body, then about 70 %–90 % of the cells in a human body are of erythroid lineage [219].
- 10^{13} diploid cells in the human body. The number of human cells in the human body is 30–37 trillion or 3×10^{13}, which we'll round down to 10^{13} [219, 36].
- 10^{24} molecules in a mole. A mole is 6.02×10^{23} molecules (Avogadro's number), which we will round up to 10^{24} molecules.
- 10^{27} meters. The observable universe is estimated to be 10^{27} meters in diameter.
- 10^{35} life expectancy, in years, of a white dwarf star. Since the universe is only 13.8 billion years old, a white dwarf is predicted to live 10^{29} times longer than the current life history of our universe.
- 10^{53} kg, the total mass of ordinary matter in the universe is thought to be about 10^{53} kg or 10^{56} grams.
- 10^{64} years. It is calculated that the required for black holes to decay by the Hawking process is 10^{64} years.
- 10^{77} atoms in the universe, excluding dark matter.
- 10^{80} cubic meters. The calculated volume of the observable universe.

5 Origin paradoxes

The only difference between a problem and a solution is that people understand the solution.

– Charles Kettering

5.1 What is a paradox?

A paradox occurs when two statements, both known to be true, contradict one another (i. e., they cannot both be true). Sometimes the apparent contradiction arises because our understanding of the statements is flawed or incomplete. Occasionally, a paradox forces us to change our understanding of reality. In every case, finding and solving paradoxes is an excellent way to sharpen our critical thinking skills. We shall see that paradoxes pop up throughout science, and the quests for their solution have led to some of the most noteworthy advances in our understanding of nature. In fact, paradoxes are prized to the extent that the individuals who first call our attention to the existence of a paradox are revered in textbooks, while the individuals who solve the paradox are soon forgotten.

Let us solve some paradoxes, focusing now on the so-called creation paradoxes, alternately known as origin paradoxes or precedence paradoxes. The general form of the creation paradox involves Object 1 whose origin requires the prior existence of Object 2, which in turn requires the prior existence of Object 1. Neither can exist without the prior existence of the other, so neither object can exist, but they do exist. The earliest precedence paradox takes us back to the creation of existence. How could the universe be created from nothingness if there is nothing pre-existing the universe with which the universe could be created?[i] [82]

5.2 The chicken and egg paradox

Let us consider the classic "chicken and egg" paradox, usually expressed as a question, "Which came first, the chicken or the egg?" The problem can be restated as a creation paradox:

1. Eggs cannot come into existence
   without hens to produce the eggs.
2. Hens cannot come into existence
   without eggs to produce the hens.
3. Hence, neither eggs nor hens can exist.

---

i In Section 3.7, "Nothing, the key to everything," we provide a plausible solution to the our universe's creation paradox. Historians may be interested in the 12th century explanation known as the Kalam argument.

https://doi.org/10.1515/9783111498690-005

The chicken and egg paradox is known to everyone, but it is often posed as an example of an absurdist question. The glib response to the chicken and egg paradox is, "We have chickens, and we have eggs, so it really doesn't make any difference which one came first." As it happens, the chicken and egg paradox is one of the most enlightening of all biological questions, and students who seriously pursue its answer will be awarded with a deeper understanding of the meaning of ancestry and life.

To begin, let us generalize the "chicken and egg" paradox, and reframe the paradox in terms that a cell biologist might appreciate. Adult animals contain two types of cells, distinguished by the method of division by which they were produced: mitotic cells and meiotic cells. The mitotic cells, also known as somatic cells, account for the vast bulk of the organism. In humans, the meiotic cells of the adult organism are the gametes: oocytes residing in the ovaries, and sperm cells inside the testes. The gamete cells are derived from primordial germ cells, and appear early in the embryonic period, from cells that divide meiotically, in the endodermal layer. Backtracking, the first mitotic cell of the conceptus (the zygote) is the product of the fusion of two meiotic cells (i. e., a sperm fertilizing an egg). We observe that two fused meiotic cells produce a mitotic cell. We also observe that mitotic cells produce meiotic cells. This poses a more general paradox: How can one class of cells produce a second class of cells that produced the first class of cells [264]?

The first eukaryotes were capable of both meiosis and mitosis. We can say this because the enzymatic apparatus for both mitosis and meiosis share many of the same steps, and the genes for mitotic and meiotic division are found in every species of eukaryotic cell today, including single-cell eukaryotic organisms that are thought to reproduce asexually [68, 207, 164, 148, 7]. Hence, the first single-celled eukaryotes were both egg and organism, at once.[ii]

When multicellular life forms (e. g., fungi, plants, and animals) evolved, they retained the ability to divide mitotically (to produce diploid cells) or meiotically (to produce haploid cells), resulting in organisms that had both components, represented by somatic cells and germ cells respectively.[iii iv v 84 85 86]

The dichotomy between haploid organism and diploid organism is best understood by examining the life of plants, particularly the life cycle of the bryophytes, an extant species of plant that flourished prior to the appearance of modern flowering plants (i. e.,

ii A fascinating variation on the haploid/diploid coexistence is found among the ciliates, an ancient class of single cell eukaryotes. Further discussion of the ciliates is provided as a note in Chapter 11.[83]

iii Somatic, from the Greek, meaning body, refers to non-germ cells (i. e., not oocytes, not spermatocytes). The somatic cells, then, consist of the differentiated cells of the body and the stem cells in their lineage. Somatic do not divide by meiosis, and, under natural conditions, cannot pass acquired mutations to progeny.

iv A gamete is a differentiated germ cell (e. g., an egg in the female or a sperm cell in the male).

v Mitotic is the adjectival form of the noun mitosis, referring to somatic cell division.

angiosperms). The bryophytes are non-vascular land plants and include the liverworts, hornworts, and mosses. The bryophyte life cycle consists of two phases, a haploid phase, consisting of a male or a female gametophyte (i. e., plant grown from a gamete), and a diploid phase, consisting of a sporophyte (i. e., plant grown from a diploid embryo). From the diploid sporophyte come the male and female gametes that grow as haploid gametophytes. From the haploid gametophytes come the sperm and eggs that fertilize one another to become a diploid embryo that grows as the sporophyte (Figure 5.1).

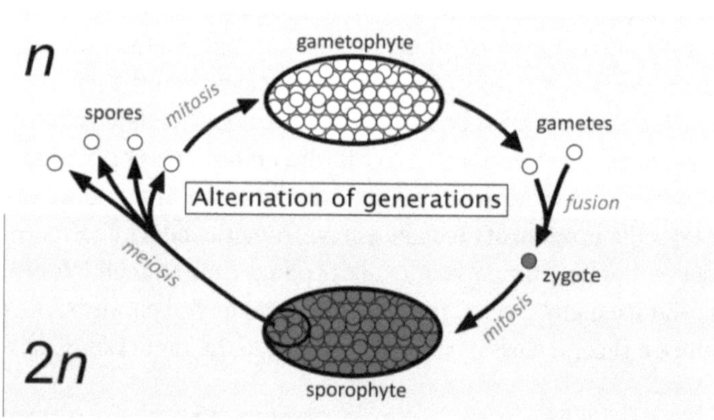

Figure 5.1: The life cycle of plants alternates between a haploid organism (one complete set of chromosomes in each cell) and a diploid organism (two complete sets of chromosomes in each cell). Source, Wikipedia, and entered into the public domain by its author, Peter Coxhead.

On first thought, we may believe that the bryophytes are unique among organisms, because they can be observed as growing plants that are either haploid (gametophytes) or diploid (sporophytes). If we think a bit more deeply, we can appreciate that all plants, and all animals have life cycles that are equivalent to that of the bryophyte, alternating between haploid organisms (derived from gametes) and diploid organisms (derived from embryos). In the case of many species of plants and animals, the haploid organism resides within the larger diploid organism (i. e., the haploid organisms of either gender are internalized in the diploid organism). In plants, the gametic organism is the fruit (female) or the pollen (male). In animals, the gametic organism is the haploid oocyte (in ovary) or the haploid sperm (in testis).[vi] [87]

---

vi Many fugi, algae, and bryophte plants can grow either as haploid or as diploid organisms, depending on their life cycle stage. The dichotomous opportunities for fungal growth have made it extremely difficult to create a satisfactory nomenclature for fungal species, a subject that is explored separately, as a Glossary item. The soma of most animals (i. e., the body of the animal, excluding its component of gametes) is diploid, but some male bees, wasps, and ants are haploid organisms because they develop from unfertilized, haploid eggs [160].

We envision meiosis and mitosis as two forms of cell division. Let us radically change our perspective, and focus on the cellular products of meiosis and mitotis — haploid organisms and diploid organisms [168]. All multicellular organisms that develop from an embryo (i. e., all members of Class Plantae and all members of Class Animalia) are chimeric life forms, with haploid and diploid components. That is to say that every animal and plant can be envisioned as two different organisms each containing its own distinctive type of cells: a haploid organism containing the genes of one gender (i. e., male or female) and the caretaker diploid organisms, containing combinations of male and female genetic material. The egg and sperm are alternate organisms that happen to reside within us, following their own biological destinies. The notion that within our diploid body, we harbor a separate, parasitic haploid organism can be unsettling; but it is true. Let us take a look at how haploid creatures differ from diploid creatures:

1. Purpose. The haploid organism's purpose is the continuation of the species. The diploid organism serves to carry or deliver haploid organisms for recombination.
2. Destiny. Haploid cells are potentially immortal. A portion of the genetic material in a haploid cell can be preserved in the gene pool of the species, and a portion of the gene pool of the species is preserved in the gene pool of descendant species. For this reason, humans share many of the same genes with animals, plants, and all eukaryotes, and these shared genes stem from haploid cells in ancestral single-celled eukaryotes. Diploid organisms have a limited lifespan.
3. Anatomy. The diploid organism has multiple organs, and the organs contain specialized cells that perform various functions intended to maintain its own health, as well as the health of the haploid organism within. The haploid organism tends to spend its life in the vicinity of a single organ (e. g., the ovary in the case of the female or the testis in the case of the male human). There are exceptions. Some plants produce autonomous haploid organisms (gametophytes) having complex body structures.
4. Cellular morphology. Diploid cells in an organism tend to all be of about the same size (with a few notable exceptions such as large neuronal ganglion cells or small red blood cells in mammals). Most diploid cells have a diameter of 30–40 microns, more or less. The sameness of size extends to nearly any eukaryote, even the single-celled eukaryotes. On the basis of size alone, the mitotic cells of diploid organisms of all eukaryotic organisms (including single-celled organisms, plants, insects, mammals) are pretty much standard, having something like a 30 micron diameter.[vii]

---

vii Exceptions are diplid organism that grow as a syncytium (i. e., having membranes that fuse), or cells that are multi-nucleated or cells that are mono-nucleate but polyploid (whose nuclei divide without cytoplasmic division). Caulerpa taxifolia, a type of seaweed, grows as a single-celled, but multi-nucleate, organism. The fronds of this organism can reach 60 centimeters in length (Figure 5.2). C. taxifolia is reputed to be the largest single-celled organism on Earth. Because we have only examined a small percentage of the extant species on our planet, such claims are almost certainly premature.

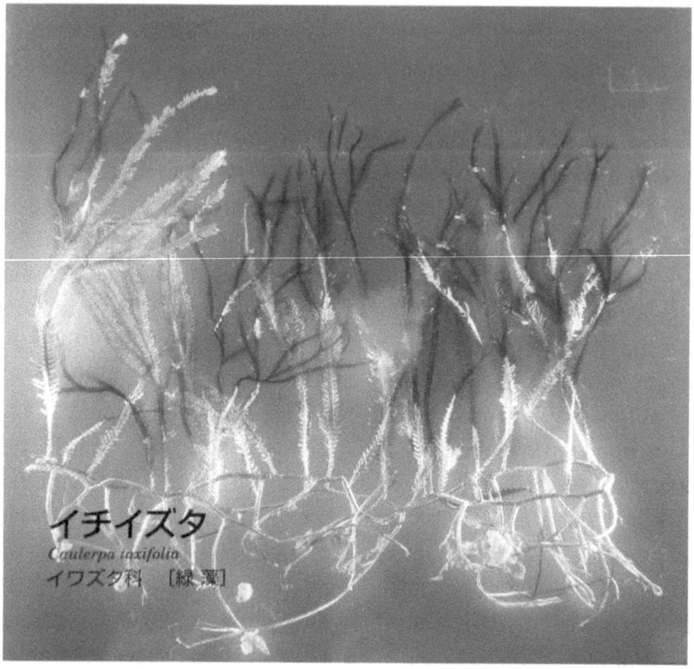

Figure 5.2: The seaweed Caulerpa taxifolia, which grows as a single cell with many nuclei. It is sometimes credited as the largest single-cell organism in existence. Source, Wikipedia, from a photograph taken at the National Museum of Nature and Science, Tokyo, Japan, and entered into the public domain by the photographer, Daderot.

It is proper to ask ourselves why it is that diploid cells of any species tend to have about the same size. It likely comes down to the most common method of cell division among eukaryotes - mitosis. During mitosis, the nuclear membrane disintegrates and the chromosomes line up along a scaffolding (the spindle) before eventually separating to opposite sides, in a carefully orchestrated process that ensures that each side of the mitotic cell receives the correct number properly positioned chromosomes, prior to splitting the cell down the middle to form two equally sized cells. This process places certain morphologic constraints on cells. For starters, the cell must be round and it must be only loosely fitted within surrounding cells. Presumably, if the cell is too big, the spindle cannot tightly control the positioning of chromosomes. If the cell is too small, it cannot provide sufficient cytoplasm for the two progeny cells. In any case, the uniformity in the size and shape of cells is observed in nearly all mitotically active eukaryotic cells. To be sure, there is some variation in the size of diploid cells among organisms, but the differences in size are less than one order of magnitude. In mammals, exceptions to uniformity occur in the post-mitotic subpopulations (i. e., cells that will never again divide), such as megakaryocytes, neurons, cartilage-producing cells, flattened keratinocytes of skin (Figure 5.3). Once again, we are what we are due to our beginnings, as single-celled eukaryotes. When mitosis evolved in the earliest diploid eukaryotic cells, all future diploid

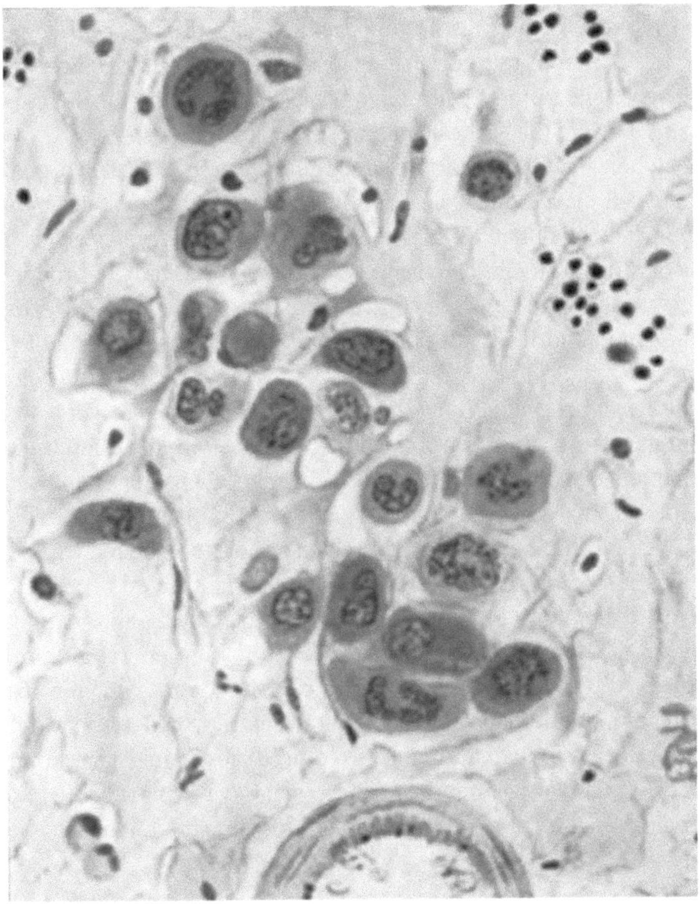

Figure 5.3: Photomicrograph of rabbit bone marrow regenerating after benzol exposure. The very large cells containing multi-lobate nuclei, are megakaryocytes (platelet producing cells). Each megakaryocyte contains a nucleus that is many times larger than the small, diploid nuclei (blue) scattered in the surrounding bone marrow tissue. The nuclei of megakaryocytes are polyploid (having more than the diploid complement of genetic material). As such, mature megakaryocytes are end-stage cells and cannot produce progeny megakaryocytes. Source, MacCallum's A Text-Book of Pathology, 2nd edition, 1921 [161].

organism were locked into the same morphologic niche. Not surprisingly, the genes coding for the machinery of mitosis are ancient and are shared among all extant eukaryotes.

Returning our attention to the morphology of haploid cells, we see that in nearly all organisms, the male haploid cells is much smaller than diploid cells or female haploid cells. There seems to be almost no upper limit to the possible sizes of female haploid cells. For example, the ostrich egg, typically weighing in at 3 pounds, is a single female haploid cell encased in a shell. The ostrich sperm, head to tail, is about 70 microns in length.

Let us examine the meaning of ancestry, as it pertains to species. When we claim that a species derives from some preceding species, we are indicating that the cells of the child species were produced by the cells of the parent species (i. e., all cells come from cells). There are no interruptions or jumps from one species to another. Hence, the cells in the first chicken were cells from its parent avian species, and so on through the lineage of living organisms, back to the earliest eukaryotic organism.

A very long time ago, when all of our eukaryotic ancestors were single-celled organisms, the single eukaryotic cell played dual roles, dividing meiotically to produce haploid cells or mitotically to produce diploid cells. When single-celled eukaryotes evolved to become multicellular organisms, meiotic division was reserved for gamete production, and mitotic division was reserved for production of somatic cells. Hence, we are the chicken and we are the egg, metaphorically speaking. *The answer to the question "Which came first, the chicken or the egg?" is that the chicken and the egg were the same cell back in the time when the chicken's direct cellular ancestor was a single-celled eukaryote.*

At this point, we can explore a favorite trope from science fiction: the alien invasion. Most space invasion plots feature bloodthirsty and rapacious aliens that bomb our cities and destroy our citizens singly or in groups using death rays. In all these scenarios, the aliens stupidly attack our diploid husks, all of which are destined to die with or without alien intervention. Furthermore, when the diploid husks of humans are killed, the remaining humans just make more humans as replacements. Movie aliens cannot grasp that they can wipe out the human race almost effortlessly, in one generation, simply by targeting our haploid cells. A gametocide, added to our air, or our water, for just a few weeks, would extinguish the human race. Our diploid selves would be unaware of their impending demise, until we tried (and failed) to reproduce. If aliens are smart enough to cross the galaxy in their spaceships, they would certainly be smart enough to cook up a potent gametocidal agent.[viii] Let us not dwell on this thought; aliens may be monitoring our libraries.

5.3 The RNA paradox

We know that RNA is synthesized from a DNA template. We also know that the DNA is replicated using a variety of proteins and nucleotide substrates, acting upon a preexisting DNA molecule. If DNA synthesis requires proteins synthesized from an RNA template and if RNA synthesis requires DNA to serve as its template, then DNA could not have preceded RNA, and RNA could not have preceded DNA.

Biologists have long inferred that RNA preceded DNA (i. e., that there were RNA cells before there were cells with DNA). The logical steps leading to this inference are:

1. RNA is the template for proteins, and proteins are the molecules that provide the structure and the metabolic pathways for cellular life.

---

viii The destruction of a species by poisoning the haploid cells contained within diploid organisms is equivalent to genocide without homicide. Haploid cells (e. g., oogonia and spermatogonia), unlike diploid forms (e. g., embryos, fetuses, and fully developed organisms), have no protected legal status.

2. RNA can be synthesized without DNA [124].
3. RNA can also serve as the mechanism by which the RNA template is translated into specific proteins (i. e., the ribosome).
4. RNA can serve as a template for the synthesis of DNA (via reverse transcriptase).
   Hence,
5. Cellular life can be sustained with RNA, amino acids and various substrates that support enzymatic activity.
   And,
6. DNA serves no purpose other than as a template for RNA.
7. RNA-based cells can exist without DNA, while there is no mechanism to expect the first DNA templates to come into existence without the participation of an RNA template.
   Hence,
8. The first cells were RNA-based, and the earliest DNA was synthesized from an RNA template.

The key premise, above, is statment number 2, "RNA can be synthesized without DNA." Montmorillonite, a soft phyllosilicate clay, can synthesize multimer RNA molecules from nucleotide substrates [124]. And that's not all. Montmorillonite also catalyzes the formation of lipid bubbles. Hence, montmorillonite clay, found near porous volcanic rocks containing water, nucleotides, amino acids, lipids, and a few other substrates, may have been the catalyst responsible for the first, lipid-delimited living cells. As an added tickler, montmorillonite deposits have been found on Mars, in abundance (Figure 5.4).

So there we have our hypothesis. RNA, created from nucleotides and a clay substrate, preceded DNA, and served as the template for the synthesis of DNA, the molecule that has become the template for RNA. Maybe. We really cannot rule out the possibility that early DNA molecules formed from mineral catalysis, or otherwise, without the participation of RNA. If this were the case, DNA may have preceded RNA. Furthermore, it is possible that both RNA and DNA may have arisen as polymers upon which spontaneously condensing amino acids (i. e., proteins) acted as templates for their own replication. Just such a scenario has been proposed [109]. In this case, the building blocks of life on Earth may have been proteins, and both RNA and DNA served as protein-encoded replication templates. There are many non-paradoxical possibilities.[88] [89]

5.4 The enzyme paradox

Consider the question, "Which came first, the enzyme or the enzyme-synthesizing machinery?" To build an enzyme, we need to have enzyme-synthesizing machinery, which itself requires the participation of enzymes. The paradox here is that the product of the enzyme-synthesizing machinery (i. e., the enzyme) must exist prior to the existence of the enzyme-synthesizing machinery (which it creates).

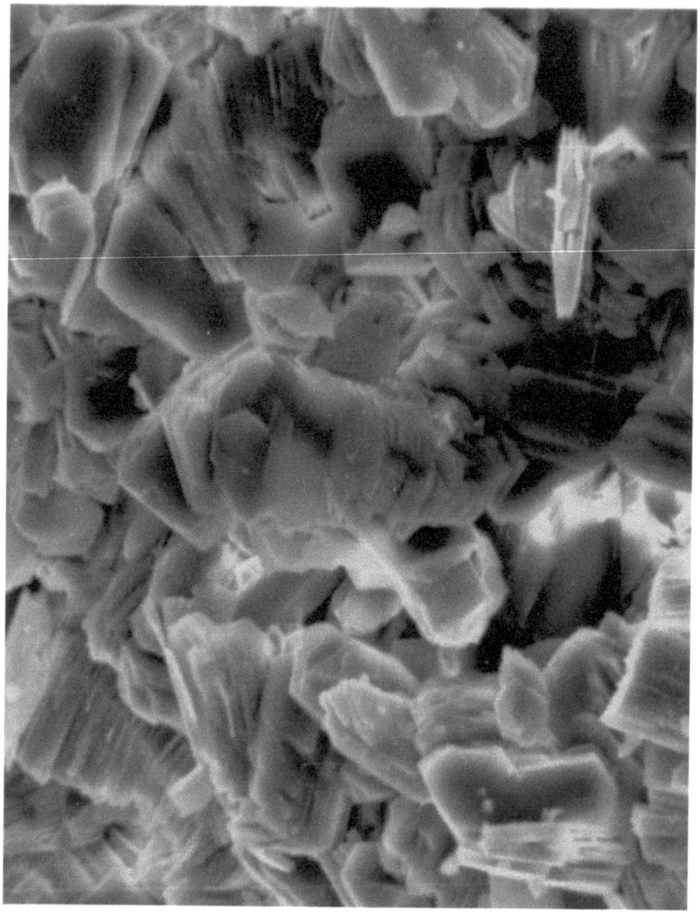

Figure 5.4: Microscopic photograph of kaolinite, a soft phyllosilicate clay much like montmorillonite. The micropores permit a large surface area for catalysis. Source, Wikipedia, from a public domain image prepared by the United States Geological Survey, an agency of the United States Department of the Interior.

To solve this paradox, we must understand a few things about catalysts. Catalysts are facilitators of reactions, and they operate by holding the substrates of a reaction in close proximity, so that the reaction can quickly proceed. Proteins make excellent catalysts, because they can evolve as structures that can custom-fit just about any substrate. Catalysts need not be proteins [251, 250, 124]. In point of fact, any type of molecule (e. g., the surfaces of rocks, various metals, RNA molecules) may serve as catalysts, but only if they happen to have the physico-chemical features that accommodate the reactants [211].

When we examine the process by which proteins are synthesized, we see that codons on a messenger RNA template molecule are sequentially processed by a ribosome to produce a protein composed of amino acids, wherein each amino acid corresponds to a specific triplet codon from the messenger RNA. A ribosome consists of proteins and specialized molecules of RNA. Among these specialized RNA molecules are the ribozyme, catalytic RNA molecules that facilitate the translation of messenger RNA into amino acid chains (i. e., proteins) [147]. It happens that ribozymes (which are not themselves enzymes) are capable of catalyzing their own synthesis. Under laboratory conditions, ribozymes can sequentially attach dozens of nucleotides to primer sequences, without the help of enzymatic proteins [129, 275, 267].[90] [91]

We can begin to see how messenger RNA and ribozymes could have catalyzed the synthesis of enzymes and other protein molecules in a world preceding the emergence of cellular life [95, 103]. The answer to our paradox, "Which came first, the enzyme or the enzyme-synthesizing machinery", seems to be that the earliest enzyme-synthesizing machinery served as both the template (messenger RNA) and as the translational machinery (ribosomal RNA). Hence, no protein-based enzymes were necessary for the creation of the first protein-based enzymes (thus evading our paradox).

5.5 The substrate paradox

When we consider the fundamental relationships between an enzyme and its substrate, we find ourselves facing another paradox of nature. A functioning enzyme requires the preexistence of a substrate that can fit into an accommodating site within the folds of the enzyme, to form a weak (easily reversible) bond, during which the substrate may react with some other chemical to produce a product that is useful to the cell. The substrate, in turn, requires the enzyme if it is to produce a useful cellular product. Substrates that have no value to the cell (i. e., that cannot participate in an enzymatic reaction) often pose a problem to the cell. We must remember that most of the identifiable substrates in enzymatic reactions are themselves the product of some prior enzymatic reaction, in a chain of enzymatic reactions. Each product of a chain of enzymatic reaction must be met with an enzyme that marks the next link in the chain. Otherwise, the final product of reaction (e. g., a newly synthesized organelle, a membrane, part of the rough endooplasmic reticulum) will be unfinished, and the intermediate product will increase in concentration, indefinitely.

We see such substrate/enzyme disconnects in a class of metabolic disorders known as storage diseases.[ix] Summarizing, enzymes must have preexisting substrates, and substrates must have pre-existing enzymes. Hence, we have a paradox.

---

ix Gaucher disease is a lysosomal storage diseases in which a defect in an enzyme of the lysosome (the organelle that scavenges non-functional cellular components) is deficient, leading to the accumulation of bloated, non-functioning lysosomes, eventually filling the cell. Dying cells in such diseases are, in turn, scavenged by macrophages (basically cells specialized to remove dead cells), whose affected lysosomes are also non-functional. The macrophages accumulate as large cells, filled with indigestible lysosomes, occupying various tissues of the body, particularly the spleen lymph nodes. Other storage diseases, particularly diseases involving long-lived cells such as neurons and hepatocytes, may fill those cells with accumulating substrate molecules producing damage at their primary site of synthesis (e. g. Niemann–

The solution to the paradox is not known with any certainty, but we have a few clues that might be useful. We know that the active sites in enzymes (that accommodate the substrate molecules), and the substrates for those enzymes, fall into a limited number of chemical classes. The presumption is that the most ancient enzymes had active sites that could be subtly altered, by mutation, to accommodate new substrates, as needed. Evidence of adaptations built on a common motif is found in ancient gene families. Gene families arose as new genes evolved from duplications of existing genes [49]. The new gene is preserved in the species gene pool if it aids the survival of the species. Because the new gene is a modification of the original gene, you might expect the two genes to have some functional similarity. Groups of genes derived from an ancestral gene are referred to as gene families, and the same gene families found in humans are found in our distant ancestral species, indicating that these gene families arose in concert with the development of metabolic pathways common to living organisms, and involving the same substrate molecules available to all cells.

The globins are an ancient gene family whose current function in humans involves distributing oxygen, through the blood, to tissues everywhere in the body [112, 228]. The genes of the globin family are so very ancient that they are present in bacteria. It is believed that the original globin molecule evolved in bacteria prior to the evolution of oxygenic photosynthesis (i. e., prior to 2.3 billion years ago) at a time when all organisms were anaerobic. Oxygen is toxic to anaerobic organisms. The purpose of the first globins in bacteria was to scavenge oxygen, thus protecting anaerobic cells from oxygen-induced toxicity. Following the advent of oxygenic photosynthesis, when the earliest aerobic organisms were evolving, the oxygen-scavenging properties of the early globins were modified to facilitate oxygen exchange. Today, a variety of hemoglobinopathies affect humans, all involving the globin gene family in one way or another [112].

One of the largest gene families is the olfactory receptor family. When we reckon the marvels of cellular existence, we tend to overlook our sense of smell. Unlike the receptors involved in vision, which only need to distinguish red, blue, and green colors, our olfactory receptors must distinguish among many thousands of chemicals. We must not allow ourselves to think that olfactory receptors evolved to detect odors. No so. Odors did not exist until our olfactory receptors, in concert with our nervous systems, created the perceptions (i. e., odors) that we happen to associate with particular chemicals. Sounds and sights exist independently of our hearing or our vision, but an odor is purely the product of our perception. Olfaction evolved in ancient organisms to identify chemicals wafting from edible plants, particulates from smoke, toxic substances, and whatever the wind blew their way. At some moment, in the distant past, the gene family diverged a bit, producing two major classes of receptors grouped by homology to our ancestors: the fish-like receptors, and the (later evolving) tetrapod-like receptors [105].

Pick disease and Tay–Sachs disease, involving the central nervous system, and glycogen storage diseases involving the liver.)

One more bit of evidence shows us that, from the very beginning, enzymes evolved in close concert with the earliest substrate (i. e., one could not precede the other). As discussed earlier, the substrate chemicals found in all cells, of all extant species, have a distinctive chiral preference. Cellular constituents of a particular kind are either left-oriented or right-oriented. For a substrate to fit into an enzyme's receptor, the receptor must have the same chirality as the substrate (a left-oriented substrate will not fit a right-oriented receptor motif). Because chirality is preserved between substrate and enzyme, throughout all of nature, we can infer that substrate and enzyme were paired at the beginning of life.

5.6 Paradox of metabolic pathways

In an evolved system, new traits do not appear in a single generation. They evolve in steps, over time, by modifying genes involved in component pathways. For example, there is simply no mechanism to evolve a feather, in one step, on an animal that had nothing previously growing from its skin. Presumably, feathers evolved from some proto-feather on an appendage that was something like a wing, that served some purpose, such as heat insulation, or protection of babies in the wing folds, or even mating attraction. We can infer backwards that there was an advantageous precursor to the precursor, and so on. That is how evolution through natural selection seems to work. Does the story of the evolution of feathers help us understand how metabolic pathways evolve?

A metabolic pathway consists of a series of enzyme-mediated chemical reactions wherein the product of one reaction serves as the substrate of the subsequent reaction, in tandem, until the series is complete. For a complete metabolic pathway to evolve, all of the steps of the pathway must evolve separately, and the final step of the pathway, which utilizes the product of the next-to-last step, must be the last step to evolve. This line of reasoning leads to paradox. When we look at a typical metabolic pathway, we really do not see much role for its intermediate products, other than serving as substrates in a chain of reactions leading to a final product. It would seem that for a metabolic pathway to evolve, each of the intermediate products must have had its own useful cellular role. Otherwise, the full pathway, including all the intermediate steps, could not have evolved. In other words, the steps of the pathway must evolve prior to the evolution of the pathway they serve; hence the paradox.

The paradox of pathway evolution has fascinated evolutionary biologists for years. One of the earliest proposed solutions was that metabolic pathways evolve backwards. The ancestral pathway begins with a single enzyme that catalyzes the transformation of the final substrate in the pathway to become the final product of the pathway. Once the last step has evolved, nature evolves another enzyme that catalyzes the next-to-the-last substrate to produce a product that serves as the substrate for last step of the pathway, and so on [123]. Thus, pathways of arbitrary length can evolve backwards, enzyme by enzyme. This mechanism simply puts the paradox at the front end of the pathway. The final product of the pathway will not be produced in any abundance until all of the steps are in place, producing the proper sequence of products.

The reason that we have fallen into a paradox is that we are relying upon three false assumptions:

1. That the mechanism of evolution of a metabolic pathway must develop the same way that a new anatomic trait (such as feathers on birds) may develop.
2. That a metabolic pathway exists in isolation of many other preexisting pathways that employ the same enzymes.
3. That each enzyme has one particular role, producing one particular product.

New metabolic pathways evolve in the midst of numerous old metabolic pathways. Presumably, a new pathway is established from contributions of pieces of existing pathways that happen to produce a product that serves some advantage to the organism. It is easy to imagine new pathways occurring from mix'n'match events. Most of these pathways will produce nothing of value and quickly fade into non-existence. Occasionally, a new pathway will appear that demonstrates great promise. Over many years, evolution will tweak individual enzymes until the pathway runs smoothly.

Just by asking ourselves how metabolic pathways came into existence, we begin to change the way we think about cellular chemistry. According to traditional thinking, a pathway was a sequence of biochemical reactions, involving a specific set of enzymes and substrates that produced a chemical product. The traditional pathway taught in biology classes is the energy-generating Krebs cycle. It was common for students to be required to calculate the output of the cycle (in moles of ATP) based on stoichiometric equations employing known amounts of substrate. Today, we know that pathways may involve the linked actions of receptors, activators, suppressors, and modifiers. The products of a pathways are often cellular actions, rather than chemical products. These pathways do not always occur within an organelle or in anatomically sequestered area of the cell. The pathways may interact with other pathways, and the direction of a pathway, and its biological consequences, may be complex and their function may vary in different cells or under different physiological conditions within one cell.

Is there evidence to support this new way of thinking about pathways? In the latter decades of the 20th century, cancer researchers were mystified by a protein that seemed to have an important role in the development of cancer. This protein, p53, was at one point considered to be the primary suppressor of human malignancy. It was thought that mutations in p53 disrupted normal cellular regulation, allowing cells to proceed down a slippery path eventually leading to cancer. In the past few decades, as more information has been obtained, cancer researchers have learned that p53 is one of many proteins that play a role in carcinogenesis, but that the function of p53 changes depending on the species, tissue type, cellular micro-environment, genetic background of the cell, and many other factors. Under one set of circumstances, p53 may participate in a

DNA repair pathway; under another set of circumstances, p53 may play a part in arresting the growth cycle [162, 278]. It is difficult to predict biological outcomes when functions of enzymes and pathways change in different cellular contexts. Various mutations in the gene that codes for the p53 protein have been linked to 11 clinically distinguishable cancer-related conditions, and there is little reason to assume that the same biological role is played in all of these 11 disorders [246].

Likewise, the Pelger–Huet anomaly and Hydrops-ectopic calcification-'moth-eaten' (HEM) are both caused by mutations of a gene, coding for the lamin B receptor. Pelger–Huet anomaly is a morphologic aberration of neutrophils wherein the normally multilobed nuclei become coffee bean-shaped, or bilobed, with abnormally clumped chromatin. The condition is called an anomaly, rather than a disease, because despite the physical abnormality, affected white cells seem to function adequately. HEM is a congenital chondrodystrophy that is characterized by hydrops fetalis (i. e., accumulations of fluid in the fetus), and skeletal abnormalities. It would be difficult to imagine any two diseases as dissimilar to one another as Pelger–Huet anomaly and HEM. How could these disparate diseases be caused by a mutation involving the same gene? As it happens, the lamin B receptor participates in two separate pathways: preserving the structure of chromatin, and serving as a sterol reductase in cholesterol synthesis [257]. These two different and biologically unrelated pathways, involving the same gene product, account for two biologically unrelated diseases.[92] [93]

Let us summarize some of the observed properties of genes and proteins:
- One gene can code for more than one protein [217].
- A single protein may have multiple functions in multiple pathways.
- A single protein with a single function may have different biological effects based on the cell type in which the protein is expressed, the stage of development in which the protein is expressed, and the cellular milieu (e. g., concentrations of substrate or protein inhibitors) for a given cell type, at a particular moment in time.

Regarding the paradox of pathway creation, we can reasonably conclude that there is no directed evolution of pathways. Metabolic pathways evolve by mixing existing enzymatic processes and selecting for helpful reaction sequences.

5.7 The computer bootstrapping paradox

When you turn on your computer, the computer "boots up", the term being a shortened form of "bootstrapping." Every time you start your computer, a bootstrapping paradox must be solved. Here is the dilemma:
1. Computer hardware requires a software operating system for any kind of functionality. Without software, a computer is just a paper weight.
2. Software operating systems cannot operate without functioning computer hardware. Without hardware, software is just a wish list.

3. When you turn on the power to your computer, the hardware cannot begin to function until it receives software instructions, but the software instructions cannot be accessed by the hardware until the hardware begins to function.
   Hence,
4. Computers cannot start up.
   But,
5. Computers do start up; and therein lies a paradox.

This human-created dilemma, which we unknowingly solve every time we turn on your computer, is referred to as the bootstrapping paradox. The term bootstrapping derives from an absurdist trope in which boys are instructed to pull themselves up by their bootstraps. Although it is certainly possible for a standing boy to pull his boots on with his bootstraps, it is impossible for a boy to gain a standing position by pulling the straps of a booted foot; no matter how hard he may pull. The term refers to a class of paradoxes in which some step in a process requires the completion of some earlier step, which itself requires the completion of a later step.

Here is how computers solve their bootstrapping paradox. At start-up, the operating system is non-functional. A few primitive instructions hardwired into the computer's processors are sufficient to call forth a somewhat more complex process from memory, and this newly activated process calls forth other processes, until the software operating system is fully up and running. The cascading re-birth of active processes takes time, and explains why booting your computer may seem to be an unnecessarily slow process.

Basically, computer scientists solved the paradox by designing a tiny start-up kernel wherein both the hardware and the software are one and the same.

5.8 The genome bootstrapping paradox

When we accept that the genome is the blueprint for the development of the organism, then we create for ourselves a paradox. How is it possible for an organisms to be assembled from a blueprint if the organism needs to be functional (i. e., assembled) in order to follow the blueprint?

The answer to this paradox is simple. The genome is not a blueprint. If the genome is not a blueprint, then what is it?

Let us back up a bit and examine what we think we know about the genome. All human life, and the lives of most animals, begins with a fertilized ovum, the zygote (from the ancient Greek word meaning joined or yoked together). The zygote contains a new and unique genome, and the genome contains all of the genetic information required to develop the an animal organism through all of its developmental stages. When we read about the genome, we are typically told that the genome holds the blueprint for the fully developed animal. We hear on countless documentaries and science shows in the popular media that medical geneticists and bioinformaticians are busy studying this blueprint, so that they can decode the process by which animals develop. Not so. Scientists who study the genome find nothing remotely resembling what we would expect to see if the genome were a blueprint. The genome does not contain identifiable "pieces" corresponding to the different organs. The genome does not contain a coded anatomic picture of the developed organism.

With just a little thought, we might see that the genome cannot be a blueprint for life, because blueprints, as everyone knows, only describe how the parts of an object fit together. Blueprints tell us nothing about the steps in which zygotes become fully developed animals. If we wanted to build an animal, starting from the zygote stage, and we had our hearts set on looking at blueprints, we would need a separate blueprint for each moment in animal development, capturing the blastula, the blastocyst, the gastrula, and so on, up through the fetal, neonatal, childhood and adult stages of the embryo. Having all those blueprints would tell us a lot about the various anatomic arrangements of cells, tissues, and organs over time, but those blueprints would not help us to create whole organisms from single cell zygotes.[x]

If we are searching for a proper metaphor for the genome, we need to be thinking in terms of a bootstrapping process, much like the bootstrapping process that brings our computer to life. The genome begins with software (the genome) and hardware (the cellular machinery contained in the oocyte). The cellular machinery of the oocyte is employed to start the most basic processes of cellular life, as directed by the genome. These would include DNA replication (i. e., replication of DNA), DNA transcription (i. e., synthesizing RNA from a DNA template), and RNA translation (i. e., creating proteins from an RNA template). Once these fundamental processes are initiated, development proceeds. But how? How does the information in the genome lead to the development of an animal organism?

The genome operates much like a recipe book, in which each successively listed recipe depends on the successful creation of the preceding recipe. In this case, there is a recipe for making a morula from a zygote, and a recipe for making a blastocyst from a morula, and a recipe for making an inner cell mass from a blastocyst, and a recipe for making a gastrula from an inner cell mass, and so on, until the animal is fully developed.

If this is the case, then how do simple genetic codes successfully describe remarkably complex living organisms? The answer is quite simple. The genetic code does not describe the organism; it merely provides the instructions that, when followed, account for the development of the organism. Just as very simple computer algorithms produce highly complex and sometimes unpredictable results that defy description (e. g., Monte Carlo simulations, fractal algorithms), we can understand why genes produce complex organisms without providing a description of the final product. Basically, instructions are not descriptions, and we should not think of the genetic code as the blueprint of life.

---

x The Bonini paradox holds that as system complexity increases, it becomes less and less feasible to describe the system. The paradox can be restated as, "the more we need a description for a system, the less we are capable of producing a description." A system as complex as an animal is beyond description.

Biologists are constantly finding metabolic pathways, coded for and controlled by the genome, that participate in the steps of development. Animals of the same developmental lineage (e. g., all mammals or all birds), pass through a closely similar sequence of embryonic phases. Aside from that, we see that all animals, from tiny insects, to large whales, develop from an embryo and share many of the same genes, despite their obvious differences in size and physiology. These observations fit with the notion that all animals use the same genomic recipe book, written as separate editions.

In summary, every organism on Earth must solve a bootstrapping paradox as it develops from a single cell. It begins as some of the machinery of the parent organism (i. e., in the oocyte's cytoplasm) activates the new genome, and it builds up to a complex life form in successive stages, using the products that the genome creates at each preceding stage, to build the succeeding stage, until the organism is complete.

5.9 The paradox of consciousness after coma

To recover from a coma, your brain must somehow arouse you, but how can your brain arouse you if it is comatose?

For those of us who learn medicine from watching soap operas, we get the impression that recovery from coma is instantaneous. According to television, a long-suffering wife sits at the bedside of her comatose husband and weeps. At the darkest hour, when all seems hopeless, the patient squeezes his wife's hand, opens his eyes, smiles warmly and calls his wife by her name. Such scenarios do not occur after deep coma.

We can imagine that the process of waking from a coma might be something like the process of booting a computer. The software (consciousness) and the hardware (live neurons connected in brain tissue) must share some sort of primitive identity wherein the physical connections among neurons have sufficient activity to begin the process that leads to consciousness. If so, we would expect recovery from coma to occur stepwise, as each increment in consciousness rallies additional neural pathways that support the next step toward consciousness. This seems to be the case.

The process of coma recovery comes gradually and may take weeks, months, or years, depending upon such factors as the degree of brain trauma and the length of time that the patient has been comatose.[xi] Coma patients may achieve something known as a minimally conscious state wherein they may track objects with their eyes, and respond to verbal or physical stimulations with blinking or vocalizing with grunts, or with small

---

xi The longest-lasting coma, as reported by the Guiness Book of world records, involved the sad case of Elaine Esposito, who, at 6 years of age, fell into a coma while under anesthesia for an appendectomy. Her coma persisted for the next 37 years, when she died, at the age of 43. The longest record for a recovery from coma occurred in the case of Terry Wallis, who became comatose following a motor vehicle accident that resulted in traumatic brain injury. He awakened 18 years later. See Benedict Carey's article, "Mute 19 years, he helps reveal brain's mysteries," in the July 4, 2006 issue of the The New York Times.

movements. Such minimally conscious states may last indefinitely, or may lead to steady cognitive improvements. Encouraging signs that recovery is progressing may include increased periods of attention (keeping the eyes open for longer periods) and increased responsiveness to touch or sound.

Attaining consciousness from a comatose state has a slow and incremental cadence, analogous to the ordered acquisition of functionality exhibited when our computers boot up. The sudden recoveries portrayed on television are staged to enhance the viewer's pleasure and to stick to the program's tight schedule; not to provide an accurate rendering of events. To be fair, it should be noted that from the frame of reference of the patient, recoveries **are** instantaneous. For recovered coma patients, the process of waking seems to occur suddenly, at the precise moment when their prior states (deep coma, light coma, and various levels of semi-consciousness) are all replaced by the fully conscious state.

Those of us who undergo surgery under anesthesia experience something akin to coma recovery. When the anesthesia is given, we lose consciousness. During surgery, we cannot be aroused as we would be if we were simply sleeping. Blood continues to flow to the brain. As far as we can tell, our neurons are not damaged in the process, but they do not support consciousness. Why do we lose consciousness when we are under anesthesia? The mechanism of action of anesthesia is poorly understood, but the various types of anesthetic agents seem to interfere, in one way or another, with the normal conduction of electrical impulses conducted through neurons (via axons and synapses). As in the recovering comatose patient, achieving consciousness after surgery occurs in steps. Patients believe that they awaken instantaneously after surgery, but the staff in the post-operative staging area, who watch and interact with patients, know that arousal after surgery is gradual.

We can speculate that the paradox of coma recovery is solved in much the same way as the paradox of the computer re-boot. There is a sequential and coordinated integration, of multiple neuronal pathways, eventually leading to the fully-conscious state. Hence, the global consciousness that we enjoy represents an assembly of processes that gradually integrate their activities. There is no need to search for the anatomic center of consciousness in our brains; it is not there.

5.10 The paradox of speech and language

It is impossible to speak without first having a language. But it is impossible to have a language without having a way of expressing that language, through speech. Therefore, it would seem impossible that we have either speech or language.

The paradox is solved when we accept that speech and language must have developed together, with each new attribute of speech (i. e., a specific and repeatable sound associated with a thing, or an action, or a description) determining a new attribute of language (i. e., the meaning of the sound incorporated into a corpus of knowledge that is used, shared, and taught by the members of a community). Hence, speech and language must develop together.

The ability to communicate through language is an inherited skill that is hard-wired into every human. From the moment of birth, humans begin to learn to communicate through one or more languages by a unique process that has very little to do with rote acquisition. Parents need not send their infants to language classes. Babies automatically make the connection between overheard words and their meanings. This holds true for the sounds of spoken language and the hand/finger motions of signed language and sounds, or words and hand motions in the case of sign languages. In a year or so, the baby begins making the sounds of words, and progress accelerates when the baby gets some reinforcing feedback from loved ones. The mother smiles when the baby says "momma", and the mother brings the bottle when the baby says "baba". By eighteen months, most babies have a small, but practical vocabulary.

Unlike the early hominids, who had to invent language and speech from scratch, societies provided newborns with a pre-existing language. All of the work of creating that language was done form them. Or so it would seem. Because babies are assigned an ancient and complex language at birth, they must re-create for themselves all the associations between sounds, language, and meanings of a highly complex system of communication. Ultimately, babies must form a conceptualization of language that is congruent with that of the other members of society. Otherwise, communication is impossible. We can begin to see that the task of learning a pre-existing complex language is as difficult as creating the sparse rudiments of a new language, if not more so.

For many of us, our mental faculties decline with age. As an early sign of cognitive impairment, we may struggle to retain short-term memories. Our thoughts may become sluggish. Eventually, we may have trouble remembering the names of friends. In extreme cases, we may forget, in the moment, the name of a close relative. The very last thing to leave us is our ability to speak. Speech and language play a large part in making us human beings. We hate to see it go.

5.11 The paradox of classification

A classification is an way of organizing reality (or some domain of reality) into individual classes composed of members who share some fundamental relationship(s) among themselves. The individual classes, in turn, are related to other classes by some rule or law, and the relationships between classes apply to every contained member of those classes.

The class structure is hierarchical, and the hierarchy of classes is established by subclassing (i. e. creating successive subclasses or child classes below a superclass or parent class). Any class may have a subclass. Inversely, any class, with the exception of the class at the very top of the hierarchy, may have a superclass (i. e., a class of which it is the subclass).

1. Each class of the hierarchy has a set of properties or rules that extend to every member of the class and to all of the subclasses of the class, to the exclusion of unrelated classes. A subclass is itself a type of class wherein the members have the defining class properties of the parent class plus some additional property(ies) specific for the subclass.
2. In a hierarchical classification, each subclass may have no more than one parent class, whereas a class may have any number of subclasses. The root (top) class has no parent class.
3. A class instance is a member of a class. For example, your copy of this book is an instance of the class of objects known as "books".
4. Every instance belongs to exactly one class.
5. Instances and classes do not change their positions in the classification (e. g., a horse cannot become a sheep and a subclass of mammals cannot become a subclass of insects)
6. The members of classes may be highly similar to one another, but their similarities result from their membership in the same class (i. e., conforming to class properties), and not the other way around (i. e., similarity alone cannot define class inclusion).

In the biological sciences, the oldest, largest, and most complete classification is the Classification of all Living Organisms [34]. Every known organism is assigned a class, and classes are arranged in a hierarchy based on evolutionary descent. Every member of a class is related to every other member by its relation to its superclass (the class from which it descended) and by distinctive morphologic or biological features that are present in every member of the class.

In the natural sciences, the most important questions are always the most fundamental questions (e. g., "What is matter?", "What is energy?", "What is time?", "What am I?"). Because classifications embody the fundamental nature of reality, we can imagine why classifications might help us understand our world. Many ground-breaking advances in science have involved establishing the class relationships among seemingly unrelated objects (e. g. humans and fish, electricity and magnetism, energy and mass, entropy and temperature, waves and particles, conservation laws and symmetry groups).[94]

Classifications, like much of our reality, are created in paradox. Here is the problem. Taxonomists (the people who build and curate classifications) cannot build a classification without some initial understanding of the relationships among objects. Therefore, every classification is created with the taxonomist's biased conception of its general structure. This bias defeats the purpose of the classification, which is intended to *reveal* relationships discovered through the process by which the classification is built.

In practice, the taxonomist begins with a root object that embodies the fundamental features of every class and every species within the classification. In the case of the taxonomy of living organisms, this root object might be "the living cell". Once the root object exists, the taxonomist can begin to create broad subclasses containing properties that are inclusive for the class and exclusive of other classes (i. e., Class Prokaryota, which lack a nucleus and Class Eukaryota, which have a nucleus). Then, based on observing the various properties of the prokaryotes, the taxonomist might define additional classes that include some organisms and exclude others. This goes on until every organism has a class, and every class is a subclass of a parent class, producing a lineage that extends backward to the root class. The root class, which contains all descendant classes and their members, embodies the classification (i. e., the first class of the classification subsumes the full classification).[95] [96]

Every thoughtful taxonomist will admit that a classification is, at its best, a self-correcting machine; not an infallible description of reality. We use the classification to create new hypotheses that can be used to see if the classification matches reality. The process of testing hypotheses may reveal that the classification is flawed; that our early assumptions were incorrect. In that case, we revise our assumptions. With a little luck, we will find that the process of testing hypotheses will re-assure us that our assumptions were consistent with new observations. By constantly testing the validity of the classification, we add to our understanding of the relations between the classes and the instances within the classification. Hence, the process of creating a classification is a manageable bootstrapping process.[97]

5.12 The paradox of evolution's origin

> Embryonic structure is no more free from the effects of new mutations, from adaptation, convergence, divergence, and the like than is structure in any other phase of the life cycle.
>
> – George Gaylord Simpson [220]

Let us look at another paradox, invoked by the following question: "Which came first, the process of evolution or the product of evolution?" If evolution (i. e., the ability of a species to evolve) is a biological trait, like height and strength, then the ability to evolve must have been acquired as a new trait at some time in our ancestral past. It is through evolution that new traits are acquired. Hence, evolution needs to exist in order for evolution to come into existence. This is a paradox.

This paradox is created from a false assumption; that evolution is a trait of living organisms that needs to be acquired. Evolution is a fundamental condition of the universe, and the law of evolution by natural selection can be restated in terms of probability. Basically, anything that happens to have a greater chance of existence (e. g., by virtue of chemical stability, or strength, or mass) is more likely to persist, or increase in number, than things that have a lower chance of existence. Evolution does not create favorable mutations in organisms; it is simply an expression of the fact that animals with favorable mutations are more likely to persist and replicate than animals that lack such mutations.

One can certainly argue that something other than evolution through natural selection contributes to the world of living organisms, but it would be very difficult to argue that evolution through natural selection does not exist. It would be equivalent to asserting that probability does not exist [31].

Where in the organism does evolution begin? This seems like a silly question. If the muscles and bones of the arms evolve to facilitate swinging from limb to limb in the forest, then surely we can infer that evolution, in this case, is something that began in the arms and ended in the arms. Nonsense. All animal evolution begins in the embryo, often expressing itself in the embryo (i. e., as improvements or modifications of the embryo), and sometimes resulting in modifications to the adult organism. We marvel at the anatomic complexity of the developed human, but a much higher level of complexity occurs in the developing embryo, which must establish the process by which adult tissues develop. The process of organismal development is best imagined as the progressive elimination of cellular properties as the first totipotent cell of the organism (i. e., the zygote), restricts the expression of genes in successive clonal generations. Ultimately, the fully differentiated cells of the adult organism are limited to one specialized functionality and morphology. The development of the adult form of an organism is a process involving many steps, that always begins in the embryo. Hence, all evolution is a consequence of changes that occur in embryonic development.

Evolution has no plan and no end-game. Changes occur in the organism, and sometimes those changes confer some survival advantage for the species. We must keep in mind, however, that increasing the survival advantage of the adult organism may confer little or no advantage to the survival of the species. For example, if there were some evolutionary change that conferred long life on human males, allowing them to live 300 years or more, such a change might have a negative effect on the survival of the species. The world would have a large number of superannuated men consuming the limited resources of the planet and wreaking wars at the drop of a hat. Population pressure might push a drop in the birth rate, ultimately endangering the human species. When it comes to evolutionary advances, natural selection always favors the fitness of the embryo, which carries forward the species gene pool, over that of the adult organism.

As discussed in Section 5.8, "The genome bootstrapping paradox", when we look at human development, from the earliest stages of the embryo to the eventual appearance of a mature adult, we see nothing remotely resembling what we would expect if the genome were a blueprint. There are no identifiable "pieces" and no indication that anything is being fastened together to form a whole organism. Instead, we see clearly that human development follows successive transformations of living stages. This process in no way resembles construction from a blueprint (Figure 5.5). A much better analogy is that the genome operates like a recipe book, in which each successive recipe depends on the successful creation of the preceding recipe. In this case, there is a recipe for making a morula from a zygote, and a recipe for making a blastocyst from a morula, and a recipe for making an inner cell mass from a blastocyst, and a recipe for making a gastrula from an inner cell mass, and so on. When we look at ancestral classes of eukaryotic organisms, we see an historical retelling of the sequential recipes for embryonic devel-

Figure 5.5: Woodcut illustration depicting the unfolding of a single beech leaf. Every leaf in a plant develops through a series of complex steps. Those steps, as well as all of the steps required to replicated and organize the specialized cells of the beech leaf, must have evolved in the embryo. The complexity of any one adult structure is trifling compared to that of the embryo, that must carefully choreograph every step in the development of every adult structure. Source, *The Natural History of Plants*, by AK von Marilaun, 1902 [248].

opment. Not surprisingly, the major milestones in the development of complex animals are marked by evolutionary advances of the embryo.

When we look at the class lineage of the metazoans (animals), we see that *each successive major class is characterized by an embryonic alteration.*

- Class Metazoa: Characterized by the formation of the blastulated embryo.
- Class Eumetazoa: Characterized by the formation of a two-layered embryo.
- Class Bilateria: Characterized by the formation of the three-layered embryo.
- Class Deuterostomia: Characterized by gastrulation and formation of an enterocoelom.
- Class Chordata: Characterized by the formation of a embryonic backbone.
- Class Amniota: Characterized by the evolution egg-laying, on land.
- Class Eutheria: Characterized by the evolution of the placenta.

Likewise, when we look at the sequential evolution of plants, we see the same pattern, with the major classes of plants being defined by the evolution of the embryo.

- Class Embryophyta: First class of plants to propagate from embryos, wherein all of the different structural parts (organs) of the plant derive from a single cell.
- Class Spermatophyta: First class of plants to nourish the embryo within a seed.
- Class Gymnosperm (also known as Acrogymnospermae): First plant to have naked seeds.
- Class Angiospermae (also known as Magnoliophyta): First plant to enclose its seed within a fruit.

- Class Monocotylodon (also known as Liliopsida): Subclass of Class Angiospermae, characterized by one embryonic leaf (i. e., one cotyledon).
- Class Dicotylodon: Subclass of Class Angiospermae characterized by two embryonic leaves (i. e., two cotyledons).
- Class Eudicotylodon: Dicotylodons with pollen grains that are tricolplate (three grooves parallel to the polar axis).

Class Spermatophyta produced the two major classes of plants that give us all of the trees and most of the smaller plants (e. g., the grasses) that we see if we are lucky enough to have an office with a view of some greenery. These classes are the Acrogymnospermae (gymnosperms or plants with naked seeds) and the Magnoliophyta (angiosperms, or plants with seeds encased in fruit) (Figure 5.6). The acrogymnosperms include the pines,

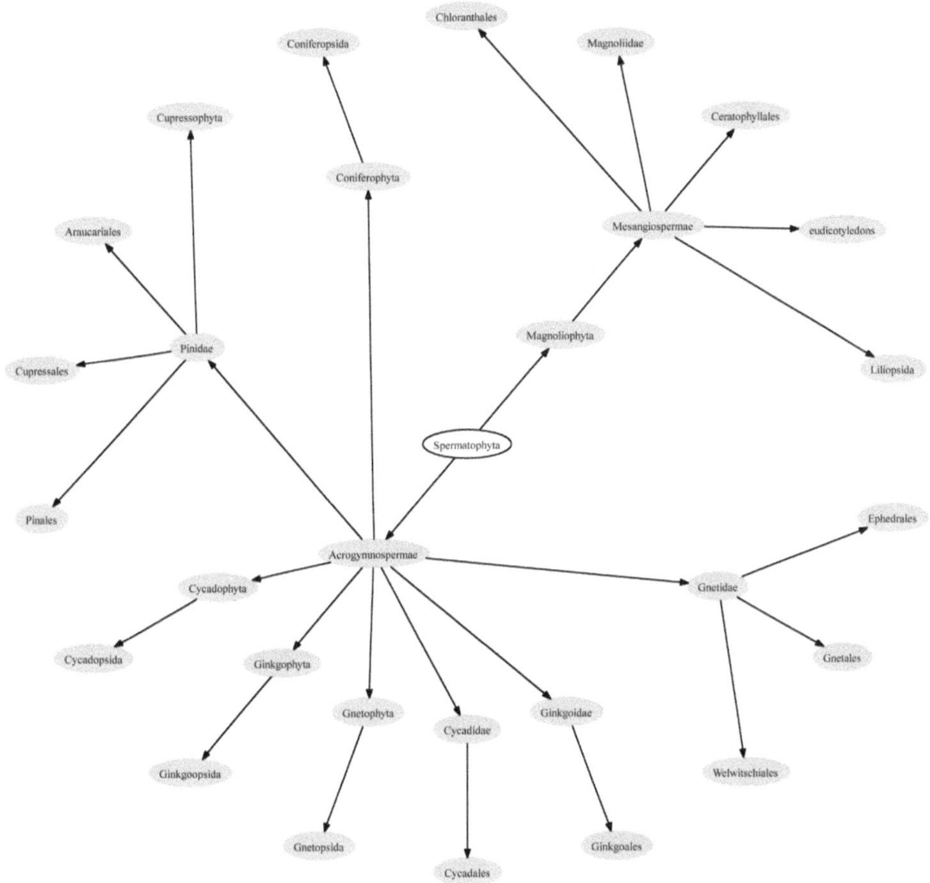

Figure 5.6: Class Spermatophyta is the first class of seed plants, in which the embryo grows from a nutritive seed. Most of what we commonly observe as structured plants (flowering plants, conifers, and grasses) are seed plants and descendants of Class. Source, author.

conifers, ginkgos, and cycads. The acrogymnosperms arose over 200 million years earlier than the angiosperms, but the number of living angiosperm species vastly outnumber the gymnosperms.[xii] The angiosperms are broadly divided into two large subclasses, the dicotylodons and the monocotylodons.

It is only through embryonic development that adult forms acquire their species-specific anatomic and physiologic features. Hence, every evolutionary feature expressed in adult organisms (e. g., feathers on birds, placentas in mammals, fins on fish) must have arisen in the embryo. These principles of embryonic supremacy call to mind a quotation from Lewis Wolpert: "It is not birth, marriage, or death, but gastrulation which is truly the most important time in your life." Wolpert emphasized that gastrulation, the embryonic process by which the primitive a cell-lined space is formed to accommodate organs developing from the endoderm, is the key event in the development of all deuterostomes (an ancestral class that includes Homo sapiens).

5.13 The paradox of self-replication

> Referring to natural selection, 'It is all about the survival of self-replicating instructions for self-replication.'
>
> – Richard Dawkins [71]

The phenomenon of replication is familiar to us. Manufacturers routinely produce replicates of all sorts of objects: toothpicks, pencils, screwdrivers, buzz-saws, automobiles, and so on. When we discuss cellular replication in school, we tend to categorize it as just another replication process, involving a tiny object. The whole trick to cellular replication is that the cell manages to perform the procedure on itself. We must remember that we never see toothpicks replicating themselves. Screwdrivers cannot divide to produce two new screwdrivers. Basically, cells are the only entities that are capable of self-replication, and they should be given due credit for this implausible achievement.

How do cells replicate? Well, they always begin with a cell. This tells us that a cell must have been present at the moment of the first cellular replication. This raises the question, how that first replicating cell came to be, if it was not itself replicated from some other cell? This is the paradox that lies at the origin of cellular self-replication.

In Section 2.2, "First cellular organism", we discussed how the first replicating cell may have evolved, in the vesicles of volcanic rock, using chemicals present in the pri-

---

xii It is estimated that the first gymnosperms appeared about 383 million years ago, and the that first angiosperms appeared about 145 million years ago. The latecoming angiosperms are more successful as a class than the earlier gymnosperms. As a rule of thumb, the newer subclasses of a superclass do better than the older subclasses. For example, there are more humans than great apes in the world. In terms of speciation, beetles are among the most successful animals on Earth. The first beetles evolved about 325 million years ago, but most of the abundant modern day species of beetles did not evolve until some time after 66 million years ago.

mordial seas. We can assume that the earliest cells were non-replicating, and that they slowly evolved the pathways of DNA replication and cell splitting that represent what we see today as mitosis. The rest is history, but our knowledge of the evolution of cellular replication tells us a great deal about the fundamental properties of organisms.

How does cellular self-replication begin? Firstly, we must recognize that cells do not replicate from DNA; they replicate from cells. A molecule of DNA, genetically endowed though it may be, has no power to do much of anything, unless it is placed into a cell. As it happens, every cell contains cellular material received from its so-called parent cell. In the case of the zygote (the oocyte fertilized by the spermatocyte), the cytoplasm comes almost entirely from the mother (sperm cells have negligible cytoplasm). The zygote (being the first cell of the new organism) inherits its DNA from the fusion of haploid maternal and fraternal chromosomes; and it inherits its cytoplasm nearly entirely from the mother. Because the product of every eukarotic cell division is a cell containing a nucleus and a cytoplasmic mitochondria, we can infer that the first embryonic cell of every eukaryotic organism contains mitochondria that comes exclusively from the mother. Because the first embryonic cell gives rise to all the cells of the body, we can say that every cell in our body contains progeny of mitochondria present in our mothers. Likewise, every cell in our mother's body contains the progeny of mitochondria of her mother, and so on. Some human populations can be tracked by their mitochondria back to their common maternal founder, and the geographic distribution of such populations tells us much about their ancient migrations. Additionally, we can infer that the very first woman of the human species must have carried forward the mitochondria found in the preceding species. Extrapolating much further backwards, through the ancestry of multicellular organisms, we can see that mitochondria, which are inherited as whole organelles from parent species to child species, have similarities that extend across all eukaryotic organisms.[xiii]

As the consequence of the maternal origin of mitochondria, we can infer that inherited disorders of mitochondrial genes are exclusively inherited from the mother. We should add that most of the inherited mitochondriopathies (disorders of mitochondria) are caused by gene mutations of the nucleus; not mutations of the mitochondria. Though mitochondria have their own genes, the mitochondrial genome codes for only 13 proteins of the mitochondrial respiratory chain. All the other proteins and structural components of the mitochondria are coded in the nucleus. Therefore, some mitochondriopathies (not involving the respiratory chain) can be transmitted by father or mother.[xiv]

xiii LUCA is the abbreviation for Last Universal Common Ancestor, also known as the cenancestor. When we ascend the hierarchical lineage of cellular species we come at last to the species that is the presumptive ancestor of every organism living on planet earth. LUCA is thought to have lived 3.5 to 3.8 billion years ago [104].

xiv It seems that as a general rule, endosymbionts (such as mitochondria) will relinquish some of their genes to the host, as they gradually lose their grasp on autonomy. As an extreme example, the symbiotic

Once self-replication begins, why does the process stop? Why do cells not continue to divide until they fill the planet, and then the universe? In the case of terrestrial cells, there are all manner of rate-limiting factors at play: limited nutrients, limited ability to survive the crush of neighboring cells, limits imposed by self-regulatory processes, and so on. Even in the case of malignant tumors, where cellular replication is less restrained than we see in normal tissues, there are practical limits to growth (e. g., inadequate blood supply, insufficient clearance of dying cells).[XV] Ultimately, growth of tumors ceases when the patient dies.

The natural limits to self-replication may not apply in our future. When we produce the first self-replicating android, the replicative process may conceivably continue indefinitely, assuming that each replicant has access to the parts required for assembly. Imagine the results of a trans-galactic voyage, wherein self-replicating androids collect energy and material en route, supplying themselves with the wherewithal to create more of themselves, and more of their transport ships, with nothing to impede the process. Maybe this is already happening, somewhere in space... heading this way.

5.14 Wave/particle duality paradox

All particles (e. g., photons, electrons, protons) have a dual nature. At times, our observations of particle physics can only be understood by modeling a particle as a wave (i. e., expressing its behavior in a wave equation). At other times, observations of the same particle behaves like a tiny ball, spinning around and careening off of other tiny little balls. How can one particle be two different things, depending on what we choose to measure? Just over 100 years ago, physicists came up with truly marvelous ways of answering this very question. We will be delving into this subject in later chapters, as we dig into quantum physics. Just for now, before we lose ourselves in the tall grass, let us conceptualize wave/particle duality as an origin paradox.

Imagine that you have a microscope that has no limit to its resolving power. Starting with a grain of stand, you examine fine particles of quartz (silicon dioxide). Then you look at individual atoms of silicon and oxygen. Then you find nuclei filled with protons and neutrons around which electrons buzz. You try to dissect the electron, but you can not. An electron is a fundamental particle, meaning that it is not composed of other particles; there can be no "stuff" inside an electron. If an electron contained "stuff", then the "stuff" would be the fundamental particle, and we would ask ourselves what is the "stuff" made from? There can be nothing "inside" an electron or any of the 25 known

---

bacterium Carsonella ruddii, found in sap feeding insections, contains only 182 genes, having transferred most of its genome to its host. See Philip Ball's article, "Smallest genome clocks in at 182 genes," published in the October 12, 2006 issue of Nature.

xv A malignant condition is one whose natural (i. e., untreated) biological course leads to the death of the patient. In common parlance, the term "malignant" is reserved for aggressive cancers.

fundamental particles. Indeed, the fundamental particles are recognized as "point" particles, having no assignable size. Insofar as all matter in the universe is composed of fundamental particles, we can honestly claim that we are composed of nothing.[xvi] Likewise, we can ask what waves are made from. Waves are propagating disturbances in a field. There is no "stuff" in a wave. What's more, we now know that waves are emitted from fundamental particles (i. e., light emitted by electrons moving from one atomic shell to another shell having a lower energy).[xvii] Alternately, waves can emit pairs of particles at particular energy levels, given enough energy [54]. This should not surprise us, as quantum events are reversible. Summarizing, we have waves that require the preexistence of particles, and we have particles that require the pre-existence of waves. It is paradoxical that we have waves, or particles, or anything, or nothing!

The wave/particle duality paradox lies in the way we think about reality. We can surmise that any universe consists only of properties. In the case of our universe, the universal properties consist of a few symmetries (such as translational symmetry, rotational symmetry, time shift symmetry) and a few constants (such Planck constant, the speed of light, the Gravitational Constant, the Cosmological Constant, and the Fine Structure Constant), giving us all we need to account for our physical laws. As we derive the laws of matter and energy, we assign more specialized properties that apply to particular types of particles (the 25 described particles of the Standard Model) and waves (electromagnetism and quantum fields) and topologies (e. g., gravity and spacetime curvature). There is a lot to learn, but it all boils down to fundamental properties from which subproperties derive. When we think of particles and waves as the properties that determine their behavior, the wave/duality paradox disappears. Basically, a particle is only known to us through its properties. A fundamental particle is nothing more than a set of properties. Regarding the wave behavior of particles, all we need know is that a particle is a wave when its properties determine that it can be observed as a wave. Generalized, a particle is whatever its properties lead us to observe.[xviii] [98]

---

xvi The notion that fundamental particles have no size was suggested as early as 1759 by Roger Joseph Boscovich in his *Philosophiae naturalis Theoria redacta ad unicam Legem Virium*. He explained that solid matter was achieved when two fundamental particles are brought so close to one another that they produced a repulsive force that prohibited any other particle from coming between them [14]. Though his theory contains errors, Boscovich correctly reasoned that it would be impossible for any fundamental particle to have size or to be composed of matter.

xvii Incidentally, we now know that fundamental particles can emit other fundamental particles. For example, the τ particle can emit other particles, such as electrons, muons, and neutrinos. Not bad for a particle having no size or mass and containing nothing!

xviii Our conceptual difficulty is due, partly, to our tendency as English speakers to confuse properties with identities. Our language is filled with idiomatic expressions that make no sense when carefully parsed. If we say "He is hungry," we are using the verb "is" to describe the relationship between "He" and "hungry". Clearly, "he" is not equal to "hungry". In Germany, one might say "Er habt Hunger (He has hunger)," correctly indicating that "Hunger" is one property that "Er (he)" happens to have at the

5.15 The paradox of electromagnetism

We are taught in elementary physics that a magnetic field is created when current moves through space. Reciprocally, movement of a magnet will produce an electric field. In electromagnetic waves (i. e., light waves of any frequency), the electric wave and the magnetic wave move in synchrony, with each component perpendicular to the other (Figure 5.7). We are taught that the electric and magnetic components generate one another, as the wave cycles. Assuming this is true, then the magnetic wave component cannot exist without the preexistence of the electric wave component, and the electric component cannot exist without the preexistence of the magnetic component. Therefore, there can be no moment when an electromagnetic wave can begin to exist.

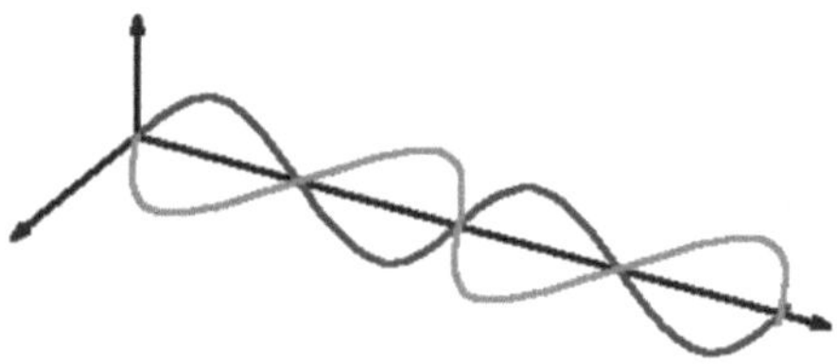

Figure 5.7: Graphic depiction of a planar electromagnetic wave. The wave has electric and a magnetic wave components, traveling in synchrony (in the same phase) but vertical from one another. Source, author.

To solve the paradox of the impossible origin of electromagnetic waves, we must eliminate a few misconceptions about the nature of light. Let us start with a thought experiment involving an observer and an electric current. Imagine that the observer moves at the same speed as the electric current. The observer uses a magnetometer to measure the magnetic field strength. In the observer's frame of reference, the current is stationary (relative to the observer). Consequently, there is no magnetic field generated. However, anyone standing to the side and watching the current as it moves along with the observer, will see that the current is indeed generating a magnetic field as it moves. The existence of the magnetic field seems to be dependent on the frame of reference of the observer.

Likewise, an observer who measures current produced by a moving magnet would be astonished to learn that if the she were moving apace with the magnet (i. e., in the magnet's stationary frame of reference), no current is measured. Again, whether there is current generated by a magnet in motion seems to depend entirely upon the frame of reference of the observer.

The problem here is that according to our understanding of relativity, all frames of reference provide equally valid observations. Assuming that relativity is a valid the-

present time. Our minds benefit when we stop thinking in terms of equivalences (e. g., a particle is this and a wave is that), and we begin to think in terms of relational properties that develop over time.

ory, the magnetic field produced by a moving current would exist for one observer and would not exist for the other observer. This is a paradox.

This paradox has a solution, and it is very simple. Magnetic fields are always generated by moving current, and moving current always generate magnetic fields, in any and all frames of reference. In the case of the observer who is moving apace with the current (or, alternatively, with the moving magnet) the relative speed of the observer is zero, and there is no net movement of current. Therefore, the observer does not cross the field lines being produced (either magnetic or electric respectively), and it is the observer's movement across a field that yields a measurable force. A generated field is present, but the measured field at the observer's location is zero.[xix]

There is yet more to learn from this paradox. Let us look at the problem from a different perspective. A stationary magnet becomes a moving magnet when the observer moves. This is easy to understand. If you run towards a magnet, it is equivalent to the magnet moving towards you. But we know that every magnet that moves creates an electric current. Therefore, when we run towards a stationary magnet, we can expect that it will generate an electric current (due to its relative movement). Therefore, in the case of the magnet that sits on a desk, there are frames of reference for which the magnet will produce both magnetic and electric fields. How can this be explained?

Again, the explanation is simple. The magnetic force and the electric force are unified when all frames of reference are considered. It is for this reason that light waves are referred to as electromagnetic waves; the two components can not be sensibly separated from one another.

How is it possible for electricity and magnetism, which we know to be two distinctly different forces, to be the same thing when viewed in different reference frames? It is best to stop thinking of electricity and magnetism separately. In the same way, space is different from time, but we now know that we cannot speak of space without considering time. For this reason, we try to consistently use the term spacetime, referring to 4-dimensional coordinates. Similarly, we could benefit by referring to electromagnetism rather than the electric force and the magnetic force. These lessons are more than a century old, but every generation of scientists must become accustomed to thinking in terms of the invariances assigned to observational frames of reference.

Now, we can return to our original paradox of electromagnetic waves. How can an electromagnetic wave begin, if the electric and magnetic components are dependent on the preexistence of each other? The paradox depends on a false premise, the presumption that electricity and magnetism are separable components. The separation of the two is just an illusion that comes from restricting our observations to particular reference frames.[xx]

---

xix We will be discussing fields in Section 6.2, "The first quantum field".

xx Einstein, in his 1905 paper *On the Electrodynamics of Moving Bodies* wrote, "The force acting on a body in a magnetic field is nothing else but an electric field," an astonishing notion at the time.

5.16 The parity paradox

Whenever a particle is created, so is its antiparticle. This assertion, based on theory, has been repeatedly verified in hadron collider experiments, wherein the birth of particles are closely monitored. As far as we can tell, every proton that has ever been created has been matched by its antiproton.[xxi] This being the case, why do we find lots and lots of protons in the universe, while antiprotons are exceedingly rare?

The proton, like its antiproton partner, is a highly stable particle with a duration estimated to be at least 1 billion years. In addition, protons and antiprotons are fermions and do not move at the speed of light. There are rocks that have been dated to have an age of 4 billion years, indicating that the protons contained in those rocks have been hanging around in the same location (relative to the movement of Earth) for 4 billion years. We would have no reason to expect that the antiprotons created at the same time and place as its paired proton, would have any ability to transport itself to a far region of the universe populated only by particles of its same charge. There should be the same number of protons as there are antiprotons everywhere in our observable universe? The overwhelming predominance of protons accounts for the parity paradox, alternately known as the baryogenesis problem.[xxii]

One possible solution posits that in the earliest moments of the universe, a large number of proton-antiprotons pairs were created; producing many more protons and antiprotons than we find in our universe today. If we allow that initial conditions may have violated charge parity, then a tiny preponderance in the number of protons, compared with antiprotons, may have prevailed. If so, mutual proton-antiproton annihilations may have left us with a universe with a surplus of protons. The surplus of protons in the early universe may account for the lopsided ratio of protons to antiprotons that we observe today.

Other solutions to the parity problem are built on the hypothesis that parity is more of a suggestion than a universal law. It was shown, back in 1957, that the fundamental symmetry between left and right (i. e., chiral parity) is not conserved by electrons emitted in beta decay of nuclei.[xxiii] Most of the emitted electrons are found to spin in the lefthanded direction [263]. The Law of Parity would demand that half of the electrons would spin to the left and an equal number would spin to the right. The same lefthanded spin dominance is observed with emitted muons. We can infer that parity does not always apply and is not a universal law.

---

xxi We note here that the parity paradox, as stated, is very much dependent upon our belief that particle collisions provide a fair assessment of reality, past and present. Conditions in the earliest moments following the Big Bang (i. e., the first 10^{-43} seconds) may have produced particles without parity. No one really knows.

xxii A baryon is a particle with mass equal or greater than that of a proton.

xxiii Beta decay occurs when unstable isotopes of elements emit particles as neutrons transform into protons or as protons transform into neutrons.

Sabine Hossenfelder suggested another very simple solution to the paradox.[xxiv] The number of particles and antiparticles provided at the time of the Big Bang is an initial condition. We have no logical reason to infer that observations made under current conditions (such as those seen within hadron colliders) would necessarily extend to the conditions that prevailed at the moment when the Big Bang occurred. Parity may simply not have existed during the Big Bang, when more protons than anti-protons were created.

It is sometimes claimed that the job of the physicist is to work from a given set of initial conditions in order to predict physical outcomes. In this case, there is no violation of proton charge parity, because our physical laws do not dictate initial conditions (initial conditions dictate our physical laws). This solution to the parity paradox side-steps the primary purpose of this book; to provide some understanding of how we received our initial conditions. Basically, we are asking, "How did our beginning begin?" In later chapters, we shall see that initial conditions, like everything else in our universe, may sometimes be explained.

5.17 A general solution to origin paradoxes

Precedence paradoxes all have the same form, namely, A cannot occur unless it is preceded by B, and B cannot occur unless it is preceded by A. Therefore, neither A nor B can exist.

The general solution to all such paradoxes is, as we have seen, fairly simple. At some point in the past, both A and B were one and the same. Whenever we encounter a precedence paradox, we should try to understand the conditions that existed before either A or B existed. Let's take a moment to think about the major consequence of our generalized solution of precedence paradoxes. In particular, we assert that over time, two identical objects may diverge from one another, in terms of appearance and properties, having codependent functions. A creation paradox seems to arise, but there shall always be clues establishing their common origin, if we look for them.

Examples of paired items that were unified, at their origin

- Consciousness and subconsciousness;
- Chicken and egg;
- RNA and enzyme;
- Haploid organism and diploid organism;

xxiv Sabine Hossenfelder is a well-known physicist, author, and educator. Her books and instructional videos are widely available.

- Particles and waves;
- Electric field and magnetic field;
- Strong force and weak force;
- Physical constants and physical laws;
- Time and space;
- Determinism (Schrödinger's wave function) and indeterminism (Heisenberg's uncertainty principle);
- Nothing and something;
- Reality and abstraction;

Throughout *How Shall We Begin?*, these seemingly mismatched pairs will be shown to be equivalent at their origins. We shall also find that paired items having a common origin tend to have many properties in common, and can be interchanged under certain conditions (e. g., the wave/particle dualities, time/space transformations, and electro/magnetic waves).

6 Initial conditions

Make everything as simple as possible, but not simpler.

– Albert Einstein

6.1 Initial conditions for a universe

We do not really know how the universe came into existence, but we can certainly surmise that any self-respecting universe would have, at its origin, some basic set of conditions to account for a non-chaotic reality. Tentatively, and without much explanation, let us suggest the following:

1. Geometry. Without geometry, there is no there there.[i]
2. Values assigned to the points in an unbounded geometric space. The values of a point in space might include the magnitude of forces and the directions of those forces. This is basically what physicists refer to when they use the term "field".
3. Some universal invariances (such as spatial homogeneity and isometry, and time-shift invariance). Invariances are operations that do not alter the objects being operated upon. Without invariances, we could not have conservation laws, and without conservation laws we could not have stable matter or energy.[ii]
4. One or more fundamental constants to set the conditions from which the properties of the universe emerge. We can think of constants as the numbers that help us to establish the relationships among the various physical constituents of reality. For example, the equation $E = h \times \nu$ establishes the relationship between energy and the frequency of light. Energy and frequency are different types of things and are expressed in different units (Joules and Hertz respectively). To legitimize the relationship between energy and light wave frequency, we need a constant that has its own set of discoverable properties (Planck's constant, h in this instance).[iii]

---

i Felix Klein (1849–1925), a German mathematician and an early proponent of group theory, asserted that "any geometry consists of a space of points and a group of transformations that move the figures in the space around while preserving the properties appropriate to that geometry." [40]. Felix Klein's definition allows us to understand geometric relationships in terms of symmetry groups and their transformations (e. g., rotations, reflections, translations) in space. We'll confront symmetry groups when we reach Section 9.2, "The "Symmetry First" universe".

ii In most cases, when we discuss invariances, we are talking about things that do not change over time. We shall soon learn why it is that invariances serve as the basis for the conservation laws, describing relationships among properties that do not change over time (e. g., conservation of energy, conservation of charge).

iii π is a universal constant, and it is fundamental, and it can be applied to all manner of physical objects and their relationships expressed as physical laws. Nonetheless, π only a number. It is not part of physical reality; it only reflects reality. Therefore, π is generally not accepted as a fundamental physical constant. The same exclusion would apply to e and to all our cherished transcendental numbers. Unfair? Perhaps.

https://doi.org/10.1515/9783111498690-006

In this chapter, we will focus on some of the most fundamental properties of matter and energy that emerge from a minimal set of ingredients that may have set the initial conditions for our particular universe.

6.2 The first quantum field

A field, quantum or otherwise, can be described as an infinite graph of spacetime coordinates, with each point in the graph assigned a value. Fields are abstractions, devised to hold information that applies to every point in spacetime. We create multiple fields when we invent the various sets of information to be held in the points of the field. For example, we can have a temperature field, with a different temperature assigned to each point in spacetime. We could have a velocity field with a speed and direction assigned to each point in spacetime. Most of the fields of physics are particle fields. The values at each point in a field may contain a simple quantity, or it may contain a vector of quantities or force descriptors, or whatever we may find useful. The values at each point may be filled with matrices, if we wished. It is convenient to restrict each field to descriptors that can be related to one another (e. g., the electromagnetic field). By doing so, physicists can predict how the values of a field will change over time (e. g., how waves move). Fields lend themselves to analysis using standard methods developed for linear algebra, and can be transformed, reverse-transformed, graphically modeled, stretched, and crushed, with impunity.

One type of field is distinguished from other types of field by the physical information assigned to each imaginary point in the imaginary field.

As such, we can emphasize the following:

1. All fields coincide within the same spacetime. Notions that fields are physically layered one atop the other can be dismissed. The locations contained in any field consists of all the locations in spacetime. When physicists entertain the notion of one unifying field theory, they are expressing the hope that all of the different fields can be related to one another through a theory, and the theory would presumably be expressed as a set of equations.
2. Because the field is a spacetime graph, the relationships of the field can help us describe how each point within the field affects other points over spacetime. Through this process, wave functions (i. e., the wavelike propagation of forces through the field) can be evaluated.
3. Point values (e. g., vectors) help us to understand how points in one field may affect points in another field (the so-called field interactions).
4. Fields are quantized when physical events are restricted to integer values, and this partly relates to the integer-dominated nature of reality (e. g., there no such thing as one third of an electron, standing waves have an integer number of nodes, exactly 6 rotations of 60° complete a circle and establish a type of symmetry).

5. Physicists interpret reality as a collection of distinguishable fields, with field-specific fundamental particles and forces, and exhibiting predictable actions between the fields.

Because field coordinates are abstractions, we need not ask when the first field coordinate system appeared in the universe, but we can certainly ask when the coordinates first attained physical values (i. e., forces with magnitude and direction). The Casimir effect (discussed in some detail in Section 3.7, "Nothing, the key to everything.") demonstrates that even in a spacetime void, shielded from all matter and energy, our spacetime coordinates are sprinkled with non-zero energy fluctuations (i. e., forces are always at play).[iv] Hence, quantum fields existed for as long as there has been spacetime.

As it happens, each quantum field is always associated with a characteristic quantum particle, and there is a simple reason why this is the case. We will learn, when we delve into particle physics, that a fundamental particle is the smallest vibration of a field. This is equivalent to saying that a fundamental particle is one quantum of its field. Whenever we have a field, we have quanta (particles). At this point, readers may be wondering how a particle made of solid matter can be equivalent to a small vibration. We'll need to wait a bit for this to make much sense. For the moment, though, we can look at one particular example for which a particular type of field creates something akin to a field particle.

For most of us, sounds are real. The notes of a musical scale are things that we can combine with other notes to create music. Our speech is filled with sounds that can be digitalized, recorded, copied, and played back to us. But what is sound, really? Sounds are oscillations of pressure that move through a field. We appreciate sound as a "thing" because our brains have a way of converting pressure oscillations into neural impulses that are somehow interpreted as distinct sounds. Pure tones are the "particles" of music. Frank Wilczek wrote, "The mathematics describing the vibratory patterns that define the states of atoms in quantum mechanics is identical to that which describes the resonance of musical instruments."[v]

Suppose we were to encounter aliens who have no natural receptors to detect small vibrations of air, and who consequently have no way of interpreting pressure oscillations as sounds. We would be hard pressed to convince these aliens that musical notes (i. e., oscillations at a particular frequency) have any particular significance. Technologically advanced aliens might have no problem measuring air pressure oscillations at every point in a volume (i. e., a field of sound), and they may be capable of formulating

iv Casimir's effect was predicted from theoretical considerations related to the Heisenberg's uncertainty principle. It holds that in the vacuum of space, wherein energy and matter have been removed, small fluctuations of energy will arise spontaneously and randomly. Such fluctuations have been confirmed by direct experimental measurement, thus establishing the Casimir's effect a physical fact.

v See Frank Wilczek's article, "On the world's numerical recipe," in the winter issue of Daedalus, 2002.

a wave equation to determine how pressure oscillations move through time and space, but they would have no special reason to think that such an air pressure field held the secret of music or notes (i. e., the "particles" of music).

Physicists are like aliens. For most of us, reality consists of silicates, hydrocarbons, concrete, and steel. However, for physicists, reality consists of fields of volumes enclosing points having properties that change through space and time, according to wave functions. The solid particles in concrete and steel are real enough for most of us, but quantum theory provides the underlying principles that account for solid reality.

6.3 All fields are derived from one root field

> But – once I bent to taste an upland spring
> And, bending, heard it whisper of its Sea.
>
> – Eccliastes

One of the most sought-after solutions in modern physics is the unification of the four forces of nature (i. e., the electromagnetic force, the weak force, the strong force, and gravity) under one consistent theoretical model.[vi] Current thinking posits that at the earliest moment of the Big Bang (i. e., in the first 10^{-43} seconds), all of the forces (and their fields) must have been unified as a single field, which soon after split into the individual fields that we observe today. This theory is based in part on calculations stemming from the Standard Model, indicating that the energies existing in the first 10^{-43} seconds of the Big Bang are so intense that the fields, as we know them, could not exist. The assumption is that there was a primordial force out of which our observable forces of nature soon emerged. There is also the belief, based upon the general trend of scientific advancement, that the operating principles of nature are simple, and the role of science is to find simpler and simpler ways of encapsulating the laws of nature. Basically, one unifying field is simpler than many fields having independent origins and operational principles. As yet, physicists have not found a model that unifies all of the forces of nature plus gravity under one model that fully accounts for our observed reality.

At this point, we might build an argument that bolsters support for a unified field theory.

---

vi The Standard Model, which deals with three of the known fields of nature, but omits gravitation, is considered to be our most accurate model of physical reality and has proven to be an excellent predictor of experimentally verifiable quantum events. Nonetheless, it has its limitations. The Standard Model fails to provide a candidate particle for dark matter. It does not explain baryon asymmetry (i. e., more protons than anti-protons in the universe). It fails to provide a satisfactory model for quantum gravity and thus does not unify the forces of nature. Sad to say, not every prediction coming forth from the Standard Model is true. A more complete discussion of the limitations of the Standard Model can be found in the Particle Data Group's Review of Particle Physics, in the journal Chinese Physics C.40, 2016, page 290.

- The properties of one field are dependent upon the properties of another field.
  Put another way, the properties of a field depend upon terms included in the equations of motion of other fields.[vii] [102] In particular, different fields operate under the same set of fundamental constants.
- All particles, regardless of their field of origin, have movement and can exchange momentum with other particles from their own field or from other fields.
  The universality of energy, movement, and momentum confirms that fundamental properties of the different fields are equivalent. It is worth noting that massive particles of a particular field do not directly interact with one another. In the case of electron-electron interactions, their like charges would repel, preventing direct collisions. In the case of other massive particles, their mutual gravitational attraction would reach infinity as the distance between particles falls to zero. Interactions between particles of one field are mediated by virtual particles of another field.[viii]
  Because field interactions are mediated by other fields, we can see that the properties of all fields are coordinated under mutual laws.
- The particles of a field may derive from the particles of other fields.
  For example, τ particles can decay into fundamental particles of different fields (electrons, neutrinos of different types, quarks, pions).
- The first three items satisfy the conditions of a precedence paradox. For two fields (A and B), field A cannot exist without the pre-existence of Field B. Likewise, field B cannot exist without the pre-existence of Field A. The general solution to any precedence paradox is that interdependent conditions must have one and the same thing the point of their origin (see Section 5.17, “A general solution to origin paradoxes”). In the case of the forces of nature, the solution of the precedence paradox is that all of the fields and their forces must have been unified as one, at some point in time.
- When we suggest that all of the observed fields in our universe came from a single field, we are equivalently suggesting that all of the symmetries of the universe derived from a single symmetry. We also know that symmetries may emerge from symmetries. Hence, the notion that our observable symmetries derived from a single symmetry seems plausible.
  As we shall discover in later sections, the laws of fields, and their forces, result from symmetries. Symmetries are subject to perturbations due to spontaneously occur-

---

vii Using terminology taken from quantum field theory, terms in the Lagrangian or the Hamiltonian depend on several fields. This concept is explained for advanced readers as a note in Chapter 11, where the Principal of Least Action, the Lagrangian, and the Euler–Lagrange equation are reviewed.[99] Proof of the Euler–Lagrangian equation requires application of the chain rule and integration by parts, both of which are proven as notes in Chapter 11.[100] [101]

viii Virtual particles are produced by small fluctuations of field energy. These fluctuations occur because a steady zero-energy state is prohibited by the Heisenberg Uncertainty Principle. Specifically, if the energy at any location were always zero, then we would be certain of the energy level, thus violating uncertainty. Virtual particles are short-lived and unobservable.

ring energy fluctuations that provide sufficient energy to "break" the existing symmetry rules for a short moment. This is the principle that made quantum tunneling microscopy a reality. When the symmetry rules are broken, it becomes possible, in theory, for a new and stable symmetry to emerge.[ix] [x] Because one symmetry can spawn additional symmetries, it seems reasonable to imagine that the symmetries responsible for the Standard Model and gravity may have all begun with one symmetry. If that were the case, and we can account for the one symmetry from which all other symmetries emerged, then we will have succeeded in explaining the fundamental nature of existence.

Of course, these points do not constitute proof, but they furnish optimism that a viable unified field theory may exist.

6.4 Forces and their mediators

The ingredients of our universe are subject to major revision at any moment. If a mathematician were to argue that the universe is just geometry and symmetries, it is unlikely that anyone could offer a rigorous counterargument. If a nihilist were to insist that reality is nothing but smoke and mirrors, then who are we to disagree? Ultimately, what we think we know about fundamentals is simply conjecture. Fortunately for physicists, many of the predictions based upon our popular conjectures have been verified. We cannot expect more from science.

Summarizing what we currently believe, our universe consists of forces (of which four have been described), particles (i. e., force mediators, of which several dozen are described in the Standard Model), and constants (whose role is to determine how the forces relate to one another). Fields are essentially vector descriptors of properties at all points throughout spacetime. Spacetime is the geometry in which forces and other properties hold sway.

The four known fundamental forces are:

1. The electromagnetic force (accounting for light).
2. The strong force (accounting for keeping protons and neutrons together in the nucleus of atoms despite the strong repulsive force between like-charged protons).
3. The weak force (accounting for nuclear radioactive decay processes).

---

ix For a detailed discussion of this subject, refer to: Dimakis N. Generation of new symmetries from explicit symmetry breaking. arXiv:2112.02776v1, Dec. 6, 2021.

x Antonio Garcia-Bellido has suggested that the evolution of species occurs through a process similar to symmetry breaking, writing "The evolution across several levels of complexity from atoms to molecules, to cells, to organs, and to organisms is an iterated symmetry break." See his manuscript, titled "Symmetries throughout organic evolution", in Proc Natl Acad Sci USA 93:14229–14232, 1996.

4. The force of gravity (accounting for the thrill of bungee jumping).[xi]

In addition to the fundamental forces, we have a number of derivative forces that emerge when the fundamental forces act within a complex system consisting of many particles and other types of forces. Examples of emergent forces are the force of friction, surface tension, and the Van der Waals force.[xii]

How do fundamental forces exert their effects? The fundamental forces are mediated by particles.[xiii] The universe's known forces are presumed to have existed since about 10^{-43} seconds after the Big Bang.[xiv] [xv]

Yet another property that seems to evade any simple definition is "mass". The law of conservation of momentum instructs us that mass is what we must overcome to accelerate an object by the use of force (i. e., $F = m \times a$). Rephrasing, mass is what resists getting our object from its initial velocity up to its final velocity. As such, mass is not a force; rather, mass is something that gets in the way of force. The notion that mass is immaterial is made clear when we examine Einstein's most famous equation:

$$E = mc^2.$$

Equivalently,

$$m = \frac{E}{c^2}.$$

Because mass can be expressed in an equation that excludes particles and their properties, we can argue that mass is not an intrinsic property of matter. In point of

xi Einstein argued that gravity is not a force; it is simply a curvature in spacetime. Even so, physicists prefer to include it with the other three fundamental forces insofar as it behaves like a force and is fundamental.

xii The Van der Waals force is the weak electric distance-dependent attraction between neutral atoms or molecules.

xiii The force carriers described by the Standard Model are photons (mediating the electromagnetic force), gluons (mediating the strong force), W and Z bosons (mediating the weak nuclear force), and the Higgs bonson (conferring mass). Outside the Standard Model, it has been suggested that gravitons exist that are the carriers of gravity.

xiv In the first 10^{-43} seconds after the Big Bang, current physical theories break down, and no one can say, with any confidence, how things may have been. Let us pretend, for now, that it does not matter.

xv Have we just created another paradox for ourselves? A force cannot be mediated without particles, but particles arise from forces. If forces cannot act without preexisting particles, and particles cannot come into existence without forces, then how can either particles or forces exist? It happens that this apparent paradox, like the origin paradoxes discussed in Chapter 5, is solved by understanding that particles and forces are dual manifestations of wave functions, operating in a quantum field. Our discussion of wave functions and fields will be saved for Section 3.2, "Nothing but waves".

fact, nearly all of the mass of the universe is accounted for by composite particles: protons and neutrons. In turn, each proton and neutron is composed of a few quarks (having themselves negligible mass), and gluons (having zero mass).[xvi] How can a massive particle (such as a proton and neutron) be composed of a few fundamental particles that have little or no mass? As it happens, the mass of the proton derives from the mass-equivalent energy that confines its constituent quarks (i. e., via the gluon-mediated strong force keeping the quarks from repelling one another) [253]. The small fraction of mass of the proton and neutron that is not accounted for by the strong force energy of confinement is attributed to the Higgs field, which impedes the passage of particles and confers mass upon them.[xvii] [103]

6.5 What is a fundamental particle?

Scientists tend to understand objects in terms of the more fundamental objects from which they are composed. For example, we say that an animal is composed of cells, cells are composed of a membrane-bound bag of fluid in which a nucleus and mitochondria various organic constituents share their space. We can go on to describe these constituents in terms of their molecular contents. A molecule of water is not a fundamental object because it is composed of two atoms of hydrogen and one atom of oxygen. An atom of oxygen is not fundamental because it is composed of neutrons and protons and electrons. Neutrons and protons are not fundamental because they are composed of various types of quarks trapped together by gluons. We continue to peel away at composite particles until we can go no further. What do we find? We find a fundamental particle that is composed of nothing. If the particle were composed of a ball of "stuff", then whatever composed the "stuff" would be more fundamental than the particle. Hence, there is no "stuff" in a fundamental particle. Furthermore, we find that our fundamental particles have no assignable size. If they had size, they would fill some region of space, and this would require some amount of "stuff", of which they have none. Therefore, fundamental particles are made of nothing, and they have no size (i. e., they are points in a field).

To summarize, the fundamental particles are not composed of other particles and do not consist of any composite matter (i. e., are not composed of anything and therefore are made of nothing). All of the fundamental particles are point particles, having

xvi The mass of a proton is about 938 MeV. The proton is composed of three quarks, two up quarks, each with mass 2.3 MeV, and one down quark with mass of 4.8 MeV, together accounting for 9.4 MeV of mass. 99 % of the proton's mass comes from confinement energy.

xvii How the Higgs field came into being is somewhat of a mystery, but it presumably arose at the same time as all the other physical fields. If not, the fundamental particles in those fields would have had no mass. But we know that fermions, present at the birth of the universe, have intrinsic mass, suggesting that the earliest universe contained the Higgs field. All field forces seem to be mediated by field-specific particles, and the celebrated mass-mediating particle of the Higgs field is the Higgs Boson.

no size. Having no size and no physical content, it seems a moot point as to whether such particles have a finite lifespan. As it happens, all fundamental particles are eternal, unless some interaction with other particles leads to their demise. The immortality of fundamental particles is discussed further as a note in Chapter 11.[xviii] [104]

If we accept these statements to be true, then we can see that the fundamental particles are only distinguishable from one another by their field properties. If this is the case, then we must ask whether our universe is an illusion. Let us not indulge our fantasies. We will add a touch of realism to this discussion when we discuss the quantal nature of fundamental particles.

If we were to think of fundamental particles as being composed of "stuff", we would be promoting the same mistake as the 19th century physicists who believed that the universe is filled with a mysterious substance, which they called "the ether." In either case, physicists could not liberate themselves from the belief that particles were obviously some kind of "stuff" and electromagnetic waves required some physical "stuff" to conduct the wave motion. Today we know with certainty that there is no invisible "stuff" that conducts waves or that mediates forces that act at a distance. Electromagnetic waves, unlike sound waves, can move through a vacuum. Similarly, the fundamental units of matter are, in fact, empty. It may be better to think of quarks and gluons and the other fundamental particles as physical properties operating in a small locality in space. When fundamental particles are combined into the familiar constituents of atoms (e. g., neutrons, protons), we see them as carriers of forces, through fields. For example, the electron is the carrier of charge. Charge is energy, and energy can be expressed as mass. Consequently, the electron has properties that we associate with matter. The properties of fundamental particles include their quantum states, and we can think of the quantum states as descriptors of points in a field. The quantum states of fundamental particles account for much of their observable behavior, and this happens to include the principle that two fermions (e. g., electrons) having the same quantum state cannot occupy the same location at the same time.[xix] This important feature of fermions accounts for what we observe as the solidity of fermionic matter (e. g., the inability of a table to slide through the floor). Altogether, we are building the argument that subatomic particles can be composed of nothing, and that what we perceive as "stuff" (e. g., refrigerators and chocolate pudding) is simply the consequence of subatomic properties playing out on a large scale.

Because fundamental particles are composed of nothing, we can accept the idea that particles can appear and disappear in accordance with universal laws of conservation, and within the limits of their own sets of particle properties (i. e., quantum restrictions).

---

xviii So-called virtual particles are extremely short-lived phantoms with subquantal energies, serving only to mediate interactions among the fundamental particles. They figure prominently in the calculations of quantum mechanics but are not part of the physically observable universe.

xix This is Pauli's exclusion principle, which we will describe later in this chapter.

If fundamental particles were made of "stuff", it might be hard to create a fundamental particle out of nothing. To preserve a net zero energy balance, we observe that each newly created electron is coupled with the creation of a positron (identical to the electron, but of opposite charge).

Does the net zero charge of the universe hold on a small scale? In particular, would it be possible to suddenly create one unit charge somewhere in space, from absolutely nothing. Would the universe really care? Physicists seem to have provided an answer to this simple but profound question. Charge conservation, on any scale, was formally proven as a consequence of a mathematically sophisticated theory, principally developed by Hermann Weyl, now known as gauge invariance [260, 192, 40].[xx] The mathematics of this theory lies beyond the scope of this book and this author, but it is possible to summarize the argument in words. Gauge theory concerns itself with certain quantitative properties in nature that cannot be directly measured. These are understood in terms of mathematical transformations that preserve symmetries (i. e., gauge invariances). One such item that cannot be directly observed is voltage. It is impossible to simply measure the voltage (electrical potential) of a point in space. We can only measure the difference in potential between points. Traditionally, the voltage at a point is measured by a comparison to the reference potential of Earth. Accepting that voltage is only measured in relative terms, then lets consider a gedanken experiment in which we imagine that charge is not locally conserved. In this alternate universe, we can create a new unit of charge or destroy a unit of charge, at will. A unit of charge would be the charge carried by a single electron. The energy carried by a unit of charge is known to be equal to the charge times the voltage:

$$\text{Energy of charge creation} = (\text{charge produced})\ x\ (\text{voltage at the point in space where the charge is produced}).$$

Hence, if we measure the energy that was spent producing a single quantum of charge, we could calculate an absolute value for the voltage. However, doing so would violate gauge invariance of voltage, which is known to be a quantity that cannot be measured absolutely. That is to say that voltage is always a relative measurement of the differences of electrical potential at two points. Thus, if charge is not conserved, then gauge invariance is violated. Gauge invariance cannot be violated. Therefore, charge is conserved, even for one lonely electron.

Let us pause a moment to think about the significance of charge conservation. Firstly, if charge is conserved, we can infer that particles that carry charge are stable

xx The gauge symmetries were developed for quantum electrodynamics. Unlike the timespace symmetries of classical physics, the gauge symmetries consist of mathematical calibrations that enforce a symmetry upon electromagnetic waves. For example, a gauge symmetry ensures that longitudinal waves produced by quantum fluctuations of electromagnetic transverse waves, will have no physical consequence.

and extremely long-lived. The charge-carrying particles are the positively charged proton and the negatively charged electron.[xxi] Protons are estimated to have a lifespan of at least 10^{34} years. Likewise, electrons are thought to have a mean lifetime exceeding 10^{24} years. Compare either with the estimated age of the observable universe, 3.8×10^9 years. As we might expect, the balance of positive and negative charges in the universe is precisely maintained. The difference in charge magnitude between a positive positron and a negative electron is equal to within a factor of 10^{-21}, and every neutral atom is evenly matched with positrons and electrons.

Given that the law of charge conservation seems to be rigidly enforced, what might happen if an electron were to suddenly vanish, in violation of the Law. Would the universe simply accept the loss and go on about its business, as usual? Or would there be hell to pay? We do not know, but it is worrisome to think that Nobel Prize winner Richard Feynman and the esteemed theoretical physicist John Wheeler had advanced the notion of a one-electron universe, in which a single electron zigzagging throughout space and time accounts for every observed occurrence of electrons [218].[xxii] If their hypothesis were true, then the decay of a single electron would result in the disappearance of every observed electron. Without electrons, atoms would cease to exist, and all physical matter, including this book, would disintegrate. Let us hope that, in this case, reality does not confirm theory.

6.6 First proof that atoms exist

It may seem difficult to credit, but the atomic theory of matter was unproven until just after the first decade of the twentieth century. As late as 1905, Ernst Mach and Wilhelm Ostwald, both eminent scientists of their time, insisted that the fundamental component of physical existence was energy, and that atoms were fictions invented as a sort of intellectual crutch to help us cope with an inscrutable physical reality [186]. Until there was proof that atoms existed, some of the greatest advances in the field of physics and chemistry were performed speculatively, on the expectation that atoms were real.

Before we delve into the modern theory of atoms, let us give some thanks to the ancient Greek philosophers. The person credited with the first atomic theory of matter is Leucippus of Miletus (5th century BCE). His thoughts on the subject were expounded by his student, Democritus of Abdera, who coined the term "atom" about 430 BCE. Atoms, according to Democritus, were uniform, solid, hard, incompressible, and indestructible, with the property of movement through empty space until stopped. How did those ancient Greeks become so smart?!

---

xxi One hydrogen atom says to another, "I've lost my electron!" The other says, "Are you certain?", and the first atom replies, "Yes, I'm positive!"

xxii See Richard Feynman's Nobel Lecture presented December 11, 1965, and published by the Nobel Foundation.

The next advances in our thinking came about 2,300 years later. In the first decade of the 19th century, John Dalton, building on the works of Antoine Lavoisier and Joseph Proust, published his own atomic theory of matter. His theory contains several assertions that are generally correct:

1. All matter is made of atoms, which are indivisible (nothing new here).
2. The atoms of any element are identical to one another.
3. Compounds are composed of combinations of atoms of elements.
4. Chemical reactions are just re-distributions of atoms among reactants (no atoms gained or lost).

Dalton knew nothing about isotopes of atoms and subatomic particles, but let us not quibble. Dalton's atomic theory was a major breakthrough, but it remained an unproven theory until the 20th century. In the interim, some of the greatest advances in the field of physics and chemistry were performed speculatively, on the expectation that atoms were real. What brought scientists to believe in the atomic theory before it was proven?

1. The polygonal shape of crystal units.
   In the 19th century, one of the best arguments, short of proof, for the existence of atoms came from observations of crystals. Because all naturally occurring crystals are composed of repeating units of a particular polygonal shape, it was concluded that discrete atoms occupied specific positions with respect to one another. Knowing that crystals are composed of the same elements found in gases and liquids and non-crystalizing solids, it seemed reasonable to infer that if crystals are composed of atoms, then all matter is composed of atoms.
2. Stoichiometric chemical reactions.
   If matter were composed of material of a non-particulate nature, then there would be no compelling reason for chemical reactions to proceed with specific proportions of constituent materials. For example, there would be no reason to have water molecules composed of exactly two parts hydrogen and one part oxygen.
3. Einstein's theoretical explanation of Brownian motion (1905), which assumed the existence of discrete atomic particles.
   Brownian motion (Figure 6.1) is the random short movements of particles suspended in water. It may seem that any particle in water should either fall as a sediment or float upwards, or lay in a still suspension. This is not the case. Suspended particles bounce around, to and fro, up and down, in short random jerky movements known as Brownian motion. Einstein reasoned (in 1905) that Brownian motion could only be caused by constant collisions between water molecules and the suspended particles. He further reasoned that such collisions could only occur if the atomic theory of matter held. Specifically, atoms of water agitated by temperature bounced constantly against suspended particles, causing them to move in rapid, random movements.

FIG. 9.

Figure 6.1: Perrin measured the perturbations of various particles suspended in water. The path of particles, studied at 30 second intervals, indicated random motion. Had Perrin reduced the time interval for observation (e. g., 10 seconds, 1 second, 1 millisecond), the picture would look much the same, but with a shorter mean path distance. Source, Figure 9 from Perrin's book "Les Atomes", published in 1913 [198].

4. Perrin's experimental confirmation of Einstein's theory of Brownian motion. In 1909, Perrin provided experimental proof of the existence of atoms based on the observed effects of Brownian motions [198].
   Perrin was able to determine Avogradro's number (the number of molecules in a mole) with great precision, for any of the types of suspended particles or solutions he cared to study. The formula used by Perrin was:

$$D = \frac{RT}{N} \times \frac{1}{6\pi a\xi},$$

   where D is the coefficient of diffusion, R is the universal gas constant, T is the absolute temperature, N is Avogadro's number (about 6×10^{23}), ξ is the viscosity of the fluid, and a is the radius of spherules (e. g. radius of a water molecule). Because

both the radius of molecules and the number of molecules could be computed from observations of Brownian motions, it seemed clear that atoms must exist.
5. X-ray crystallography. X-ray crystallographic analysis (1912) demonstrated the atomic nature of crystals.

By 1912, the atomic theory became the atomic fact. Of course, now that we have abundant proof of the existence of atoms, physicists feel obliged to challenge their own conclusions (that is what good scientists do!). What we believe to be solid particles can be thought of as wave packets in fields. The field properties of matter, which include the symmetry conditions of the various fields, dictate the observed behavior of wave packets. Today we have come nearly full circle, and what we think of as particles made of matter ore just oscillating waves in fields, obeying symmetry transformations.

6.7 The first moment of time

Obscurum per obscurius, ignotum per ignotius.
(Explanation more obscure than the thing it explains.)

– Latin admonition for instructors

When contemplating the origin of time, it is easy to take the easy way out and to simply say time is infinite, having no beginning and no end. At some moment in the past, the universe may have been completely devoid of matter and energy, but time was always passing, just the same. If there comes a time when there is no matter and no energy in the universe, then time will continue through the void, without end. The notion of eternal time, tempting as it may seem, is actually quite problematic. Let us look at two of the conceptual difficulties posed by the "eternal time" hypothesis.

1. If time were infinite, then space must also be infinite (because space and time cannot be separated, and every point in space occupies a point in time). The condition of an infinite spacetime violates the law of energy conservation. The reason why this is so relates to the vacuum energy of empty space. It has been predicted on theoretical grounds, and proven to be true experimentally, that empty space (devoid of matter or forces of any kind) contains fluctuations of energy (as described by the Casimir's effect, to be discussed in Section 3.7, "Nothing, the key to everything"). In an infinite spacetime universe, these fluctuations could add to yield a universe with infinite energy (positive or negative energy). If the universe had infinite energy, then there would be no law of energy conservation.
2. For those fundamental particles that travel at light-speed, time does not exist.[xxiii] For example, the photons, like all fundamental particles, have no size. In addition,

xxiii Particles that travel at the speed of light are called bosons, and include photons, gluons, the Z and W bosons, and the Higgs boson.

> they are timeless in the sense that they travel at light speed, wherein the passage of events cannot occur — they are stuck in the "eternal now".[xxiv] The reason for this is that all events depend on the movement of light for their observation. Light cannot move past a photon because photons move at the speed of light. Hence, for a photon, time stands still. When we accept that we live in a universe populated with particles for which time does not apply, then we can begin to doubt that time is a fundamental condition of existence.[xxv]

We feel comfortable inventing infinities (e. g., infinite distance, infinite time), without much justification. Perhaps we do so because we are short-lived, impatient creatures with a greatly exaggerated sense of duration ("Why is this turkey taking FOREVER to cook!?). Space and time are constructs that deal with changes; changes in the position of an object in a field, over time. We do not really need to have an infinite amount of spacetime to accommodate our calculations; we only need some way of understanding relative changes in spacetime position. Mathematicians have a very simple way of dealing with continuous changes; it is called calculus. With calculus, we can shrink time to a single moment that continuously transforms into the next moment, proceeding indefinitely to produce what seems like an infinite duration. Viewed this way, there is no beginning of time. There is only the continuous "now" that carries forward all of the matter, energy, and properties of the universe from moment to moment.

6.8 The Big Bang, muffled

Let us quickly dispense with the Big Bang, one of the few topics that can spark interest in cosmologists and cosmetologists alike. So much has been written about the Big Bang that it would be redundant to repeat the story here. Let us just summarize by asserting that about 13.8 billion years ago, there was a cosmological event that has been likened to the starting pistol that set the universe in motion. The Big Bang, or something akin to it, is an historical event. We base our understanding of the Big Bang on cosmological observations of distant space, which provide a backwards look at our universe as it developed over time (Figure 6.2). To review just a few of the findings that support the Big Bang theory, consider the following:

1. Wherever we look in the skies, galaxies are receding from us. This observation, discovered by Vesto M. Slipher in 1912, was confirmed by Edwin Hubble (1929), who further showed that the speed at which galaxies were moving away from earth was

---

xxiv To the best of my knowledge, Karl Pearson, in *The Grammar of Science*, first published in 1892, introduced the concept of the eternal now. Pearson speculated further, suggesting that an observer traveling faster than light would see time in reversal [197].

xxv Time is basically a property of fermions (particles traveling less that the speed of light), with non-integer spin. As such, time functions to ensure that fermionic events do not all happen at once.

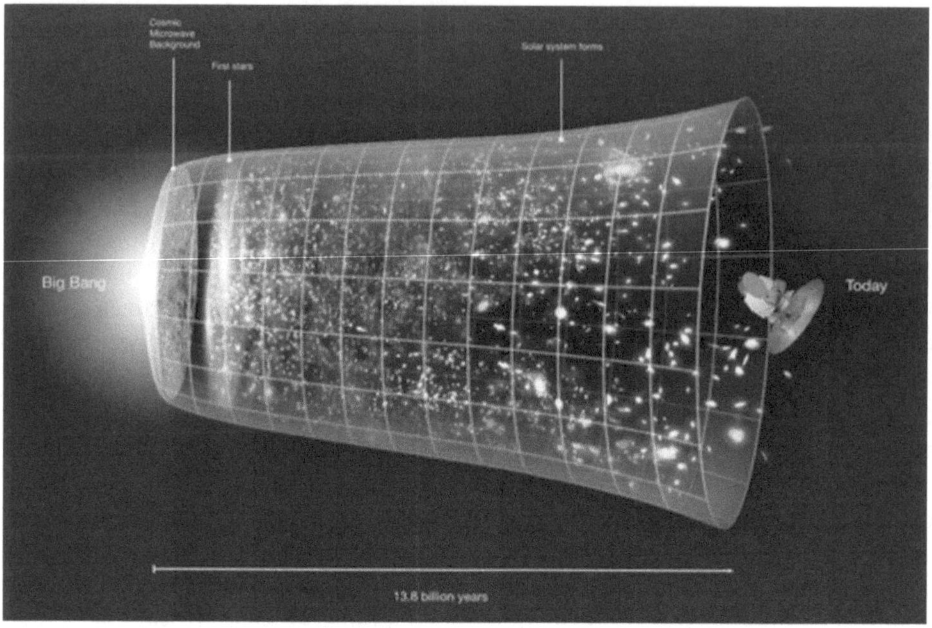

Figure 6.2: Timeline of the universe, from the moment of the Big Bang, until the present. Galaxies and satellites are not actual size. Source, U.S. government public domain document, provided by the NASA/WMAP science team.

directly related to the distance from earth (i. e., the further the galaxy, the greater its speed). Early observations led Alexander Friedmann (1922) to suggest that the universe is expanding and led Georges Lemaître (1927) to suggest that expansion began at a moment in the distant past when all matter was confined to a single point; then suddenly released.

2. Hydrogen and helium account for nearly all the atoms in the universe. This suggests that there was a moment when matter existed as a condensed hot plasma of protons, neutrons, and electrons.
3. There is a cosmic background of radiation that is present in all directions, and this background looks about the same everywhere. This suggests that the aforementioned hot plasma cooled to the point where electrons were captured by light nuclei (protons or proton-neutron combined), releasing confined photons at once, filling the universe with a background radiation with an energy distribution corresponding to a black body spectrum corresponding, at that moment, with the prevailing temperature of the plasma. Because the cosmic background radiation is the same wherever we look, we can assume that the event filled the early universe, from a single location and moment.
4. Space telescopes are basically machines that view the past, permitting us to view the development of the observable universe, beginning with Big Bang. When as-

tronomers announce that they have detected a star that is located 5 billion light years from Earth, we are really saying that we are currently receiving starlight that has traveled 5 billion years after being emitted by a star. What the astronomers see is what the star looked like 5 billion years ago. The current location of the star, and whether the star even exists at the current moment, is unknown to us. By viewing stars at increasing distances from Earth, we can reconstruct a sequence of events moving backwards in time. Consequently, astronomers physicists have a fairly good understanding of how the universe changed over time, leading back to a moment when the universe was hot and dense and tiny.

5. Despite our best efforts, we cannot find stars whose light traveled more than about 13.5 billion years before arriving here on earth. The absence of older starlight strongly suggests that stars did not exist more than 13.5 billion years ago. Allowing some time for the first stars to form following a Big Bang event, this would suggest that the Big Bang probably occurred about 13.8 billion years in the past. Conversely, if we found a star that formed, say, 20 billion years ago, then we would have proof that the Big Bang, as currently conceived, must not have occurred. No such star has been found.

All five observations from which the Big Bang theory emerged have the advantage of hindsight. Retrospective analysis is easier and more reliable than prediction insofar as we can be more certain about the past than the future. Scientists have used observational data to trace a cosmic timeline that leads back to a hot event occurring suddenly, in a tiny spot, about 14 billion years ago.

Let us not get stuck in the weeds. Allowing that the Big Bang actually occurred, we don't really know where the Big Bang lies in the sequence of events that litter the space-time continuum (i. e., our universe may have existed long before the Big Bang). We do not know why the Big Bang occurred. The very first moment of the Big Bang (preceding the first 10^{-13} seconds of so-called epoch time) lies in mystery. Physicists have suggested in the very first 10^{-43} seconds of epoch time, the laws of quantum physics could not apply.[xxvi] We have no proof that the Big Bang had anything to do with the beginning of the universe, and it is only a matter of speculation that the universe had a beginning.[xxvii] There is a great deal of conjecture regarding the origin of the Big Bang. For the discussions to follow, we shall stipulate that the Big Bang was a cosmological event that

---

xxvi On a simplistic level, Werner Heisenberg's uncertainty principle, a bulwark of quantum physics, is violated when the diameter of the universe is smaller than a Planck distance.

xxvii The smallest unit of time that has physical meaning is the so-called Planck time, which corresponds to 10^{-43} seconds. The Heisenberg uncertainty principle would suggest that a time interval less than the Planck time is incompatible with physical law, as we know it. Therefore, assuming that the universe began at the moment of the Big Bang, we can infer that the laws of physics could not apply in the first 10^{-43} seconds. This implies that our physical laws came into existence moments after the first 10^{-43} seconds of the Big Bang epoch.

occurred about 13.8 billion years ago. In the aftermath, one thing led to another until, ultimately, this amazing book materialized!

6.9 Making a point with identicality

We live in a universe where unique objects (such as you and me) lose their uniqueness when parsed at the subatomic level. Every seemingly unique object is composed of a few varieties of perfectly identical (i. e., non-unique) fundamental particles. Because of identicality, we can infer that a fundamental particle does not change with time — if a fundamental particle were to change over time, then it would become non-identical to other particles of the same kind. Because fundamental particles do not change, we can infer that stable particles present at the time of the Big Bang are still present today.[xxviii]

All of the non-identical items that exist today (e. g., cream-filled donuts, stars, and black holes) arose as emergent forms resulting from the interactions of multiple identical particles. We can use our highest precision tools to produce two identical ball bearings, but we know that our best efforts are imperfect. Likewise, "identical" twins are never truly identical. Geneticists inform us that identical twins are born with the same exact genetic sequence in their respective genomes, but we know that this is not strictly true. Gene replication is never flawless, and a host of epigenomic influences, distinguish one twin from another well before birth. Identical twins are at best an approximation of one another. It is only the fundamental particles of nature that are truly identical.

We find evidence of identicality only in atoms and their subatomic constituents. Let us look at some examples, leading up to an experimental proof that particles of a kind are identical to one another.[xxix]

1. Distinct elements.

   It would be easy to imagine that chemists in the 19th century may have supposed that elements, like everything else they observed, were composed of similar (not identical) units, and that some of the units of one element might be indistinguishable from the units of another element having similar chemical properties. In this case, the collection of elements could have been viewed as a kind of spectrum, where one element blends into the next element, with only arbitrary cut points to distinguish where one element ends and the next begins. If such were the case, it would be impossible to build the periodic table of the elements, because there would be no sensible way of assigning atoms to elemental classes.

---

xxviii Fundamental particles do not change with time, but unstable or short-lived "virtual" particles can drop in and out of existence.

xxix There is more to come. We will see in Chapter 9, "The Mathematical Universe," that the identicality of like particles is a fundamental property of spacetime, resulting from the symmetries that determine the conservation laws upon which our universe operates.

When Mendeleev submitted his first draft of the Periodic Table, back in 1869, he relied heavily upon Dalton's atomic theory of matter, first published in 1808. Dalton asserted that all atoms of a given element are identical to one another, in size, mass, and every measurable property of matter. It was Dalton's concept of the identicality of atoms of any particular kind that provided Mendeleev with a rationale for creating and enumerating classes of atoms (i. e., elements). Of course, we now know that there are different isotopic forms of every element, which differ from one another according to the number of neutrons in the nucleus. Nonetheless, for any particular isotope of any particular element, one atom is identical to the others.

Although Dalton asserted that the atoms of any element are identical to one another, and Mendeleev had accepted Dalton's assertion in the best of faith, the proof of Dalton's theory did not come until late in the 19th century, well after Mendeleev had proposed his Periodic Table and before the arrival of quantum physics.

2. The regularity of crystals.

   Crystals are formed by atoms fitting into an orderly structure composed of repeating units (Figure 6.3). If atoms were composed of non-identical particles, there could be no crystals. All solid matter would be lumpy.

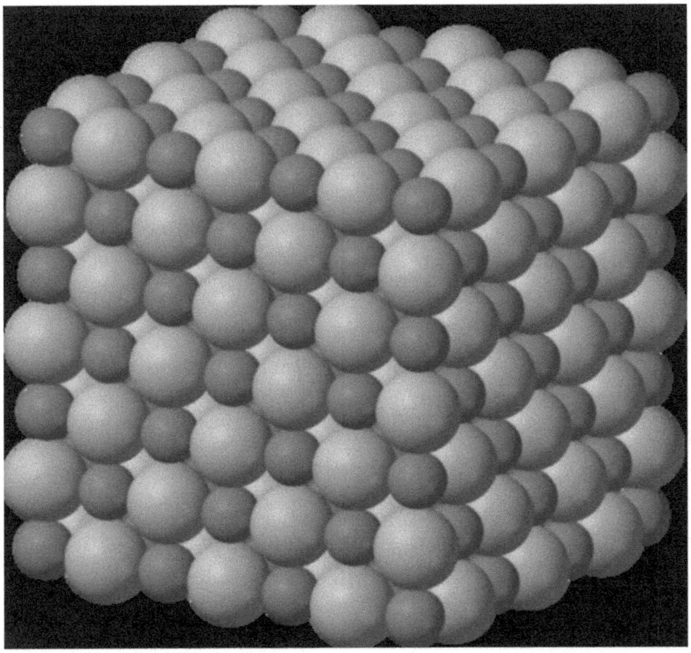

Figure 6.3: A graphic representation of a sodium chloride crystal, with large spheres representing sodium cations and small spheres representing chloride anions. The regularity of the aggregate crystal is achieved only if it is composed of identical sodium atoms and identical chlorine atoms, both having characteristic size, and occupying their own specific positions in the crystal. Source, Wikipedia, and entered into the public domain by its author, Benjah-bmm27.

3. Charge balance.
   If two particles are identical, we would expect their properties to be equal. All electrons have the same exact electric charge. As it happens, the charge of an electron is exactly balanced by the charge of a proton.[xxx] If the charge on different electrons (or different protons) were even the tiniest bit unequal, then matter would carry an excess electric charge (i. e., the sum of the differences in charge between electrons and protons would not be zero). In that case, matter would be repulsive to matter (because like charges repel). Once again, the universe as we know it could not exist without identicality.
4. Particle-antiparticle annihilation.
   When particles encounter their antiparticles (proton/antiproton, electron/positron, neutrino/antineutrino), they self-annihilate. We would only expect this might happen when each particle is exactly balanced by its antiparticle.

Casual observations lead us to presume that particles of a kind are identical to particles of the same kind. Is there any experimental evidence leading to the same conclusion? Yes, indeed. The experiment-based argument for the identicality of fundamental particles rests on the second law of thermodynamics, which holds that the entropy of isolated systems left to spontaneous evolution always achieves a state of thermodynamic equilibrium for which the entropy is maximal (i. e., maximal particle randomness). A strange phenomenon occurs, defying simple explanation, whenever we mix different types of gases; the so-called mixing paradox. If there are two different types of gases, held under the same conditions (volume, pressure, temperature, and number of atoms), the total entropy of the system, when the two different gases are mixed, is greater than the total entropy if we had mixed two volumes of the same gas. The lack of entropy change observed when gases of the same type are mixed is an apparent violation of the second law of thermodynamics (hence, a paradox). [102]. Explaining the mixing paradox has been an ongoing labor involving contributions from the fields of quantum chemistry, quantum thermodynamics, and classical entropy physics. For whatever reason, the mixing paradox, also known as the Gibbs paradox, seems to be an empiric fact. Because gases of the same type, when put together, do not increase their entropy, we infer that the individual atoms of the gas must be identical to one another. Because the atoms in gases are composed of fixed numbers of electrons, protons, and neutrons, we can infer that each of these fundamental particles is identical to all fundamental particles of the same type (e. g., one electron is completely identical and interchangeable with every other electron).[xxxi 105]

---

xxx Experimentally, the charges of electrons and protons are equal, within a factor of 10^{-21} [196].

xxxi For readers who take an information theory approach to physics, we can think of the "mixing" paradox as a demonstration of entropy. When we mix identical particles, there is no change in the information needed to describe the system, no change in entropy, and no change in heat. When we mix

The identicality of particles of a given type reinforces our notion that the fundamental particles are made from nothing. If the fundamental particles were made from matter, then particles of a type would not be exactly identical to one another. We might expect to have one particle to contain just a bit more or less "stuff" than the next electron. Or a particle might lose some of its "stuff" over time. Or relativistic effects would cause the particle to change its size as it moved. And we would no longer be able to claim that a fundamental particle has no size.[xxxii]

If fundamental particles are not made from matter, then what are they? Every fundamental particle is a point particle (i. e., having no size) that consists of nothing other than its properties. Put differently, a fundamental particle is the set of properties that determine what we observe as the physical nature of the particle. This definition is actually quite useful in that we need not ask ourselves what is inside a fundamental particle (nothing material is inside), and we need account for the size of a particle in our calculations (it has no size). We need only concern ourselves with its particle-specific properties, which are experimentally observable. Functionally, fundamental particles carry or mediate forces, according to their type properties. The forces mediated by fundamental particles seem to be limited to the forces described in the Standard Model (electromagnetism, the strong nuclear force, and the weak nuclear force) plus gravity.

Hold on. If the fundamental particles are just properties and are not composed of "stuff", then how is it possible for one fundamental particle to decay into several different particles? For example, we know that one τ particles can decay into several fundamental particles (electrons, neutrinos of different types, quarks, pions). Only particle physicists seem capable of keeping track of all the options. If a fundamental particle is composed of other particles, then it cannot be composed of "nothing". Well, no. When a particle decays, it does not decay into its constituent parts; it transformes into other particles in accordance with its properties. We cannot open up a τ particle and expect a neutrino and a pion to pop out![xxxiii]

different things, we increase the information needed to specify the system, with a consequent increase in entropy and heat.

xxxii We will come to understand that a fundamental particle is a field's characteristic wave packet. As such, the fundamental particle has no size, but it occupies a space circumscribed by its possible locations (basically, the spread of the wave packet in the field, at any point of time.)

xxxiii If we wished, we could argue that τ decay can only occur if τ and all of its possible combinations of decay products, are composed of nothing. If a τ particle (all of which are identical to one another) contained sub-particles, then we would expect to find only one possible set of τ decay products, consisting of those particular objects held inside the τ. Because there are many options for τ decay, including non-overlapping sets of particle products, we can surmise that τ could not have contained any particular set of sub-particles. Particles suddenly decay and other particles suddenly appear, according to the field properties.

6.10 Identicality and Pauli's exclusion principle

In this section, we will be discussing Pauli's exclusion principle, a scientific advancement on par with Einstein's special theory of relativity. The exclusion principle can be stated in a few unpromising words: "Two or more identical particles with half-integer spins cannot occupy the same quantum state within a quantum system simultaneously." Do not be fooled, when Wolfgang Pauli formulated his exclusion principle, in 1925, he provided the solution to several of the most important puzzles facing physicists of his day.

After the periodic table of the elements was established as a reasonable way to classify atoms, Niels Bohr set about providing a physical justification for the periodicity of properties displayed by atoms. Bohr suggested that the electrons are arranged in shells that orbited the nucleus. Each shell had a certain number of allowed electrons, and when the allowance was met, the shell was complete. Any additional electrons would need to occupy a higher orbit. The properties of atoms were determined largely by the total number of electrons of the atom, and these electrons were distributed into the different orbitals within shells. As more was learned about quantum theory, Bohr saw that electrons could transition between available spaces in shells by absorbing or releasing quanta of energy. The spectrum of light produced by a heated element consisted of a series of discrete lines corresponding to the characteristic quanta of energy absorbed when electrons shifted orbitals.

The known observations of the elements in the periodic table, fit well with Bohr's atomic theory, except for a few loose ends, but these loose ends grew to dominate the new physics that was born at the dawn of the twentieth century: quantum physics. For particle physicists the loose ends were largely solved by Pauli's exclusion principle.

1. Why is matter, which we know to consist largely of the empty spaces through which electrons orbit, solid? As a specific example, why don't chairs sink through the floor, as their atoms fall through the spaces of the ground beneath?
2. Why do the electrons fill the shells of atoms up to a defined maximum number for each shell, and according to a periodic rule?
3. What is the physical basis of the anomalous Zeeman effect (to be discussed forthwith), a phenomena that appears to defy any sensible understanding of the physical world?
4. Why is it that some types of fundamental particles (such as photons) can superpose upon other fundamental particles of the same kinds, while other particles (such as electrons and protons and neutrons) cannot? What strange forces might be involved in this process?
5. Why don't orbiting electrons spiral into the nucleus, when they lose energy? This problem was solved using the Schrödinger's wave equation, but we can add that

another way to solve this same problem is provided by the Pauli exclusion principle.[xxxiv]

Yes, Pauli managed to answer all of these questions with one principle, and he based that principle entirely on the fact that fundamental particles of any type (e. g., electrons, photons, quarks, neutrinos) are all completely identical to all other particles of the same type.[xxxv] We notice that Pauli's breakthrough is aptly referred to as a "principle", not a formula or an equation. Like Einstein and his theory of special relativity, the brilliance of Einstein was largely conceptual. Einstein saw the relationship (between time, space, and the invariance of the speed of light) that others failed to see. Similarly, Pauli saw the relationship between identical particles that others missed. Because Pauli's exclusion principle accounts for initial conditions following the Big Bang, and because it explains much about our physical universe, we will devote some time to its discussion. Furthermore, because the Pauli Exclusion Principle is conceptual, we can describe its proof without resorting to a long mathematical derivation.

On a certain level, Pauli's exclusion principle is trivial. Our everyday experience informs us that a teacup placed on a table will not sink through the molecules of the table, falling to the floor, and onwards, to the center of the Earth. When we examine the teacup as a collection of particles having no size, and made of nothing (i. e., no constitutive matter), it becomes much less obvious that a teacup cannot simply fall through the spaces of a table, and pop out on the other side. Pauli's exclusion principle assures us that teacups cannot fall through tables, but how do we know that the exclusion principle is valid?

For background, we must understand the concept of symmetric and antisymmetric function. A function is symmetric if its value stays the same when the variable that it acts upon changes its sign. A function is antisymmetric when the value of a function changes its sign when its variable changes its sign.[xxxvi]

$$f(x) = f(-x) \quad \text{Symmetric function,}$$
$$f(x) = -f(-x) \quad \text{antisymmetric function.}$$

---

xxxiv Wolfgang Pauli applied his exclusion principle to a vexing problem unjustly ignored in standard physics books. When asked, "Why did the chicken cross the road?" Pauli deftly applied his exclusion principle replying, "Because there already was a chicken on that side of the road!"

xxxv Readers may notice that we use the terms "fundamental particle" and "point particle" interchangeably. We can do so because all of the fundamental particles are made of nothing (but their properties), have no physical size, and can be represented geometrically as points. Non-fundamental particles, such as protons and neutrons, have a size determined by a containment space.

xxxvi Mathematicians manage to derive many useful theorems that distinguish the properties of symmetric and antisymmetric functions (also known as even and odd functions). Several of these theorems determine how the Fourier transform, named for Joseph Fourier, is calculated for symmetric and antisymmetric functions, and these theorems are frequently applied in the realm of particle physics.

The function that describes our knowledge of a quantum system is its wave equation, Ψ, and the absolute value of the square of a wave function tells us the probability that a particular quantum state occurs.[xxxvii]

$$|\psi(x)|^2 \quad \text{Wave probability function.}$$

The wave probability function for a quantum system containing two identical particles must not change when we switch one particle with another (that is what identicality means).

$$|\psi(x_{a,b})|^2 = |\psi(x_{b,a})|^2 \quad \text{Wave probability function.}$$

This condition holds true under two circumstances:
When we switch a and b and the wave function does not change,

$$\psi(x_{a,b}) = \psi(x_{b,a}),$$

or, when we switch a and b, and the wave function changes its sign

$$\psi(x)_{a,b} = -\psi(x)_{b,a}.$$

Thus, both symmetric and antisymmetric forms of the wave function satisfy the quantum probability wave function (and are equally possible and valid).

Let us imagine a situation in which we have two identical particles (particle1 and particle2) that are occupying the same quantum state (i. e., have the same quantum properties, such as energy level and position). In this case, their wave equation will be unchanged when we exchange one particle for another.

$$\psi(x_{\text{particle1,particle2}}) = \psi(x_{\text{particle2,particle1}}).$$

The value of the wave equation does not change when particles are switched because particle1 and particle2 are identical and indistinguishable in every way, including their quantum state.

This seems well and good until we remind ourselves that the wave probability function has two solutions: a symmetric solution (where the sign of ψ does not change when particles are switched) and an antisymmetric solution (when the sign of ψ changes when particles are exchanged). But we have just shown that the sign of ψ cannot change when identical particles occupying the same quantum state are exchanged! Therefore, there must be some restraint on the quantum state of two identical particles such that when

---

xxxvii If we were to sum the wave probability function over all possible quantum states, the probability must add up to 1, as in $\int_{-\infty}^{\infty} |\psi(x)|^2 = 1$, a simple trick that we will use when we derive the Heisenberg's uncertainty principle.

the identical particles are exchanged, the value ψ changes its sign and we have an antisymmetric solution. In other words, in order to produce our antisymmetric solution when we switch identical particles, we cannot allow those particles to both occupy the same quantum state. Amazingly, by just playing a bit with the meaning of identicality, we have arrived at the root concept of the Pauli exclusion principle!

That was just too easy. What's the catch? The catch is that we have just invented a universe in which there is a class of particles for which the wave equation is symmetric, and another class of particles for which the wave equation is antisymmetric. Two identical particles satisfying the symmetric wave equation can occupy the same quantum state, and identical particles satisfying the antisymmetric wave equation cannot both occupy the same quantum state. What in the world are these newly invented particles?!

Physicists have given the name bosons to the particles that satisfy the symmetric wave equation. Identical bosons can occupy the same quantum state simultaneously. The photon is an example of a boson that we are all familiar with. As described earlier, light waves, as with all waves, have the property of superposition (and the related property of wave additivity). Any number of light waves can occupy the same place at the same time. We can shine one beam of light, consisting of photons, through another beam of light, and they both emerge from the intersection unchanged by the encounter.

Physicists have given the name fermion to the particles that satisfy the antisymmetric wave equation. No two fermions can occupy the same quantum state simultaneously, and everything that we know as solid matter (atoms and molecules) are composed of fermions (i. e., protons, neutrons, and electrons). No two identical fermionic particles can occupy the same quantum state at the same time.

How does the quantum state of bosonic particles differ from the quantum state of fermionic particles, to account for their symmetric and antisymmetric (respectively) solutions of the wave equation? Pauli (and other physicists) added a new parameter to the description of the quantum states of particles, and he called this the spin of the particles, because it happens to conserve the particle's angular momentum. Bosons have an integer spin, and two identical bosons occupying the same quantum state may have spins of 1 and −1 (for example), and this allows the wave equation to have a symmetric solution. Fermions have a half-intiger spin, and when we assign spins of 1/2 and −1/2 to two electrons occupying an otherwise identical quantum state, we get an antisymmetric solution to the wave equation.

It turns out that Pauli's exclusion principle provided solutions to previously listed "loose ends" of the Bohr theory of the atom.

An important objection to Bohr's theory was the mysterious anomalous Zeeman's effect. In the 1890s, Pieter Zeeman had shown that when excited atoms were placed in a magnetic field, the spectral lines produced when electrons transitioned to lower orbitals were increased in number. Bohr and Summerfield launched a new theory to explain this phenomenon. In the new theory, each electron is provided with a set of 3 quantum numbers (n = orbital shell number, k = the shape of the orbital shell, and m = the direction in which the orbit is pointing). The quantum of electromagnetic energy released during an orbital shift is determined uniquely by the set of these three quantum numbers. In the presence of a magnetic field, the m value changes (i. e., the direction in which the orbit points changes), with 6 allowable values for m. Thus, according to the Bohr–Summerfeld theory, the spectral lines produced by transitions between orbitals can increase 6-fold under the influence of a magnetic field.

To everyone's relief, the periodic table and the emission spectra under a magnetic field seemed to be self-consistent. Soon, however, very fine analyses of emission spectral lines revealed even more spectral lines than were predicted by the Bohr–Summerfeld theory. Quantum theory was again in jeopardy. Was another quantum number needed?

We can now bring back Wolfgang Pauli, who provided further explanation that, as far as anyone can determine, has withstood every experimental challenge. Let us expand upon what Pauli suggested, and what it tells us about the behavior of matter.

Physicists in 1925 introduced the fourth quantum number (in addition to n, k, and m), which came to be known as spin and abbreviated as "s". The spin quanta are characterized by one of two numbers: spin plus one-half or spin minus one-half. History credits Ralph Kronig, George Uhlenbeck, Samuel Goudsmit and Wolfgang Pauli for their nearly synchronous contributions to this important breakthrough in quantum physics. In the same year, Pauli asserted that every electron in an atom is characterized by all four quantum numbers, and that no two electrons at the same location are characterized by the same four quantum numbers. This assertion applies to all particles with half-integer spin (i. e., fermions) and excludes all particles with integer spin (i. e., bosons). Because Pauli's exclusion principle does not apply to bosons (which includes photons), it would be theoretically possible to superposition any number of boson particles at the same location.

At this point, the new quantum number, spin, seemed to account for the anomalous Zeeman's effect and satisfied the Pauli exclusion principle. The extra spectrographic lines emitted by atoms in a magnetic field were the result of spin (one line for transitions of the spin 1/2 electron in an orbital, and another line for transitions by the spin −1/2 electron in the same orbital). How does the exclusion principle relate to our understanding of the periodic table? It turns out that when spin is taken into account, the permissible quantal states match the filling complement of shell electrons represented in the rows of the periodic table. Hence, spin number reconciled the Bohr's early atomic theory with the allowable quantum states predicted by Pauli.

When we speak of two particles occupying the same place, Pauli's exclusion principle moves us away from the traditional idea of a "place" being determined by an x, y, and z coordinate, and replaces it with the idea of a set of quantum numbers that refer to a probability or wave configuration. We no longer pretend to know the exact coordinates of a particle. Instead, we use a quantum configuration characterized by a wave function. Two quantum particles are thought to be the same place if they have the same quantum numbers and the same positional likelihood. Prior to quantum theory and Pauli's exclusion principle, it was known that matter is filled with emptiness; the emptiness lying between orbiting electrons and the nucleus, and the emptiness between atoms. Physicists asked themselves why a table does not descend through the floor, as its atoms ooze through the copious spaces provided within the matter below. Today, we accept the idea that electrons cannot occupy the wave function of another electron (having the same quantum numbers), which occupies a probabilistic space larger than a single point. Hence, atoms and subatomic particles carve out their own private quantum space that no force in nature can violate. Accordingly, we cannot walk through solid walls, we do not sink to the center of the planet, and our material world exudes a sense of solidity.[xxxviii] Wolfgang Pauli, in one blow, solidified two tables: the table that sits in my dining room, and the periodic table of the elements.[xxxix]

When Wolfgang Pauli provided a mathematical description of quantum spin, he thought he was creating particles that make no physical sense; and this troubled him deeply. Electrons are point particles (having no size), just like any other fundamental particle. Their only function is to carry charge. How can a particle with no size spin? Additionally, in the case of electrons, the mathematics would indicate that the two allowable quantum states for spin could be either $\pm\frac{1}{2}$. Pauli's own calculations indicated that an electron would need to spin around twice, just to get back to its starting point (a seeming impossibility). Pauli is quoted as saying, "I have done a terrible thing. I have postulated a particle that cannot be detected."

As it happens, electron half-spin is easily detectable. In fact, electron spin was detected in 1922, three years prior to Pauli's mathematical description of spin, and one century before this book was published. The corroborating experiment, known today as the Stern–Gerlach experiment, demonstrated that quantum spin was real, and that fermions (e. g., electrons) and bosons (e. g., photons) behave much as their symmetry group (mathematically speaking, the irreducible representations of the double cover of the Poincaré group) would predict. Specifically, fermions have spin $\frac{1}{2}$ and photons have spin 1.

The experiment, conceived by Otto Stern and conducted by Walther Gerlach, involves passing a beam of electrons, through a magnetic field. Doing so splits the beam in two; one beam deflected upwards, and the other beam deflected downwards, and each beam composed of the same number of electrons. The reason that the electron beam

xxxviii The mutual resistance by which two identical fermionic particles resist superpositioning is known as the degeneracy pressure. This pressure is of a purely quantum origin, arising from Pauli's exclusion Principle.

xxxix The allowable states of orbiting electrons in atoms are determined by Pauli's exclusion principle and the probability density functions of electrons in orbitals are calculated with the spherical harmonic representation of the Schrödinger's wave equation. We will not delve into the mathematics of spherical harmonics here, but a note on the subject is included in Chapter 11.[106]

was split into two equal beams is that the original electron beam contained equal populations of electrons that were alike in every way, but with half the electrons having spin $\frac{1}{2}$ and the other half having and equal and opposite spin of $-\frac{1}{2}$, the two allowable spin states described by Pauli. The spin $\frac{1}{2}$ electrons moved in one direction under a magnetic field, and the spin $-\frac{1}{2}$ electrons moved in the other direction. This phenomenon seemed to echo the Zeeman effect, wherein spectrographic emission lines split into two closely-spaced lines when atoms were subjected to a magnetic field.

By blocking the electrons of one or the other of the split beam, a beam of electrons having "polarized" spin of exclusively one spin is produced, and the experimental system acts as a spin directional filter. If two such filters are placed in tandem, and rotated with respect to one another, the doubly-filtered output beam is observed to vary in intensity from 100 % (when the filters are aligned) to 0 % when the filters are oriented at 180 degrees.

If we duplicate this experiment, using a beam of photons (i. e., a light beam) instead of an electron beam, and using a polarizing filter instead of magnets, we get similar results, with the a filtered beam that can be doubly-filtered by a second polarizing filter. As in the case of an electron beam, we find that we can vary the intensity of the doubly filtered light beam by changing the orientation of the polarizing filters with respect to each other. When the polarizing filters are aligned, the beam strength is at maximal intensity. When we rotate the filters, the beam strength drops to 0 % when the filters are oriented at 90 degrees.

Wait a minute! In the electron beam experiment, the doubly-filtered beam strength drops to zero when the filters are 180 degrees out of alignment. In the light beam experiment, the doubly-filtered beam strength drops to zero when the filters are 90 degrees out of alignment. How is this possible? Electrons (a type of fermion) have spin 1/2, while photons (a type of bosons) have spin 1. The electron must spin twice to come to its original configuration, while the photon needs to spin only once. This explains the two-fold difference in the effect of orientation angle on the intensity of the filtered electron and light beams.

Pauli thought that spin was an abstract property, that could not be interpreted in a classically physical manner. He was wrong. Stern and Gerlach's physical observations fit perfectly with the strange, abstract notion that an electron must spin around twice, to come back to its original configuration!

Almost immediately following Wolfgang Pauli's announcement of his exclusion principle, cosmologists applied the principle to explain white dwarf gravitational degeneracy. When stars collapse, they reduce their size to something comparable to the diameter of Earth, achieving enormous densities of plasma consisting of bare nuclei and electrons. Why does the collapse of neutrons and electrons stop when it does, instead of collapsing even further, to the extent that there is no space whatsoever between particles, or even to the point where particles superpose upon themselves (i. e., where two particles occupy the same space)? This problem had been vexing cosmologists, but Ralph Fowler, in 1926, applied Pauli's exclusion principle and showed that at a certain point, when the space between neutrons nearly vanishes, Pauli's exclusion principle prohibits any further crowding, and the collapse of a white dwarf must cease [99, 137].[xl]

The application of Pauli's exclusion principle to cosmology serves a an example wherein fundamental advances in one field (quantum physics in this case) often lead to the solutions of seemingly intractable puzzles in other fields (such as cosmology and chemistry). In tandem with the Schrödinger's wave equation, it provides an explanation for how the atomic orbits are filled with electrons of permissible quantum states, the numbers of electrons in each valence shell and their subshellls, and the structure of the periodic table of the elements.

Now we can return to one of the original questions that challenged the validity of the atomic theory. If atoms are characterized by a positively charged nucleus and encircled by negatively charged electrons, then why don't the electrons, attracted by an opposite charge, fall into the nucleus? Again, Pauli's exclusion principle provides an answer. The electron and the proton each behave according to their respective properties (enshrined within the Schrödinger wave equation[52]). Because protons and electrons are different from one another, their interactions are not encumbered by Pauli's exclusion principle. A proton's probability space can (and does) intersect with the probability space of its orbiting electrons. Ground state electrons overlap the same probability space as the nucleus, but the electron does not spiral into the nucleus because it has attained its lowest energy state in the inner orbit (not within the nucleus). When all else fails, we can avoid becoming confused by the Bohr model of the atom if we think of electrons and protons as waves whose so-called orbits are described by wave functions.

xl Readers may be wondering why collapsed stars yield a product composed purely of neutrons. How did all the protons and electrons, present in the atoms of stars, disappear? As stars collapse, increasing gravitational pressure causes electrons to combine with protons to become neutrons, via a process known as electron capture. Because atoms have an equal number of oppositely charged electrons and protons, the process yields a collection of neutrally charged neutrons.

7 Fundamental constants

Infinitely sharp constants are failed clocks. They imply complete delocalization in time.

– João Magueijo [163]

7.1 Fundamental constants of nature

We have all witnessed constants at work. They are the numbers that we plug into mathematical formulas to balance equations. There are many constants of physics and chemistry, as seen in Table 7.1 below.

Table 7.1: Physical constants. Source, National Institute of Standards and Technology table provided as a public domain work of the U.S. Government.

Quantity	Symbol	Value	Unit	Relative std. uncert. u_r
speed of light in vacuum	c	299792458	$\mathrm{m\,s^{-1}}$	exact
vacuum magnetic permeability $4\pi\alpha\hbar/c^2c$	μ_0	$1.25663706212(19) \times 10^{-6}$	$\mathrm{N\,A^{-2}}$	1.5×10^{-10}
$\mu_0/(4\pi \times 10^{-7})$		1.00000000055(15)	$\mathrm{N\,A^{-2}}$	1.5×10^{-10}
vacuum electric permittivity $1/\mu_0 c^2$	ϵ_0	$8.8541878128(13) \times 10^{-12}$	$\mathrm{F\,m^{-1}}$	1.5×10^{-10}
characteristic impedance of vacuum $\mu_0 c$	Z_0	376.730313668(57)	Ω	1.5×10^{-10}
Newtonian constant of gravitation	G	$6.67430(15) \times 10^{-11}$	$\mathrm{m^3\,kg^{-1}\,s^{-2}}$	2.2×10^{-5}
	G/he	$6.70883(15) \times 10^{-39}$	$(\mathrm{GeV}/c^2)^{-2}$	2.2×10^{-5}
Planck constant*	h	$6.62607015 \times 10^{-34}$	$\mathrm{J\,Hz^{-1}}$	exact
		$4.135667696\ldots \times 10^{-15}$	$\mathrm{eV\,Hz^{-1}}$	exact
	$\hbar$	$1.054571817\ldots \times 10^{-34}$	J s	exact
		$6.582119569\ldots \times 10^{-16}$	eV s	exact
	hc	$197.3269804\ldots$	MeV fm	exact
Planck mass $(\mathrm{hc}/G)^{1/2}$	m_P	$2.176434(24) \times 10^{-8}$	kg	1.1×10^{-5}
energy equivalent	$m_P c^2$	$1.220890(14) \times 10^{19}$	GeV	1.1×10^{-5}
Planck temperature $(\hbar c^5/G)^{1/2}/k$	T_P	$1.416784(16) \times 10^{32}$	K	1.1×10^{-5}
Planck length $\hbar/m_P c = (\hbar G/c^3)^{1/2}$	l_P	$1.616255(18) \times 10^{-35}$	m	1.1×10^{-5}
Planck time $l_P/c = (\hbar G/c^5)^{1/2}$	t_P	$5.391247(60) \times 10^{-44}$	s	1.1×10^{-5}

What exactly qualifies a number as a fundamental constant of nature? When and how did fundamental constants enter existence? Why do the fundamental constants all have precise and unchanging values? In this chapter, we will try to answer these questions, and we shall see how certain numbers are fundamental building blocks of our universe.

A constant, by definition, never changes its value. A fundamental constant is distinguished from ordinary constants in that it cannot be derived from other constants.

https://doi.org/10.1515/9783111498690-007

Logic would dictate that if a constant cannot be derived, then our only way of learning its magnitude is by experimental measurement. Our experimental measurements are all based upon having known units of measurement (e. g., length, duration, mass, charge), but all of these units have a derivation, either observational or theoretical. Summarizing, our fundamental constants, like so much else in our universe, begin as the following paradox:

1. A fundamental constant, being fundamental, cannot have components (i. e., it must be irreducible)
2. But a fundamental constant can only be determined by measurement.
3. The measurements are expressed as units, and the units are the expression of constant quantities.
4. Therefore, a fundamental constant must be reducible, and cannot be fundamental.
5. Therefore, fundamental constants cannot exist. But they do exist!

The fundamental constants are commonly treated as though they were given to us by the universe. This concept makes no sense. If we simply accept fundamental constants as things given to physicists by the universe, then we are basically accepting them as magical objects. As scientists, we must not succumb to such nonsense. We should try our utmost to predict how the constants of nature arose from the same fundamental relationships that gave rise to physical properties. Let us begin our efforts by listing the issues that relate to the physical meaning of the fundamental constants. These are:

1. The dimensionality problem. Do we have an objective and consistent way of assigning units to measurements?
2. The measurement interpretation problem. What we know about the universe comes from observation and measurement. If we cannot trust our own measurements, we cannot be certain about anything that we observe.
3. The inclusion problem. Do we really need dozens of fundamental constants? If a constant can be derived as a value computed from other constants, then can we just drop it from the list? What are the true fundamental constants that we would need to describe our universe (if any)?
4. The precision of measurement problem. How can we attain the precise value of constants when there are practical limits to the accuracy of our measurements? If we cannot know the value of a constant, then how do we know that its value is truly constant.
5. The origin problem. The fundamental constants of nature are more than mathematical abstractions that balance equations. The fundamental constants are physically manifest quantities that control the operating system of the universe. We cannot understand reality if we do not understand how the fundamental constants came into existence?
6. The "constancy" of constants problem. We assume that the universe is homogeneous and that the laws of physics and the fundamental constants do not change from place to place and time to time. This assumption may be false in general or false

under specific sets of circumstances. If this were the case, we would need to explore the consequences of such physical lapses.

Let us take a few moments to look at each of these problems individually. We will learn that much of the confusion we have concerning the nature of our universe comes from the difficulties encountered when we examine the fundamental constants of nature.

7.2 The dimensionality problem

"Only theory can tell us what to measure and how to interpret it." – Albert Einstein

Whenever we measure anything (atmospheric pressure, temperature, shoe size), our measurements are made in units. Units are intimately associated with the objects being measured, so much so that the object and its unit of measurement can be substituted, one for the other. We think of a British pound as currency, but it's name derives from its value in unit (pound) weight of sterling silver. We speak of yard goods to indicate textiles that are sold in sheets of yard width. Our cavalier use of language has transformed units of measurement into physical objects.

When the industrial age came into full swing, there was a strong commercial incentive to develop a measurement for the power of steam engines. It wouldn't do to let customers think that all steam engines were of equal power, and industrialists were reluctant to allow any one company's engine serve as the standard for the other companies to follow. It was soon decided that steam engines would be forever measured against horses. One horsepower is a measure of weight lifted by a horse per unit second. Potential buyers were not interested in purchasing a steam engine that could not compete favorably against a horse (they could always buy a horse). Of course, some horses have more power than others. In addition, the proposed unit cannot be scaled, because a horse that lifts a certain weight in one unit of time may not necessarily lift twice that weight in two units of time. This being the case, what would 2 horsepowers represent? Despite practical and theoretical deficiencies, the horsepower settled upon 550 foot-pounds per second (the equivalent of lifting 550 pounds to a height of one foot in one second). To be marketable, steam engines would need to do better than one horsepower. Risible as it may seem, the horsepower unit has withstood the test of time.

The origin of horsepower as a unit of measurement tells us something about the inherent problems of measuring constants in terms of units that are artifactual and arbitrary. Moving to present day science, when we measure the speed of light in terms of meters per second, we are matching a true constant (the speed of light) against two units whose values are relative. Time and distance, as Einstein has shown, are relative values that change in different frames of reference. The speed of light is invariant in all frames of reference. Just about any form of physical measurement (e. g., clocks, rulers, scales) cannot provide us with units that independent observers can agree upon. Basically, it is impossible to measure a constant with units that are not themselves universal and constant.

As a remedy, some physicists argue that the laws of physics cannot be dependent upon anyone's choice of units. In that case, only dimensionless constants can be considered fundamental, all other constants are derived from the units that are anything but fundamental.

Let us look more closely at the speed of light through a vacuum, c. Because c is does not change in any frame of reference, it is generally considered a fundamental constant of the universe. When we begin to think about our measurement of the speed of light, we find that we must determine its value by measuring items in spacetime that are neither fundamental nor constant. We express the speed of light in terms of meters per second.

$$c = 299,792,458 \text{ meters/second}.$$

The length of a meter ruler and the time between ticks of a clock change, relative to velocity, though their ratio remains the same. Are there alternative ways with which we can define the speed of light? Can we find absolute spacetime measurements? We can certainly do better than defining c in terms of meters and seconds. For example, as a constant unit of distance, we can choose the Bohr radius a_0, this being the distance between the nucleus and its electron, in the ground state hydrogen atom, as shown:

$$\mathrm{a}_0 = \frac{\hbar}{m_e c a},$$

- $\hbar$ is the reduced PLanck constant;
- m_e is the mass of an electron;
- a is the fine-structure constant and is dimensionless.

It happens that the fine-structure constant is a fundamental constant of the universe, and it is dimensionless (i. e., it is a number)

$$\frac{1}{a} = 137.035999174.$$

Instead of using seconds as our unit of time, we can use the Bohr time,

$$\text{Bohr time} = \frac{\hbar}{E_h},$$

where E_h is:

$$E_h = \frac{\hbar^2}{m_e^2}.$$

Now we have the units of distance and time expressed in terms of physical constants found in quantum physics. It's beginning to look as though the speed of light is a true universal constant, but is it really? To be considered fundamental, a constant cannot be derived from anything else. When we look at our expression of c in terms of atomic units, we find that c is ultimately derived from the fine structure constant α and is therefore not fundamental. Confusing? There is more to come.

On the subject of dimensionless constants, can we include the list of numeric constants that pop up everywhere in equations that describe physical realities? For example, are π and e, and i universal physical constants? The abstract constants of mathematics, such as π, e, and i, exist independently of the physical universe. Nonetheless, abstract constants can represent physical shapes and forces, and they appear in the equations that describe our reality (e. g., the Schrödinger wave equation). Without them, our physical reality would be incomprehensible.[i]

In 1960, Eugene Wigner wrote a fascinating essay entitled, "The Unreasonable Effectiveness of Mathematics in the Natural Sciences." The thesis of this essay is that the most fundamental and abstract concepts in mathematics seem to play crucial roles in almost every aspect of the sciences [261]. For myself, the most unreasonably effective equation in mathematics is Euler's identity.

$$e^{i\pi} + 1 = 0 \quad \text{Euler identity}$$
$$e^{i\omega t} = \cos(\omega t) + i\sin(\omega t) \quad \text{from Euler identity}$$

Euler's identity, considered one of the most beautiful equations in mathematics, demonstrates the relationships among 5 of the most important quantities in the mathematical universe: i, pi, 1, 0, and e. Using Euler's identity, a novice mathematician can quickly come to grasp the DeMoivre's identity, the general wave function, Gaussian distributions, the Fourier transform, signal analysis, and combinatorics. A proof of Euler's identify is provided in Chapter 11.[ii] [107]

The constants in Euler's identity are abstractions that have somehow come to preside over our physical reality. It would seem that the much of the universe is organized by a few numbers that only exist in our imaginations.[iii]

---

i The number i is imaginary (that is what the "i" stands for), and not real. Yet, i serves a very real purpose, providing solutions to algebraic equations that could not be solved with only real numbers. In addition, i makes an appearance in a variety of fundamental laws and in the formulation of several physical constants. Shouldn't a real universe be describable without a resorting to a wholly imaginary number? Could we, perhaps, substitute cosines and sines into our equations and produce equivalent results? According to Renou and his coworkers, the answer is "no" [209]. A sound quantum theory that cannot be falsified must contain complex numbers (i. e., numbers in the form $a + bi$ where a and b are real numbers). Therefore, our reality must contain i, a number that is definitively "unreal".

ii Word to the wise. Do not forget the Euler identity. It will come in handy when we discuss wave equations.

iii Beauty, as found in the mathematical literature, applies whenever something of profound meaning can be expressed in a minimum of symbols. Euler's identity, relating e, i, pi, 0, and 1 in a simple equation, is held as an example of beauty in mathematics.

Physicists have suggested that we are very lucky that the value of $\frac{1}{\alpha}$ is 137.035999174. Had it been off by just a few percentage points, our universe would contain no galaxies! This prediction has actually been used to argue that the universe must have been precisely created by a benevolent and wise entity who carefully determined the best possible value of α to supply us with a multitude of galaxies in which to live and pray. Does this line of thinking make any sense? Isn't it no different from saying that if π had been created just a bit smaller, then there would be no circles because their circumferences would not close. It seems best to think that our fundamental constants are baked into the fabric of reality and cannot be conceived as an emergent properties arising from an existing physical system.

7.3 The measurement interpretation problem

Measurements require dependable observations of particular quantities (length, time, volume, mass). Our observations require some form of energy, typically the energy of light or electrons with which we view events. The energy of measurement alters the things we measure. The sad truth is that we do not have a firm grip on any of the necessary components of measurement — energy, space, or physical quantities. We will see that at a fundamental level, we never fully understand what our measurements mean.

No measurement is more confusing to physicists than the quantum measurements of waves. We refer here to the so-called collapse of the Schrödinger's wave function.[iv] The situation is the following: When we pass light through two slits, a wave diffraction appears on the screen behind the slits, making visible periodic wave peaks and wave troughs. The reason we see these peaks and troughs is that the edges of a slit cause the light (at the edge) to bend and spread out, producing multiple spreading wave patterns that, when combined on the screen, will reinforce or cancel periodically. The result is called a wave interference pattern insofar as the peaks and troughs of the two diffracted outputs cancel out, in repetition. These observations are what we would expect from any self-respecting wave (Figure 7.1).

If we were to modify the experiment slightly to measure the wave as it approaches one of the slits, then the light passing to the slit no longer behaves like a wave. Instead, it behaves like a beam of photons. The screen behind the two slits measures exactly two columns of impacted particles. The placement of a detector between the light source and the slits was the only condition that changed in the two experiments (Figures 7.2 and 7.3). Apparently, the act of measurement instantly transformed our wave into a particle beam! Moreover, the instantaneous transformation occurs at both slits, including the slit at which no detector was placed! How did the particles at the second slit know

---

iv The derivation of the Schrödinger's wave function is provided as a note in Chapter 11.[52]

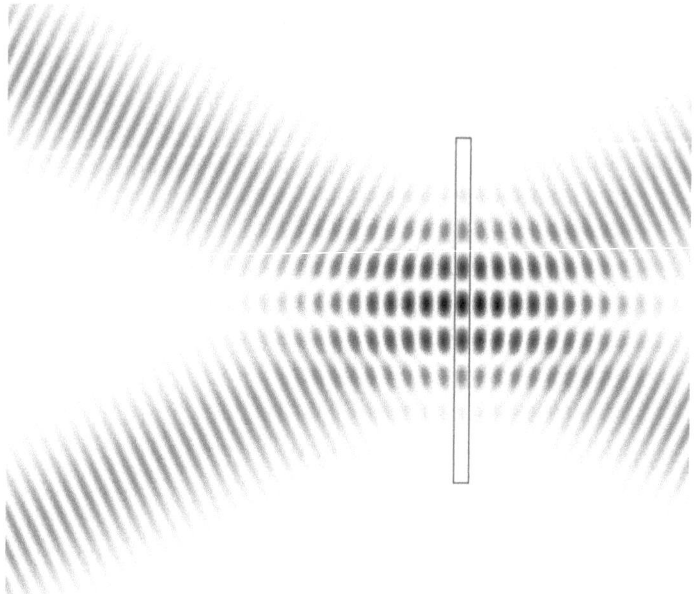

Figure 7.1: An interference pattern created when two planar waves meet. The pattern produced when the waves "collide" is the so-called diffraction pattern or wave interference pattern. If we were to confine our attention to a screen placed at the intersection plane (the thin rectangle at the center of the image), we would see a succession of dark and light blocks of light corresponding to the locations where the two waves reinforce one another (the dark areas), separated by the areas where the two waves cancel out. Source, an image placed into the public domain by its creator.

that the particles at the first slit were being sampled and measured? Was this an example of action-at-a-distance (i. e., an effect produced at a point in space without a causal force located at the point that acts at the moment in time when the effect is produced).[v]

In the quantum world, the behavior of particles, from one moment in time to the next, is determined by the Schrödinger wave function. The wave function models particles as harmonic oscillators, and the position of the particle at any given moment is anywhere within a range of probable locations. The particle is said to occupy all allowable quantum states simultaneously. When the wave is measured, the Schrödinger wave function ceases to hold, and the function is said to collapse into one specific quantum state. Hence, observation altered the observed object![vi] [vii]

---

v Einstein referred to this phenomenon as "spooky action at a distance", an apparent violation of causality.

vi Physicists offer a variety of ways in which particles can be perceived, including: the quantum excitation of a field, an irreducible representation of a Poincaré group, and a vibrating string that is one Planck unit in length.

vii Just for fun, let's add another questionable hypothesis into the mix. Let us imagine that the Schrödinger's wave function collapses to an impulse function (a function that is always zero except at

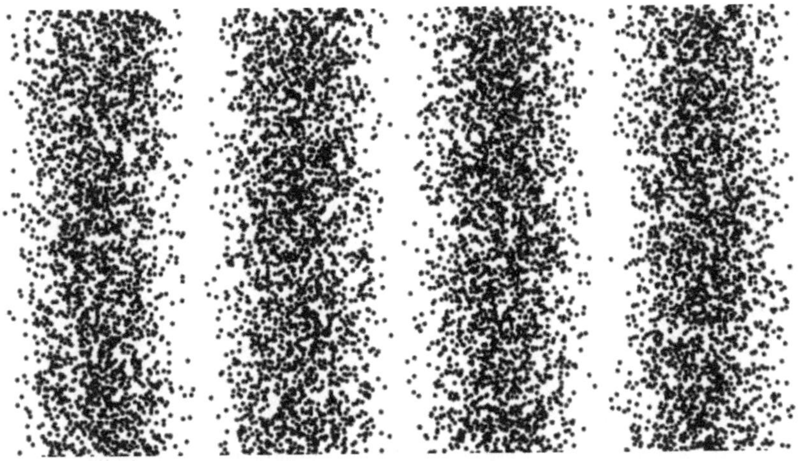

Figure 7.2: When we send a beam of electron particles through two slits, and we place a screen at the focus where the two beams meet, we see the same kind of interference that we always see with light waves. There is a repeating pattern of dark areas (wave reinforcement) separated by light areas (where the waves cancel out). This finding indicates that particles are waves. Source, Wikipedia, from a graphic provided to the public domain by Thierry Dugnolle.

Figure 7.3: When the electron particles moving through either slit are observed with a particle detector, the waves (from both slits) instantly lose their wave properties, behaving much like a stream of tiny marbles shot through both slits, and streaming out as two (non-repeating) columns of particles impacting on the screen. The endlessly repeating wave interference pattern is suddenly gone, replaced by two split beams! This phenomenon, known as the collapse of the wave function, has been an enduring topic of speculation among physicists. Source, Wikipedia, modified by author to appear as an expected outcome, from a graphic provided to the public domain by Thierry Dugnolle.

7.4 The inclusion problem

We find that several constants that tell us a great deal about the nature of our physical reality are, in fact, derived from other constants. Should such constants be stricken from our list of fundamentals? At the same time, we find that some of our wholly derived constants (that cannot be measured directly) seem to be telling us something fundamental about our universe. Should these constants be added to our list? Let us spend a moment to see what we may learn from the Planck units?

As originally conceived by Planck, the Planck units were a way of representing length, mass, temperature, frequency, and time in units derived from a few accepted and precise fundamental constants (h, G, c). Somewhat later, the Planck temperature was added to the list, employing the Boltzmann constant, k.[viii] Planck hoped that these derived units would establish international standards, would simplify the representation of such units in mathematical formulas, and would eliminate the necessity of using somewhat arbitrary and nonstandard methods for measuring unit values.

Here is a list of the Planck units:

- Reduced Planck constant, $\hbar$:

$$\hbar = \frac{h}{2\pi}$$

- Planck length, $\ell_P = 1.61625510^{-35} m$

$$\ell_P = \sqrt{\frac{\hbar G}{c^3}}$$

- Planck mass, $m_\mathrm{P} = 2.176434 \times 10^{-8}\mathrm{kg}$

$$m_\mathrm{P} = \sqrt{\frac{\hbar c}{G}}$$

- Planck time, $t_\mathrm{P} = 5.391247 \times 10^{-44} s$

$$t_\mathrm{P} = \sqrt{\frac{\hbar G}{c^5}}$$

one point, where the impulse resides). Impulse functions can be decomposed as a collection of waves that can be added together to create a single spike. In this sense, the collapse of the wave function can be interpreted as a transformation into the impulse function, whose physical counterpart is a particle moving through spacetime. We will visit the topic of impulse functions later in this chapter.

viii The Boltzmann constant, named for Ludwig Boltzmann (1844–1906) relates the average kinetic energy of particles with temperature and pressure. It is an emergent quantity whose meaning only makes sense when we observe lots of particles interacting. As an emergent quantity, the Boltzmann constant is not strictly fundamental, and constants derived from the Boltzmann constant may not be fundamental, as well.

- Planck temperature, $T_P = 1.416784 \times 10^{32°} K$

$$T_P = \sqrt{\frac{\hbar c^5}{G k_B^2}}$$

- Planck frequency, $F_p = 1.855 \times 10^{43}$Hz

$$F_p = \sqrt{\frac{c^5}{\hbar G}} \approx 1.855 \times 10^{43} \text{ Hz}.$$

At the time that Planck published these new units, and for decades thereafter, no particular significance was assigned to each quantity. In 1959, C. Alden Mead tried to publish a paper in which he asserted that the Planck length and the Planck time had important physical meanings [176]. Famously, his paper was not received favorably by reviewers. Over the next few decades, Mead's idea, refined by others, has become conventional wisdom. Nevertheless, every physical interpretation of the various Planck units has its detractors, in no small part because they either contain assertions about unknowns (e. g., the consequences to spacetime that may occur when the accepted theories of physics do not apply), or they include claims of predicted phenomena that cannot be confirmed (e. g., the creation of a black hole occurring when the frequency of light exceeds that associated with the Planck wavelength). Regardless, it seems that the Planck units have much to tell us about the limits of what we might expect to find in our universe.

Physical significance of Planck units

- Planck length — corresponds to the Planck wavelength, which is the shortest possible observable wavelength of light.
- Planck time — the time following the first moment of the Big Bang during which our current understanding of physics does not apply.
- Planck mass — the mass whose quantum wavelength is equal to the Planck length. In one manuscript, Stephen Hawking suggested that the minimum mass required to initiate black hole formation was roughly equivalent to the Planck mass [115].
- Planck temperature — the theoretically highest possible temperature in the universe.
- Planck frequency — the theoretically highest possible frequency of a wave.

The fundamental constants are nearly always defined as being measurable quantities that cannot be derived from other constants. Essentially, they are "given" to us by our beneficent universe, no questions asked. We should be thankful for the fundamental constants, because without them the universe would be chaotic and have no stable form.

When we look closely at the fundamental constants, we begin to find that the relationships among the constants imply that some of the fundamental constants can be derived from other fundamental constants.

Let us look at α, the fine structure constant. As to its physical significance, it has a multitude of meanings that we need not examine.[ix] [x] For our purposes, we only need to know that the Fine Structure Constant is dimensionless (i. e., a number without units) and that it is typically expressed as its inverse:[xi] [108]

$$\frac{1}{\alpha} = 137.035999174$$

In addition, α can be computed from any of numerous equivalent mathematical relationships:

$$\begin{aligned}\alpha &= i\left(\frac{e}{q_{\mathrm{P}}}\right)^2\\ &= i\frac{e^2 c\mu_0}{2h} = \frac{1}{4\pi\varepsilon_0}\frac{e^2}{\hbar c}\\ &= i\frac{k_{\mathrm{e}}e^2}{\hbar c} = \frac{e^2}{2\varepsilon_0 ch}\\ &= i\frac{c\mu_0}{2R_{\mathrm{K}}} = \frac{e^2 Z_0}{2h}\\ &= i\frac{e^2 Z_0}{4\pi\hbar} = \frac{Z_0}{2R_K}\end{aligned}$$

---

ix There are multiple interpretations for the fine structure constant, α. Here is a partial listing: 1) the energy to overcome the electrostatic repulsion between two electrons a distance of d apart divided by the energy of a single photon having angular wave d; 2) The ratio of the velocity of the electron in the first circular orbit of the Bohr model of the atom to the speed of light in vacuum; 3) α^2 as the ratio of the potential energy of the electron in the first circular orbit of the Bohr model of the atom divided by the energy equivalent to the mass of an electron; 4) any of several simple ratios involving three constant lengths (the classical electron radius, the Compton wavelength of an electron, and the Bohr radius; 5) a direct relation to the coupling constant for the strength of the interaction between electrons and photons; 6) one fourth the product of the characteristic impedance of free space, and the conductance quantum; 7) as its reciprocal (i. e., $\frac{1}{\alpha}$, the limiting value for the atomic number of an element); 8) as the square of the elementary charge to the Planck charge.

x Clever physicists refuse to be bound by the petty dictates of space, time, or quanta. A recent paper by Tejinder P. Singh constructs pre-spacetime, pre-quantum derivation of the fine structure constant [221].

xi It has been suggested that if the value of the fine structure constant were altered by just a tiny amount, there would be no stars in our universe. Of course, this speculation makes no sense; it is analogous to declaring that if the value of π were just a bit smaller, then there would be no circles because the two ends of a circumference would not touch one another.

If α is the product of other constants, and if it has multiple different physical meanings, then do we really need to include it in our list of constants?

Likewise, the maximum speed of light is often characterized as one of the defining characteristics of our universe. Does it really qualify as a constant?

As mentioned earlier, the speed of light can be computed from other constants; specifically as a inverse of the square root of the products of the permittivity (ε_0) and the permeability (μ_0) of space.[xii]

$$c = \frac{1}{\sqrt{\varepsilon_0 \mu_0}} .$$

This being the case, it seems that the speed of light is a derived number. Because the speed of light in a vacuum represents an absolute maximum speed for matter (and both the maximum and minimum speeds for light traveling in a vacuum), would it not be wiser to simply refer to the speed of light as 1. Something moving at half the speed of light would have speed one-half. A stationary object (in its own frame of reference) would have a speed of zero. If this were the case, what units would we use for the speed of light? The units would be "speed of light" units (e. g. 1 speed of light, one-half speed of light, zero speed of light).[xiii]

7.5 The precision of measurement problem

What are the practical limits of precision measurements? Let us stretch our imaginations, for a moment, and pretend that we have just found an artifact left by an alien race known throughout the galaxy for its prowess in the science of measurement. As a sort of time capsule for the universe, their top scientists decide to collect the history of their civilization, encoded in binary. Their story might look something like "001011011101000..." extended to about 5 million places. Rather than print the sequence out on a piece of paper or a computer disc, these aliens simply converted the sequence to a decimal length (i. e., .001011011101000...") and marked the length on a bar composed of a substance that would never change its size. To decode the bar, and recover the history of the alien race, one would simply need to have a highly precise measuring instrument, that would yield the original binary sequence. Computational linguists could translate the sequence to

xii When we refer to the speed of light, we are really indicating the speed of all electromagnetic waves. The term "light" is often reserved for electromagnetic waves of the visible spectrum. Nonetheless, no one bothers to write "speed of electromagnetic waves", and throughout this book, we will use the term "light," "electromagnetic waves," "electromagnetic radiation," and "electromagnetic energy" interchangeably.

xiii Fans of Star Trek will note that we would be using the same scale as that used by the Federation, which refers to Warp 8 and Warp 9 to signify 8 or 9 times the speed of light, with units unencumbered by measurements of time or distance.

text, and the recorded history of the alien race would be revealed! Of course, the whole concept is built on an impossible premise. Nothing can be measured accurately to the 5 millionth decimal place.[xiv] [109]

Aside from the inherent imprecision of our measuring instruments, there are things that we cannot measure, no matter how hard we try. In a universe that expands faster than the speed of light, objects at great distance from us will recede beyond our detection inasmuch that the light they emit will never reach us.[xv] [xvi]

We run into strict limitations when we attempt to directly measure quantum objects. As an example, let's imagine what would be involved if we wanted to directly measure an electromagnetic wave whose frequency equals the Planck frequency. The Planck frequency is considered to the highest frequency that can possibly be attained in our universe (i. e., a wave that repeats with a period equal to one Planck time) and is derived mathematically. Suppose we wished to directly measure the Planck frequency. To do so, we would need to have a measuring device that could observe the Planck frequency wave with light that had a higher frequency than the Planck frequency (because the length of a wave can only be measured from peak to peak using optics that are smaller than the peak-to-peak distance). Aside from being logically impossible (we have stipulated that the Planck frequency is the highest possible frequency), it is also physically impossible. The energy associated with a light wave having a frequency higher than the Planck frequency (remember, $E = h \times \nu$) approaches infinity as the unit volume shrinks to sub-Planck size. Such energy could conceivably produce a singularity spawning a new universe. We will have none of that![xvii]

---

xiv The most precise measurement achieved by modern science has been in the realm of atomic clocks, where accuracy of 18 decimal places has been claimed [38]. Nonetheless, many scientific disasters are caused by our ignorance of our own limitations, and our persistent gullibility, leading us to believe that precision claimed is precision obtained [30].

xv If an object were greater than 14 billion light years from Earth, then its emitted light would require more than 14 billion years to reach us. But the Big Bang occurred just under 14 billion years ago. Therefore, the light reaching us today from an object at a distance exceeding 14 billion light years from Earth must have begun its journey before the observable universe came into existence. Therefore, objects more distant that 14 billion light years from Earth are unobservable.

xvi For an easy way to see the problem, imagine that you shout at a man standing at the caboose of a moving train. The sound of your voice travels at 1,000 feet per second. If the train is moving away from you at 1,001 feet per second, then the man will never hear what you have said. He is moving from you at a speed exceeding the sound of your voice.

xvii For this reason, so-called virtual particles (short-lived particles resulting from fluctuations in vacuum energy), having sub-Planck size, are always unmeasurable.

7.6 The origin problem

A constant is only fundamental as a matter of convention.

– Eite Tiesinga and coworkers [238]

In the absence of any understanding of the origin of the fundamental constants, we can speculate with impunity upon their origin and their significance.

To most of us, physical constants, such as the speed of light, the fine structure constant, the cosmological constant, the gravitational constant, and the Boltzmann constant, are "given" quantities that come bundled with our universe. There is no reason to try to understand their origin; natural constants are just measured descriptors of physical reality. Yet, when we give these constants a moment's thought, we can see that they are not simply descriptors; they are participants in the laws of physics that account for reality. That being so, the universal constants must have been created.

It is difficult to imagine what goes into the creation of a fundamental constant. A constant is not a form of energy, or a field, or anything that sets it apart from any other number. Yet the fundamental constants enter every law of physics, unifying the relationships among fields and their forces. By what mechanism can a fundamental constant be created? We could simply shrug our shoulders and say that the constants are initial conditions; something "given" to us at the start of existence. Doing so would be tantamount to believing that the universe is magical. Let us agree to insist that our initial conditions, like everything else in existence, are legitimate subjects of investigation.

To understand something about the conditions that produced exact fundamental constants, we shall shift our focus, for a moment, on what it means to have a sharp wave impulse. Imagine what happens when you hear a single clap of your hands. You know that a signal has been generated at a certain frequency, but you cannot determine its pitch (i. e., you cannot guess the note produced by the clap). When we think about the meaning of a musical note, we see that a note represents a vibration at some characteristic frequency. Every note has its own characteristic frequency. Those of us who are lucky enough to have perfect pitch can accurately determine a note when they hear it, and there are tables that identify the characteristic wavelength that define the note. This being the case, why is it impossible to identify the pitch of a clap, even when the wave frequency of the clap would seem to have the characteristic frequency of some particular note?

When we listen to a single note, we receive multiple oscillations of a wave having one particular frequency. As time passes, the amplitude of the wave diminishes until it is zero. The note is a smoothly declining set of wave peaks, of a characteristic frequency. A clap represents a short pulse. An clap (i. e., a pulse of sound having a duration shorter than the duration of the frequencies that our ears recognize as sounds) has no specific frequency because its brevity is shorter than any wave cycle.

Given that short wave pulses have no discernible frequency, it would seem pulses of waves contain no useful information. As it happens, such is not the case. An instantaneous wave pulse in the time domain is equivalent to the collection of an infinite spectrum consisting of all possible waves when transformed to the frequency domain. Essentially, an instantaneous wave pulse carries **all** of the waves that can be generated of any particular type (i. e., a sound pulse carries all sound wave frequencies, an electromagnetic pulse carries all possible electromagnetic wave frequencies). We explain why this amazing assertion is true in Section 3.4, "Converting waves into digital signals", where we cover convolution. Just For now, content yourself with knowing that instantaneous wave pulses are described by the Dirac delta function and provide much of the utility of the Fourier transform, the Laplace transform, and for the convolution operation that underpins a good chunk of quantum physics and nearly all of modern digital signal processing (Figure 7.4). In point of fact, what we have discussed so far is a central dictum of digital signal processing, namely, when the time domain is compressed until it becomes an impulse, the frequency domain is expanded until it becomes a constant amplitude, for all possible frequencies [223] (Figure 7.5).

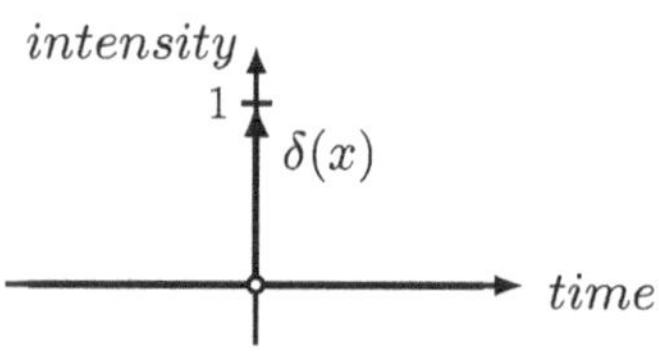

Figure 7.4: The Dirac delta function is a wave pulse of zero time duration, delivered at a particular time. All values of the pulse are zero at every time outside the pulse. Notice that the wave has an amplitude at a moment in time, but we cannot assign any specific wave frequency to the pulse. The pulse is just not wide enough to discern periodic oscillations in its waveform. Source, author, generated in Latex and modified from a short Tex script written by Izaak Neutelings.

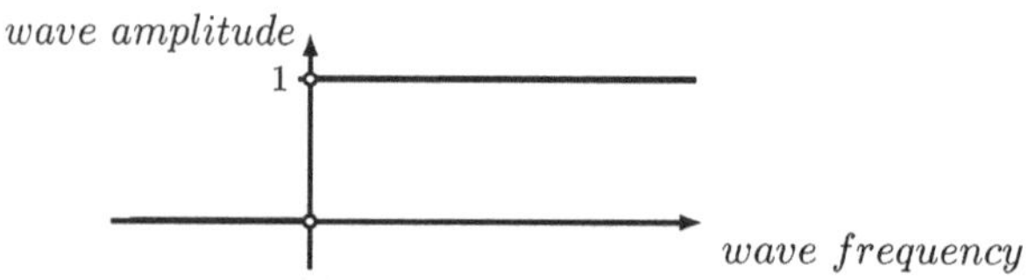

Figure 7.5: The delta function expressed as its Fourier transform. Notice that the delta pulse viewed in the frequency domain yields a constant amplitude for all possible wave frequencies (i. e., an infinite spread of frequencies all having the identical amplitude). Source, author, as an image generated in Latex and modified from a short Tex script written by Izaak Neutelings.

How does the topic of instantaneous wave pulses relate to the fundamental constants of nature? Let us return to some of the questions that we have been chewing over regarding the fundamental constants. They seem to be disembodied numbers representing some physical measurement that is constant throughout the universe, and precise, and having some kind of physical influence without having any physicality. Do the fundamental constants have **any** conceivable explanation? At this point, we must be prepared to consider all proposals.

Just for fun, let us consider what would happen if, at the beginning of time, short (instantaneous or perhaps less than Planck time) pulses of one or several types of waves were generated — perhaps as the result of the same set of circumstances that accounted for the creation of our universe. Each pulse would have sent out waves of every frequency of its kind (e. g., a gravity pulse wold have sent out gravity waves of every frequency of gravity, expanding through the universe). In one short clap, an instantaneous pulse at the beginning of time and space accounts for all of the properties associated with a fundamental constant:

- Origin. A wave pulse of zero time duration, delivered at the beginning of time;
- Timelessness. The timelessness of waves moving at light speed (remember, photons live in an “eternal now”).
- Omnipresence. A wave spreads radially outward, eventually filling the universe.
- Constancy. A wave has the same frequency wherever we choose to make our measurement.
- Physicality. Waves have physicality manifested by frequency, wavelength, movement, and energy.

Even if we were to accept the theory that the fundamental constants were created by wave pulses emitted at the beginning of time, we still have not addressed the problem of how those pulses were actually created. We take refuge in recognizing that the Big Bang was nothing other than a Big Pulse.

7.7 The “constancy” of constants problem

We have reason to assume that the values of the fundamental constants will stay the same over time. Here is why:

1. The observable universe exhibits time-shift invariance.
   This means that an experiment performed today will yield the same results when repeated a billion years from today and conserved entities such as energy and momentum will be the same at all past and future times.
2. To maintain time-shift invariance, the laws of physics must be timeless (i. e., equally valid at all times)
3. The laws of physics express the fundamental constants.
4. Therefore, the value of the fundamental constants cannot change over time.

Though we have some tentative trust that the values of the fundamental constants do not change over time, we can be much less certain about our ability to validate our trust. Gravity is a property of space that changes in strength from time to time and place to place. We can shield our measuring devices from most forces, but we cannot shield our measuring devices from the effects of gravity [184]. Therefore, there is no way to evaluate the "constancy" of constants when using measuring devices that can be influenced by gravity (and all measurements can be influenced by gravity).

7.8 Metaphysical issues raised by the fundamental constants

The deeper we delve into the meaning of the fundamental constants, the harder it becomes to understand their nature. Consider the following points:

1. The fundamental constants of nature (such as the speed of light, the fine structure constant, the Planck's constant, the gravitational constant, and the cosmological constant) describe the physical properties of our universe. As such, the universal constants do not apply to "nothingness"; they have no meaning outside of our observable universe. But the universe requires the pre-existence of fundamental constants to describe the initial conditions the determine the behavior of the universe. Hence, neither the universe nor the fundamental constants can exist prior to the other. This is a classic origin paradox, and it would suggest that the fundamental constants and the universe itself must have a common origin. But what could that be?
2. We measure in units of space and time, both of which are relative quantities and unsuitable as units for the fundamental constants. To sidestep the issue, physicists have suggested that the only valid fundamental constants are the dimensionless constants (such as the fine structure constant, α). In this case, the constant is a number, no different from any other number, and this would imply that the fundamental constants are mathematical abstractions, with no physical meaning. The laws of the universe are determined by the fundamental constants. Therefore, we could suggest that reality is a mathematical abstraction, with no physical meaning.
3. The plot thickens. When we begin to think of all physical constants as derivatives of physical things, then we need to consider that the constants (fundamental or not) could not have existed prior to the existence of the physical things from which the constants were derived. This being the case, we need to ask whether there was a time when the constants did not exist. If so, then the constants existing today must have had a beginning. Exactly when did our constants come into existence?

How did we get to this impasse? Maybe we misunderstood the term "derivation". Let us take a step back and consider quarks, gluons, electrons, and photons; all of which are considered fundamental particles. These particles all came from something, and the current thinking on the subject is that they came from symmetries of the universe. The symmetry properties of the universe account for all of the conservation laws and determine the field properties that fully account for all of the fundamental particles. The fundamental particles derive from symmetries (to be discussed in Chapter 9, "The Mathematical Universe"), but they do not derive from other particles (i. e., a fundamental particle does not contain any other fundamental particle).[xviii] Likewise, the fundamental constants derive from symmetries (typically expressed as conservation laws), but they do not derive from other fundamental constants. We solve the paradox of fundamental constants by recognizing that symmetries "create" fundamental constants, just as they create fundamental particles. We will be examining symmetries in Section 9.2, "The 'symmetry first' universe".

7.9 Unifying the constants

If we can mathematically relate one fundamental constant with another, then we can legitimately ask whether all of the fundamental constants can be unified under one theory. The hope that a single theory could account for, and predict, all of the fundamental constants, has been artfully described as "the landscape of modern physics". The constants filling the landscape of physics are: the cosmological constant, Λ; the fine structure constant, α, the gravitational constant, G, and Planck's constant, h. Could these constants be unified under one theory, much as the fundamental forces (the strong force, the weak force, the electromagnetic force, and excluding gravity) are unified under the Standard Model?

As it happens, physicists in 1980 came up with a brand new theory intended to predict the values of the fundamental constants, from first principles. This theory, affectionately as SuSy (for supersymmetry, or supersymmetric string theory) should have given us a rational understanding for the constants of the so-called string theory landscape. Despite four decades of work by scores of earnest physicists, theoretical predictions could not be confirmed [62].[xix]

Our best models of physical reality (i. e., the Standard Model, string theory, and the closely related supersymmetry theory) have failed to derive our constants from theory. This failure has been a source of great embarrassment and frustration for theoretical physicists and an indicator that our cherished models may be wrong. The universe is still waiting for a theory from which the fundamental constants could be determined and unified.

xviii If a particle were to contain another type of particle, it would be divisible and hence not fundamental.

xix Much the same could be said for string theory. Many physicists contributed to the early work in string theory, but credit for being the very first string theorist usually goes to Gabriele Veneziano, whose 1969 paper is considered a defining milestone. Experiments designed to justify string theory have failed to yield encouraging results. Physicists seem to be abandoning string theory and focusing their attention on more fruitful problems.

7.10 Fundamental constants save the ether

"Is there any point to which you would wish to draw my attention?"
"To the curious incident of the dog in the night-time."
"The dog did nothing in the night-time."
"That was the curious incident," remarked Sherlock Holmes.

– from The Memoirs of Sherlock Holmes, (1893)

Because the fundamental constants make command appearances in every measurement of every interaction that occurs in the universe, and because we must assume that fundamental constants are an intrinsic property of the universe, it would seem reasonable to infer that the fundamental constants must tell us something highly profound about the nature of our universe. But what does it tell us? When we survey our own educations, we can find no hint whatsoever that the fundamental constants represent anything other than fudge factors that make our equations fit our measurements — duct tape for physicists. In this section, we will see that the existence of fundamental constants gives credence to the "ether", a long-discredited concept that deserves re-evalutaion in light of recent developments.

First, let us review the long and illustrious history of the ether theory, and the events leading to its demise, at the start of the 20th century. For millennia, scientists believed that there were four visible elements: Earth, air, fire and water. In addition there was believed to be a fifth element, shrouded in mystery, referred to as the fifth "essence" or, with Latin prefix, the quintessence. The ether was the quintessence, the invisible and impalpable material through which every other essence moved. To this day, we persist in applying the word "quintessential" to indescribable but powerful properties of nature. The fifth element was considered to be the stuff of which the heavens were composed.[xx]

Plato (circa 424 BCE – 348 BCE) imposed a geometric framework to the five elements, noting that there were exactly five regular polyhedral solids, each corresponding to one of the 5 elements of nature, from which all transformations of matter and energy ensued (Figure 7.6). Plato and his colleagues were wrong, of course, but it is fascinating that a genius such as Plato had touched upon a deep truth, even when working in error. He recognized that geometric symmetries, such as those displayed in the five regular solids, were somehow related to our laws of physics, an insight that would serve as an early clue to how group symmetries underly all of quantum physics (much more on this topic in Chapter 9).

To physicists in the mid 19th century, nothing could be more obvious than the ether; the quintessence of the universe whose primary function was to conduct the passage of light. At the time, it was believed that all waves (e. g., water waves, sound waves, light waves, gravity waves) were periodic disturbances of a medium (e. g., water, air, ether).

---

xx Scientists in the 21st century occasionally apply the word "quintessence" to dark energy, thus restoring an ancient belief in an invisible but essential component of existence. Why not?

Figure 7.6: The five Platonic solids: cube, octahedron, tetrahedron, icosahedron, and dodecahedron, respectively corresponding to Earth, air, fire, water, and ether. All the faces, edges, and vertices of a Platonic solid are identical to one another, supporting various symmetry operations in space. In 3-dimensional space, there are exactly 5 convex regular polyhedra. In 4-dimensional space, there are six regular polytopes. In any dimensional space greater than 4, there are always 3 such solids; the *n*-dimensional analogs of the tetrahedron, the cube, and the octahedron. Source, author, as a PovRay rendering from modifications of source code originally contributed by S. R. Hall.

In the absence of a medium, there can be no waves. For example, if you jingle bells in a vacuum jar, no sound is produced. Hence, light waves must have a medium through which they travel, even if that medium cannot be seen or felt. We called that medium "the ether", or, to indicate its purpose, "the luminiferous ether".

Nineteenth-century scientists were confident that light was carried by an invisible substance, known as the ether. Just to be on the safe side, two physicists working near the end of the 19th century conducted a clever experiment that designed to detect the presence of the ether, conclusively. Their methodology was based on the following line of thinking:

1. If there is an ether, then the Earth must move through it as it circles the sun.
2. As the Earth moves through the ether, it must produce a sort of "drag" or "ether wind" diminishing the speed of light that happens to travel against the direction of the wind.
3. Two light beams, originating at the same moment, but traveling at right angles to one another, will experience different levels of ether wind.
4. Hence, the two beams will travel at different speeds.[110]
5. The difference in speeds of the two beams will become apparent in a precision interferometer, even when the differences in speed are quite small (Figure 7.7).

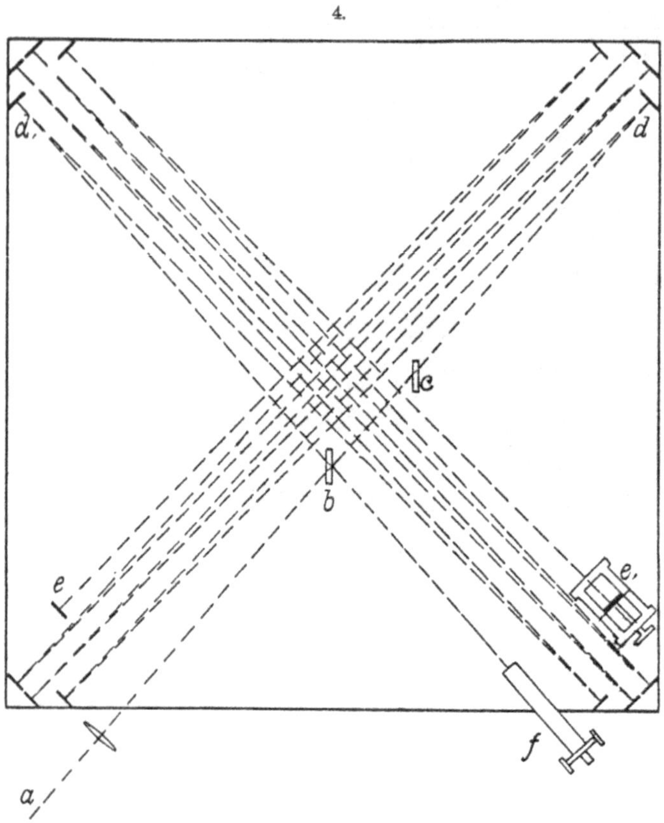

Figure 7.7: Sketch of the interferometer employed to detect the difference in the speed of light as it travels in one direction, or in a direction at a 90 degree from that direction. Source, Michelson and Morley's publication, "On the relative motion of the Earth and the luminiferous ether," published in 1887 [178].

6. The finding that light travels at different speeds, depending on direction of drag, produced by the Earth's movement through the ether, would serve as proof that the ether exists.

Famously, Michelson and Morley found no difference in the speed of the two light beams, regardless of how the interferometer table was positioned. Their experiment, published in 1887 [178, 179], challenged the prevailing notion of the ether, and promoted the idea that the universe operated under fundamental constants, such as the speed of light.

Aside from the Michelson–Morley experiment, a clever 19th century physicist may have dismissed the idea of a universal ether on theoretical grounds, based on the observed homogeneity of space.[xxi] If one segment of empty space is no different than any

---

xxi Space is said to be homogeneous if an experiment done under a specific set of conditions produces the same result regardless of the location in space in which the experiment is performed.

other segment of empty space, then we can enlarge or reduce the size of space without observing any differences in the properties of space, or in the electromagnetic waves moving through space. If, however, we believe in the presence of an ether permeating space, then the ether would be diluted as we enlarge a segment of space, and this would affect the passage of light in that region (something that just does not happen in our currently expanding universe). Hence, there is no ether.

After the Michelson–Morley experiment discredited the notion of the ether (1887), scientists were unable to fully explain reality. Without the ether, there was nothing to carry forces such as electromagnetic energy and gravity. Physicists at the time had no credible way of thinking about forces acting at a distance with nothing in between to deliver the effect. As a result, physicists were quick to accept a remedy in the forms of quantum physics and general relativity, both of which brilliantly addressed the seemingly intractable problem of forces acting at a distance. The services of the ether were no longer required.

Einstein was more than ready to drop the notion of the ether as the carrying medium for light, but he insisted that the universe has properties, and that the ether represents the physical qualities of the universe, which operate outside of matter and energy. He remarked, in a 1920 lecture, that without the ether there would be "no possibility of existence for standards of space and time (measuring-rods and clocks), nor therefore any space-time intervals in the physical sense."[xxii] Einstein understood that without the ether there would be no universal fundamental constants, all of which specify physical qualities of the universe and are not produced by forces or particles.[111]

To quote Einstein,

> To deny the ether is ultimately to assume that empty space has no physical qualities whatever.
>
> – Albert Einstein, from 1920 lecture delivered at the University of Leiden, titled "Ether and the Theory of Relativity" (translated from the German)

The Michelson–Morely experiment disproved the existence of an ether wind, believed at the time to carry light waves through the vacuum of space. The Michelson–Morely experiment by no means proved that the vacuum of space is devoid of any physical properties. In Chapter 9, "The Mathematical Universe," we will show that Einstein was right, as usual, and we will demonstrate that the nothingness of space is endowed with a collection of curious properties.

xxii From Einstein's 1920 lecture delivered at the University of Leiden, titled "Ether and the Theory of Relativity" (and translated from the German).

8 What's new? Not much

There's nothing new under the sun, but there are lots of old things we don't know.

– Ambrose Bierce

8.1 New things are modified old things

We all like to think that at the moment of our birth, a new and unique being arrived on Earth, someone who had never before existed, and who will never again exist. Perhaps we are mistakenly conflating newness and uniqueness. Our genome is simply a fusion of two existing haploid genomes, with a bit of scrambling in the process. Our bodies are stamped into the mold of our species. Our species developed through the tiniest of modifications of its immediate ancestral species, and up through a lineage that ascended to the first living cells. Everything about our development has occurred through steady modifications of pre-existing forms. The resulting organism has an ancient origin.

Just as living organisms are derivative forms, so is our data. For scientists, "new" data is data that has been collected prospectively, under experimental protocols, or through observations made independently of any prior observations. As a general rule, prospectively collected data is thought to have more value than 'old" data that had been sitting around in some existing dataset of dubious validity. We are fooling ourselves. New data is always dependent upon the "old" data that preceded it. The air temperature that we measure today is very much dependent upon conditions that prevailed yesterday. Yesterday's weather was determined largely by conditions of the day before, and so it goes back to the beginning of time. This point is discussed further as a note in Chapter 11.[112]

Likewise, new ideas are modifications of old ideas. In the sciences, it would be difficult, if not impossible, to find a recent concept that was not built upon a firm foundation of older concepts. The same applies to engineering. For example, every component of the modern automobile (e. g., engine with spark plugs and pistons, chassis with drive train, wheel with brake) is the product of continuous modifications of prior works. For computer programmers, every "new" line of code is a modification of an old line of code, and within every "new" program dwells the ghost of algorithms and modules taken from older programs.

One of the most important lessons of modern genomics has been that we humans contain genes inherited from ancient organisms. The overwhelming majority of genes in extant species are ancient, arising hundreds of millions of years ago, sometimes billions of years ago. Consider the following observations:

1. There are over 500 core genes that are present in all extant metazoan (i. e., animal) species [100].
2. Not only do we find homologs of human genes in single-celled eukaryotic organisms, but we also find homologs that extend all the way up our ancestral lineage

https://doi.org/10.1515/9783111498690-008

and into the bacterial kingdom (as in the case of an actin homolog in the proteobacteria Haliangium ochraceum [271]).
3. Approximately 60 % of the annotated protein coding genes in the mouse genome originate from prokaryotic and basal eukaryotic ancestors [185], and we infer that much the same applies to humans.
4. Nearly 75 % of human disease-causing genes are believed to have a functional homolog in the housefly [193].
5. The human genome and the chimpanzee genome are between 97 % and 92 % alike, indicating that closely related species have nearly the same set of genes [259].
6. Genes extracted from one species can be inserted in another species, and retain their functionality. For example, Pax6 is a regulatory gene that controls eye development. This gene, like so many other basic regulatory genes, has been strictly conserved throughout metazoan evolution. Remarkably, Pax6 in mice is so similar to its homolog in insects that the corresponding genes from either species can be interchanged and function properly [85].
7. Nearly identical gene families are found in many different classes of animals.
8. The human genome is packed with relic sequences from ancient ancestral organisms. About 8 % of our genome is derived from sequences with homology to known infectious retroviruses, and these sequences can usually be recognized by their subsequences that contain viral genes (e. g., gag, pol and env genes) and by the presence of long terminal repeats. The viral sequences, remnants of ancient retroviral infections and the occasional non-retroviral infection, were branded into our DNA and subsequently amplified [110, 121, 195].

Because all animals have similar sets of genes, as amply demonstrated in recently collected genomic databases, we can conclude that the genomes of all animals must contain many old genes passed down from species to subspecies, and generation to generation. Had we lacked these recent findings, from modern gene-sequencing studies, we could have arrived at the same conclusion, based on the following reasons:
1. When we look at various animals, particularly the eutherians (i. e., placental mammals), there is little difference from one species to the next. As noted earlier, all eutherian species have essentially the same skeletal structure and contain the same range of cell types. The liver cells in the mouse look like and function like the liver cells in the human, and these same likenesses apply to all organs and all cells in all eutherians. To have these close equivalences, we correctly infer that the genes of the mouse must be roughly the same as the genes of the human, or of any other placental mammal. As it happens, many of the cells and tissues and body features of humans are also found in non-eutherian animals (e. g., marsupials, birds, and even fish).
2. If each species evolved its own genes from scratch, then we would see chemical incongruities among living organisms. Enzymes with similar functions in different animals might be expected to have active sites that prefer substrates of one particular isomer or another (left- or right-handedness). In that case, various vitamins and

food constituents might have nutritional value in one species but not in another. We do not see any evidence of structural differences in enzymes or their substrates among the different species of animals.

3. There is no known mechanism by which new genes could evolve *ab initio* (i. e., from the beginning and starting with nothing) within a single species. All evidence suggests that genes evolve through changes in pre-existing functional genes or through alterations of pre-existing stretches of non-coding genetic material carried in the genome.

   We can assume that the first organisms on Earth must have begun with small proteins, several amino acids in length. Through mutations and natural selection over time (perhaps hundreds of millions of years), the chain of amino acids may have grown to several hundred amino acids in length. No organisms has the wherewithal to create a large protein hundreds of amino acids in length, from scratch. Not to put too fine a point on it, but evolution works best when it operates on something that has previously evolved.[113]

8.2 How are organisms modified?

> A molecular biologist can be certain that a gene has an ancestral homologue, but a paleontologist can only hope that a fossil has a descendant species.
>
> – Anon

If we insist that all new things develop from old things, then through what mechanism do such modifications take place? In the case of evolving organisms, the knee-jerk answer that is drilled into us through grade school and college-level courses, is that the chemical driver of evolution is genetic mutation. This is seldom the case. If mutation were the chemical driving force of evolution, the differences between the species of animals, in respect to gene sequences, would be profound. But it is not the case. The sequences of functional genes are closely conserved and change very little among different species of animals. In point of fact, the differences in gene sequences between the chimpanzees and humans are not significantly greater than the differences in gene sequence that we would expect to find from one human to another. The reason that chimpanzees look and behave differently from humans is largely determined by subtle alterations in the regulation of gene expression (i. e., the manner in which genes are transcribed, translated, chemically modified, and subsequently utilized in cellular pathways). Put simply, evolution operates largely by selecting for variant forms of regulatory sequences and genes that are waiting and available within the species gene pool. Any mutations that are favored during the natural selection process may have been present in the species gene pool for many millions of years and ubiquitous among related species. Such mutations are best thought of as expected population variants.

Minor alterations in endocrine balance can account for major changes in the form and function of organisms. Consider the differences between a bull and a cow. Both genders have nearly the same genes, but male hormones account for the size, strength, and disposition of the bull. The variety of mechanisms by which gene expression can be modified is enormous. Many of these mechanisms are executed through the cell's epigenome, a collection of chemical alterations to chromosomes that does not alter the sequence of genes (i. e., the alterations are non-mutational).

Mutation does play a role in the evolution of new proteins, and we have previously discussed some of the ways that mutation, in concert with natural selection, may operate (see Chapter 4, "Origin Paradoxes"). Of the various methods by which new genes arise from old genes, the most significant seems to be gene duplication followed by mutation of one of the duplicate genes. Notes in Chapter 11 provide a detailed explanation of the mechanisms whereby new genes are acquired.[114] [115] [116]

8.3 New species; old species lost

George Gaylord Simpson, the eminent evolutionary biologist and taxonomist, wrote, in 1945, "One of the most fundamental goals of modern biological research is comprehension of the way in which species arise." [220]. It is intriguing that one year after Simpson remarked on the importance of understanding speciation, the science fiction author, Ray Bradbury, found the answer. In "The Million-Year Picnic," one of the stories in *The Martian Chronicles*, Bradbury relates how a man takes his family from Earth, to settle on Mars. Once there, they burn the rocket that transported them, so that they can never return to Earth. One evening, the father tells his children that he is going to take them to see the Martians. The family walk to a canal, and the father shows his children their own reflections in the water. The family had become, by virtue of leaving the Earth behind, a new species; a species of Martians.

The family depicted by Ray Bradbury had separated itself from the gene pool of Earthlings, and had established, along with the other Martian colonists, an isolated gene pool that would evolve in its own manner, over time. From the moment that the earthlings arrived on Mars, they became a new species. As natural selection operated in the Martian environment, the new species would become visibly distinguishable from its parent species on earth. The Martians might develop a horny coat to protect against cosmic radiation, or any number of modifications to cope with the toxic Martian atmosphere. Regardless, we can be certain that the gene pool available to the colonizing martians would evolve differently than the evolving gene pool of Earth-bound humans.

Can a single member of a species account for a diverse collection of subspecies? Richard Dawkins has explained the process quite clearly in his book "The Greatest Show on Earth: The Evidence for Evolution," [71]. We suspect that it happens frequently when animals of one species travel away from their group and find themselves on islands or in geographic areas isolated by mountains, rivers, or habitat barriers.

As an example, imagine a female primate of Class Lemuriformes desperately clinging to a tree limb as she is is swept to sea, only to wash up on the shore of Madagascar. After a few days, she assesses her situation and determines that she is the only strepsirrhine (a subclass of Class Lemuriformes) on the island. Adding to the drama, she discovers that she is pregnant.

If she and her brood survive, and if the children successfully breed, they will establish a new, small gene pool derived from a sampling from the much larger gene pool of the parent species. A new species will evolve from the new gene pool. In fact, a scenario something like the one described has been suggested for the origin of all the species of lemur on Madagascar. The reasoning is simple: lemurs are found only on the island of Madagascar and on the Comoros Islands, just northwest of Madagascar. Hence, they must have arisen on the island from an ancestral species that lived only on the island. The non-lemur members of Class Lemuriformes, the galagos and lorisids, live in Africa and Asia. Hence, the founding parent of the lemurs of Madagascar most likely came from Africa.

The important point here is that the member of Class Lemuriformes who landed in Madagascar was nothing out of the ordinary. Her only claim to fame was that she contributed to an isolated gene pool. This gene pool evolved over time, and the process of evolution required nothing more than natural selection of the available pooled genes. Eventually, all the different species of lemur now living in Madagascar came to be. Had there been a different pregnant strepsirrhine, arriving in Madagascar under the same set of circumstances, we would have no lemurs as such, but we might have a different set of species in their place.

Evolutionary biologists suspect that just as one strepsirrhine primate may account for all the different types of lemur on Madagascar, one caviate rodent may have been responsible for all the extant rodents of South American. New species are just old species that have lost their way and established their own evolving gene pool. The initial gene pool of the new species is a subset of the gene pool of the parent species, and the first generation of the new species are indistinguishable from the members of the parent species. Hence, the creation of the new species does not require the acquisition of new genetic mutations. Over time the gene pool of the new species will evolve, accumulating new variants of genes, and serving as the genetic material for individuals who will look less and less than the parent species. The greatest threat to the survival of a new species is cross-mating between members of the parent species and the child species; this results in a mixing of the genes, and we would no longer have two separately evolving gene pools. Consequently, we often find a profusion of new species in locations that are not easily accessible to the parent species (e. g., islands like Madagascar and the Galapagos).[i]

i Species are not distinguished from one another by any single gene, or by any specific gene combination. The identity of a species is determined by the species gene pool; the aggregate collection of genomes occurring in the members of the species, and available for procreation. Humans have their gene pool.

8.4 A deterministic universe

In a deterministic universe each moment of the future has been determined by physical laws operating upon the conditions of the present. There is nothing new in the universe; there is only something old that is being continually updated according to physical laws.

The law of conservation of Information holds that the total amount of information in the universe that describes the positions, velocities and quantum states of everything in the universe, is conserved. This law arises as a consequence of determinism. The reasoning that leads to the law of information conservation is as follows:

1. Everything in the universe behaves from moment to moment, ad infinitum, according to known physical laws.
2. The future of the universe, from any moment to the next, can be accurately predicted by applying our physical laws to the contents of the universe (e. g., particles, waves), at their locations, and with knowledge of their status (e. g., velocities and quantum states). This is just another way of asserting that the universe is deterministic.
3. Although we could never know all this information, and although we could never perform the mathematical operations required to advance the universe from moment to moment, we can be confident that all of the information must exist. Otherwise, there would be insufficient information to advance the universe from moment to moment, and the universe would be chaotic (which it is not).
4. The assertion that all the information describing the universe must exist from moment to moment is equivalent to asserting that all information in the universe is conserved (because the information at any one moment of time provides all the information describing the next moment of time).

The law of conservation of Information has been invoked to explain the motion of objects that follow a trajectory. The trajectory of a launched object is determined by the set of information pertaining to the object, at the moment of its launch, and the observation that one, and only one, trajectory is possible implies that each point along its path holds all the information needed to move the object to the next predetermined point. It is as though the universe has total knowledge of reality, and uses its omniscience to determine the object's trajectory, a condition that could only apply if the information at all points along the path is conserved.

It takes some mental accommodation to accept that information exists that we cannot attain, in any practical sense. We certainly cannot know the quantum state of every particle in the universe, but the deterministic model of our universe depends upon

Gorillas have their gene pool. There is never any need to search for a "species gene" that separates one species from all others.

"playing out" the destiny of particles from their known initial quantum states followed throughout all of time.

There is an easy way to make sense of the law of conservation of information. Throughout this book, we have maintained that quantum particles are nothing other than their quantum properties. This is equivalent to asserting that quantum particles are just information (i. e., the information that describes their properties). Because all matter and energy are composed of quantum particles, we can infer that matter and energy are composed of information. Hence, the law of conservation of mass/energy is equivalent to the law of conservation of information.

Yet another rationale for the law of conservation of Information can be developed from Schrödinger's wave function, which captures all the information of a quantum object. Because all the universe is composed of quantum objects, and because all wave functions are linear and additive (i. e. can exist in superposition), we can conclude that the composite Schrödinger's wave function contains all the information of the universe and all the information required to determine the state of the universe from moment to moment, for eternity. Hence, the Schrödinger's wave function embodies the law of conservation of information.[ii]

Physicists, in their overly clever manner, have come up with a strange rebuttal to determinism, and it is based upon the law of conservation of information. Brace yourselves, and imagine that you are residing within a spaceship that is fast approaching the event horizon of a black hole. The moment that you cross its horizon, you disappear forever into the black hole, along with all of the information related to you (i.e, the positions and quantum states of every particle in your body are no longer accessible to the universe outside the black hole). All your information, so far as we can tell, now exists in black hole limbo.

Steven Hawking tells us that black holes are not eternal. They slowly shrink as they expel energy, into space, in the form of black-body radiation.[iii] Black body radiation is a continuous spectrum whose peak amplitude correlates to a particular temperature. Because the emission spectrum of a black body is continuous (i. e., not radiated as discrete emission lines as we might see with atomic emissions and without any relation to the information that had been engulfed by the black hole), we infer that black hole radiation contains no information. Over time, the black hole shrinks into nothingness, at which time all of its contained information is lost (i. e., not conserved).[iv] Summarizing,

---

ii Experimentalists will be pleased to learn that the law of conservation of Information seems to work at the quantum level. Samuel Braunstein and Arun Pati demonstrated that when a qubit is randomized, its information is not lost but is distributed to other qubits in the environment [46].

iii First described by Stephen Hawking, in 1974, and now known eponymously as Hawking radiation [116].

iv A black hole of one solar mass might persist for 10^{64} years, while a super-massive black hole, such as those found in the center of galaxies, might persist for 10^{100} years or more.

black holes lose their contained information and replace that information with random radiation that influences the universe in ways that could not have been known at the moment that the black hole's event horizon was crossed.

From here, the logic gets a little tricky. Let us see:

1. If information is not conserved forever by black holes, then the law of conservation of information is violated.
2. If the law of conservation of information is violated, then the universe does not hold all of the information required to move the universe from one moment to the next.
3. If the universe lacks information on its state at any given moment, then it cannot determine how events unfold throughout the universe from one moment to the next.
4. If the universe cannot determine outcomes from moment to moment, then the universe must be nondeterministic.

More than a few physicists have devoted a sizable portion of their careers trying to reconcile black hole information loss with the law of conservation of information. Such efforts require a careful examination of the assumptions underlying the Standard Model of quantum mechanics, the laws of general relativity and of gravity, and the physics of black holes. No doubt, solutions will be proffered, but will any of these theoretical solutions be testable? Let us wait and see.

Determinism is incompatible with free will. Not to put too fine a point on this, but in a deterministic universe, we humans are nothing but automatons, fulfilling a predetermined course, much like the rock thrown into the air. In a deterministic universe, every decision we make, no matter how trivial, is predetermined by an inscrutable universe. That saddens us. Most of us would prefer to believe that we have free will and that there are many different possible paths open to the universe. Essentially, we would prefer to think that the universe is a work in progress, and that we are its architects.

One factor that may help us to distinguish a deterministic universe from a probabilistic universe is the notion of randomness (i. e., unpredictability). In the computational realm, there is general agreement that no algorithms are capable of producing a truly random sequence of characters. Eventually, even with the most clever "random" number generators, the output sequence repeats itself. For this reason, computer algorithms intended to produce a random sequence of characters are known as "pseudorandom generators." Computer scientist know their limits.

What about the real world? Are there naturally occurring phenomena that occur probabilistically, whose outcomes cannot be predetermined? Let us look at a few candidates for purely random behavior.

- Radioactive emissions. These are weak force nuclear decay events, such as those sensed by a Geiger counter. We know that they occur, but we cannot predict when any single event will happen.
- White noise. Natural signals are never pure. There is always a low level of noise, and the noise is characterized by a seemingly random sequence of pulses.

- Brownian motions. Molecules bounce around in every possible direction, colliding with other molecules, with unpredictable paths.
- Black body radiation. At any give temperature, objects emit a continuous spectrum of light waves, independent of the source, and thus carrying no information about the source of emission.
- Vacuum energy/field fluctuations. When the energy and matter is removed from a volume of space, we see apparently random small fluctuations of energy. Because vacuum state fluctuations occur spontaneously, without an energy source or a preceding event that accounts for their existence, they are thought to occur randomly.

So far as we can tell, the aforementioned list of natural phenomena are truly random. A random event is, by definition, an event that cannot be predetermined. Therefore, it would seem that events occur that are not deterministic, and this would imply that the universe is not deterministic.

Maybe not. A determined determinist might argue that events that are random to observers are not random to the universe. For the determinist, the test for determinism is based on reversibility. If an event, even a random event, is reversible, then it is deterministic. Reversibility means that we can go backwards and forwards knowing both outcomes with certainty. Because quantum events are reversible, we can infer that non-quantum events, being nothing other than amalgams of quantum events, must also be reversible.[v]

Those who champion a non-deterministic universe have another arrow in their quiver; an observable and repeatable phenomenon known as symmetry breaking.[vi] Symmetry breaking results when physical events occurs in apparent violation of a known symmetry. Various metaphoric scenarios have been used to illustrate the meaning of a symmetry break, the most common being the Mexican hat example. Imagine that a marble has bounced around, obeying the laws of physics as best it can, until it comes to rest at the very top of a Mexican hat, exactly at its center. Perched prettily atop the hat, there are no forces that would tend to push the ball off-kilter, to slide down towards the hat's brim. Nonetheless, if we wait long enough, the ball teeters, and we soon have a new equilibrium position for the ball, in the large hat's gutter. The ball has broken the physical symmetry of the system.

v There is, perhaps, a fly in the ointment. Yakir Aharonov has proposed that seemingly random quantum events are determined by events happening in the future. If this were the case, there could be no true randomness in nature. The proof of Aharanov's theory is elusive insofar as it requires some form of measurement, and measurements abolish coherence (i.e, the coherence between a future event and a current apparently random event would be destroyed). Nonetheless, Aharonov and his colleagues claim to have a proof [5].

vi We will be discussing symmetry in Chapter 9. For now, we can think of symmetry as the physical status quo provided by conservation laws.

Though the Mexican hat metaphor is vivid and memorable, it lacks some important subtleties. A better way of thinking about symmetry breaking is as an accommodation for small fluctuations around a balance point. The symmetry holds for the system, but when transient fluctuations occur at a point of perfect balance (of forces), we might expect to see the system behave in a manner that is not predictable. For example, if we have a symmetry involving rotation, what would happen if we lowered the forces to an extent such that shifts in translation occur at distances smaller than the Planck length. As we shall discuss in Chapter 9, “The Mathematical Universe”, rotational symmetry accounts for the law of conservation of angular momentum. At sub-Planck length distances, tiny fluctuations in the system may set conditions that operate outside the rotational symmetry model, essentially breaking symmetry. For all of the physical symmetries, there may be local fluctuations for which the symmetry may not fully apply, and when such conditions prevail, an outcome may arise that resets reality. The formal analyses of symmetry breaking involve perturbation theories wherein the possible effects of small deviations within a system are analyzed. We will not delve into the physical manifestations of broken symmetries. Suffice it to say that symmetry breaking is an observed phenomenon that accounts for various aspects of superconductivity, antiferromagnetism, ferroelectrics, liquid crystals [8], and certain types of phase transitions of matter.[vii]

Physicist and Nobel laureate Philip Anderson, referring to the new realities produced by symmetry breaking, wrote, “The actual universe is the consequence of layer upon layer of emergence.” All well and good, but how does symmetry breaking relate to the issue of determinism? Put simply, broken symmetries yield outcomes that are nondeterministic (i. e., outcomes that could not be determined by the law of conservation of information).[viii]

8.5 New light and old light

Electromagnetic radiation is the currency of the universe. The universe’s energy is waves, represented by the simple equation, $E = h\nu$, where h is the Planck constant and ν is the energy’s characteristic wavelength. The matter in our universe is waves, described succinctly by the Schrödinger’s wave equation. All motion in the universe produces electromagnetic waves. When electrons orbiting atoms transition between energy levels, photons of light are either absorbed or emitted. Vibrational transitions of

---

vii Symmetry breaking may also account for the bulk of the mass of the matter (atoms) in our universe. Nucleons (protons and neutrons) are composed of a few quarks and gluons whose combined mass represents about one percent of the mass of a nucleon. Most of a nucleon’s mass arises from a type of containment energy that is the putative result of symmetry breaking [262].

viii Our treatment here of determinism, and of the Law of Conservation of Information, and of symmetry breaking will certainly raise the hackles of physicists. These topics are famously controversial. My interpretation makes perfect sense to me, but, like other facile explanations, it is likely to be wrong.

molecular bonds emit infrared radiation [152]. Even gas phase molecules have quantized rotational levels, and when transitions of these levels occur, electromagnetic radiation, typically in the microwave range, is emitted or absorbed. What was the origin of the universe's allotment of electromagnetic radiation? As it happens, most of the light traveling through the universe today arose from one astonishing event that occurred about 380,000 years after the Big Bang (i. e., most of our light has been continuously shining for close to 13.8 billion years).

Some moments after the Big Bang, subatomic particles formed, and these included photons, electrons, quarks and gluons. (along with their antiparticles doppelgangers). Quarks and gluons aggregated to form protons and neutrons. For a long time, the universe was much too hot and dense for atoms to form from electrons, protons and neutrons, as heat energy greatly exceeded the ionization energy of orbiting electrons (specifically, the energy needed to break an electron from its orbit in a neutral Hydrogen atom). During this period of the universe, photons moved the very short distances from particle to particle, being repeatedly absorbed and emitted, but never traveling freely through space.

At about 380,000 years following the Big Bang, the universe cooled down to a comfortable 3,000° Kelvin, at which time the temperature of the universe could not ionize the electrons in hydrogen atoms, and every positively charged hydrogen and helium nucleus (these were the only two atomic nuclei synthesized at the time) quickly bonded with one or two negatively charged electrons (one electron for hydrogen and two for helium). Suddenly, photons had a clear path by which they could move through the expanding universe. At this moment, the universe became transparent and the cosmic background radiation of the universe was released![ix]

When we look into the night sky, we see tiny specks of light, coming from stars. We do not see the cosmic background radiation because, as the universe has expanded, this primordial light has diminished in wavelength, dropping into the invisible microwave range.[x] When we aim our Earth-bound telescopes in any direction, we find that all of space is saturated with microwave radiation. Because this radiation has just about the same intensity wherever we look in the universe, we deduce that this is the cosmic background radiation that emanated through the cosmos 380,000 years following the

---

ix Electromagnetic waves tell us little about the universe preceeding the release of the cosmic background radiation. In contrast, gravity waves, in theory, are unaffected by absorption or scattering by primordial plasmas. In principle, by studying gravity waves emitted at the time of the Big Bang, we can study the earliest moments of our universe, gaining insights unattainable with electromagnetic waves.

x The cosmic background radiation currently peaks at a wavelength of about 1.9 mm and a frequency of 160.2 GHz. The distribution of the wavelength of the cosmic background radiation conforms to a blackbody radiation curve, such as we would expect with a non-reflective body heated to a temperature of 2.7° Kelvin. This happens to represent the average ambient temperature of our universe, indicating that the cosmic background radiation is primarily responsible for heating the universe to its current temperature.

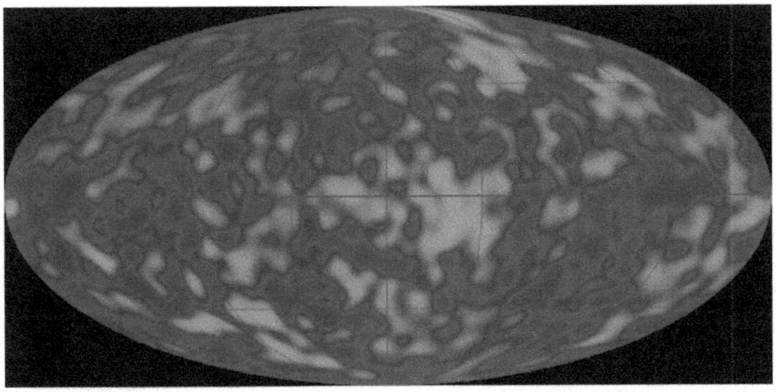

Figure 8.1: Cosmic microwave background radiation (measured as temperature) for the entire visible universe, corrected to remove the contribution from our galaxy. The differences in color represent minor differences in temperature. There are no areas devoid of cosmic background radiation. The cosmic microwave background permeates all of visible space with light that originated 380,000 years after following the Big Bang. The cosmic background remains the largest contributor of electromagnetic radiation in our universe. Source, Public domain document produced by the U.S. National Aeronautics and Space Administration.

Big Bang (Figure 8.1). To this very day, the majority of the light in our universe consists of the cosmic background radiation.[xi]

When we gaze into space at night, we see tiny specks of twinkling light coming from distant stars. The night sky is beautiful, but it does not compare in intensity to the universal afterglow produced by the cosmic background radiation. Although we cannot detect microwaves with the naked eye, our televisions can assist us. Readers who have not discarded their analog (cathode ray tube) televisions can watch the cosmic background radiation day or night, from any location on Earth. Turn on your television and detune the receiver (by turning to a channel that does not receive broadcast signals, or by disconnecting your antenna). Your picture tube will display "snow", a cacophony of flickering black and white dots completely filling the screen, produced in part by microwaves. It has been estimated that a few percent of the snow on analog TV sets comes from the cosmic microwave background.

At the same moment that the cosmic background radiation was unleased upon the universe, the first atoms of hydrogen and helium formed. Whatever happened to all those primordial atoms? Those original atoms are still with us. Today, about 70 % of the atoms of the universe are hydrogen and 28 % are helium. We are taught that stars (particularly massive stars such as red giants and neutron stars) are the crucibles for nucleosynthesis (the creation of atoms). This is true, and but nucleosynthesis is the process by which heavy elements are produced from lighter elements. The atoms heavier than helium account for only 2 % of the atoms in our universe. A good portion of the

xi The cosmic background radiation is remarkably uniform throughout space. Variations in its intensity are small and only the most sensitive instruments can detect local fluctuations.

remaining 98 % of matter (hydrogen and helium) has been floating in the vast expanse of space for nearly 14 billion years.

8.6 Nothing is new, not even our universe

Notwithstanding the Big Bang, which only accounts for the universe that we observe today, we cannot say much about what the universe was like prior to the Big Bang. For all we know, the universe stretches back to infinity, and has no beginning. If the universe has no beginning, it seems reasonable to guess that the universe has no ending. Drawing a simplistic analogy, we have geometric shapes that have neither beginning nor endings nor edges. The Klein bottle comes immediately to mind (Figure 8.2). It has volume, but lacks a discernible inside or outside. It seems to be an open container, but from any point on the bottle, we can follow a path indefinitely, never emerging, like a genie trapped in a bottle for all of eternity. If we can imagine the universe as a multidimensional Klein bottle, then we should be able accept that the object may have no definable beginning or ending.

If, on the other hand, we can imagine that our universe has a beginning, then it seems only fair to try to provide some sort of ending. What would the ending of the universe look like? Would it be empty spacetime, without matter or energy?

If the universe is prepared to wait a very long time, its destiny may tell us something about the physical meaning of nothingness. The late-stage of stellar evolution seems to be mainly limited to just brown dwarfs, white dwarfs, neutron stars, and black holes [2]. White dwarfs represent the road's end for most main-sequence stars and can persist for about 10^{35} years. As main-series stars, like our sun, reach their destinies, they leave behind a universe filled with dead-end objects, predominantly white dwarf stars. The Milky Way is thought to have already accumulated 10 billion white dwarfs, and the number will only increase. At some point in the future, illuminated galaxies such as ours will be gone, and the universe will be a celestial nursing home filled with superannuated dwarf stars. While this tedious process proceeds, black holes will play an expanded role as the universe's grim reaper. Black holes, with a predicted life-span of 10^{64} years seem to have no size limit. They will engulf any matter that reaches their event horizons, sucking in other black holes and white dwarfs in their vicinity. There seems to be no theoretical reason to think that at some point in the far distant future, a single black hole will have absorbed all of the matter of the universe. Eventually, when the universe consists of nothing but engorged black holes, these behemoths will slowly disappear, as they emit black body radiation that carries no information.[xii] All that shall remain are meaningless bosons traveling in a sparkling clean void.

---

xii The evaporation of black holes via Hawking radiation is discussed in Section 8.4, "A deterministic universe".

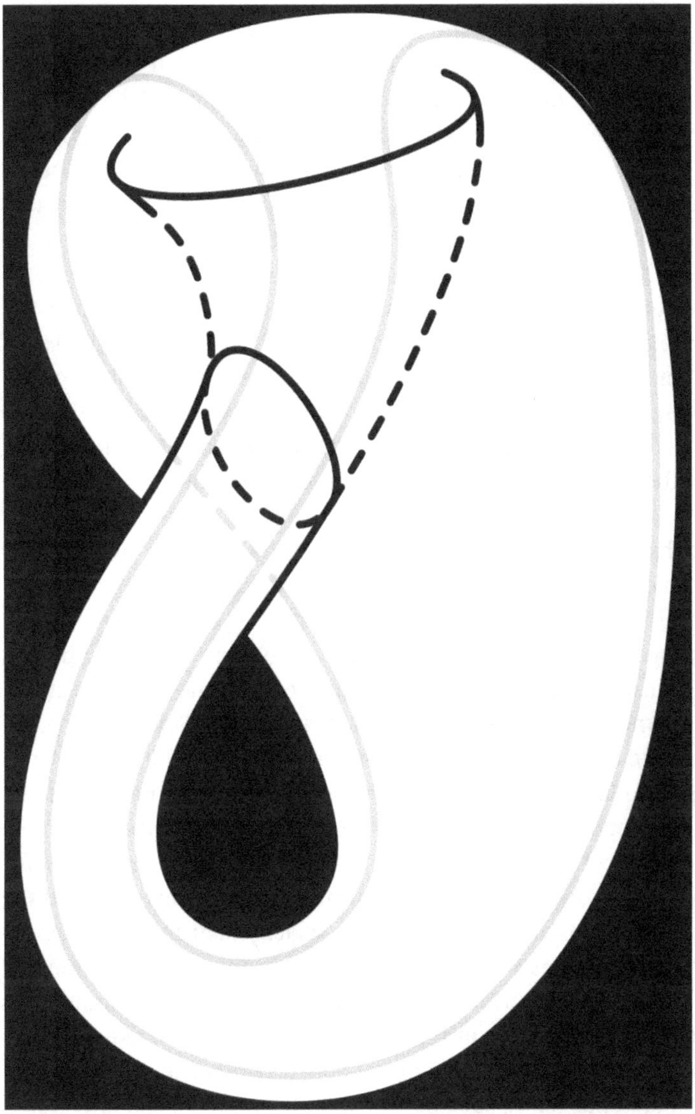

Figure 8.2: Drawing of a Klein bottle, a topologic oddity in which the outside cannot be distinguished from the inside. Source, Wikipedia, and released to the public domain by its author, Vierkantswortel2.

Just for fun, let us consider one more irresponsible and wildly speculative scenario by which our universe may devolve into nothingness. Let us return to the notion of the cosmic microwave background, the light that was emitted about 380,000 years after the first moment of the Big Bang, and which fills our entire universe. From this moment onward, the CMB has accounted for nearly all of the directly observable energy of our universe. As the universe has expanded, the cosmological background has kept apace of the expansion by stretching its wavelength. At present, the wavelength of the CMB, throughout the universe, is about 1 mm, accounting for a black-body temperature of 2.7° Kelvin, this being the temperature of empty space. At some point, if the universe continues to expand, the wavelength of the CMB will stretch to the point that it is length is so long that it no longer has observable energy (basically, when its energy falls below the level needed to create a quantum of light (i. e., a photon), and its "wave" becomes indistinguishable from a straight line). At this juncture, the universe will cease to have any physical existence, other than the physicality of nothingness.

Our morbid fascination with the end of things is predicated upon the condition that the laws of our universe shall remain static. In turn, our faith in the steadfastness of physical laws is based on measurements of the present and the past. Based on observations of distant objects, as they existed in the early universe, we can say with some degree of confidence that our physical laws have not changed in the past 15 billion years. We can extrapolate that those laws will not change any time in the future. But we cannot be certain.

Let us just assume that there comes a time when all of the matter of the universe is absorbed into black holes, and all of the black holes have radiated away into bosons (e. g., photons). At this point, time stops because bosons travelling at the speed of light, cannot observe the passage of time. Furthermore, there is no limit to how far bosons (e. g., photons) can travel in a vacuum. In the boson's frame of reference, it exists everywhere at every time or nowhere at no time. It's all the same to a boson! Summarizing, the universe seems to be headed towards a moment when it consists of nothing but bosons that are everywhere at once, in the bosonic frame of reference, which happens to be the only frame of reference remaining in the universe (because everything else is gone). At this point, the total amount of information held in the universe will be zero, insofar as the black holes will have consumed all of the information of the universe, and the radiated bosons have no information, because we know exactly their location (i. e., everywhere at once). The end-times described here consist of bosons outside of time and without information or entropy (entropy being a state described by information). This being the case, the far-distant future of the universe shapes up to be much like the imagined beginning of the universe (i. e., nothing but bare-boned symmetries). This kind of analysis, or something like it, has led to speculation that the ending of the universe is the beginning of the universe, unfolding as a perpetual cycle. My goodness!

9 The Mathematical universe

All things are number.

– Pythagoras of Samos (570–495 BCE)

9.1 Conservation laws and invariances

All the universal laws of science are simple relationships that apply to the fundamental building blocks of our universe: matter and its constituent particles, and energy and its various forces. Let us review the features of scientific laws.

1. Testable and validated.
   A law does not mean much if you cannot put it to the test. If a law passes every test that scientists can devise, we often say that it has been validated (i. e., shown to be true, as far as anyone can determine).
2. Universally applicable.
   One of the very nice things about our universe is that it provides us with laws that apply everywhere. We can determine that the light (electromagnetic waves) received from galaxies billions of miles away obeys the same scientific laws as the light emanating from our smartphone screens. We can observe that orbiting stars in distant galaxies obey the same laws of gravitation that cause hammers to fall here on Earth. Furthermore, our laws seem to be stable over time. Nothing much seems to have changed since the Big Bang.
3. Simple.
   Most scientific laws can be described in short equations, and their simplicity has been an enduring puzzle among scientists and non-scientists alike. We will soon see that scientific laws are almost always based on invariances (i. e., situations in which things do not vary under transformations). When we look at things that do not vary, we might expect that the equations that describe relationships would be simple; nothing really changes. This happens to be the case, and every student of physics should be grateful that the most significant scientific laws can fit comfortably on a single index card.

As mentioned in point 3, scientific laws often describe simple relationships among fundamental things (i. e., matter and energy) that depend on some universal invariance. An invariance, as its name implies, is anything that does not change when some specified condition is altered. If an object has the same properties when moved through space, then we say that the properties are invariant through translation. If a circle looks the same, when flipped across any axis that passes through its center, then we say that the circle is invariant under rotation, indicating a geometric invariance.

Conserved quantities, such as the conservation of energy or the conservation of charge, are both based upon invariances over time. If the quantity being measured does

https://doi.org/10.1515/9783111498690-009

not change over time, then we say that the quantity has been conserved. Hence, conservation laws can be thought of as the subset of invariance laws in which the invariance is observed over time.

We tend to take invariances for granted, as though they are all so obvious that they could have no significance and could not possibly be worthy of our attention. In this section, we will look at some invariances that have profound influence on our understanding of the universe, and how the laws derived from these invariances took humans thousands of years to fully grasp.

Let us start with one of the simplest invariances that is taught to every middle school student: the law of inertia.

– The law of inertia, also known as Newton's first law, states that if a body is at rest or moving at a constant speed in a straight line, it will remain at rest or keep moving in a straight line at constant speed unless it is acted upon by a force.

In accordance with Newton's first law of physics, if a rocket ship were to coast through empty space at 50,000 miles per hour, it would maintain that speed forever, without requiring any additional fuel, and would eventually traverse the entire universe, if it could spare the time. This is quite an amazing property of the universe, providing us with the ability to travel for eternity, at constant speed, without any investment of energy. It would seem that the law of inertia is one of the energy conservation laws because the total energy of the system does not change, and there is definitely something conserved. Specifically, the momentum of the rocket ship, calculated as the mass of the ship times its velocity, remains constant and is, thus conserved. More fundamentally, the law of inertia tells us something about the invariance of spacetime. Because the speed of the rocket ship stays the same, regardless if where it is translated (i. e., movement in a straight line) and when it is observed, we can conclude that space is homogeneous and that momentum is spacetime invariant.

What about that part of the law of inertia that deals with masses at rest staying at rest? The homogeneity of spacetime implies that one frame of reference is just as valid as any other because the universe does not change when positions are translated through space and time. We can see now that the mass moving at constant velocity, when observed in a frame of reference moving at the same velocity, is equivalent to a mass at rest. So, of course, the mass at rest stays at rest.

An object can be stationary in space, but it cannot be stationary in time. Time is always flowing in one direction, and it is for this reason that poets liken time to a river. Poets have a tougher time finding the words to describe the traditional three-dimensional coordinates, which are disdainfully described as whatever happens to lie between point A and point B.

Similarly, time seems to move forward inexorably. Any actions, occurring at any given moment of time, can be repeated at later moments, producing the same outcome. This signifies that the available energy necessary for the action does not change as time moves forward. Therefore, energy does not change over time. Hence, the shift symmetry of time accounts for the law of conservation of energy. That being settled, we must now ask ourselves, "What is energy?" Very few textbooks seriously attempt to define energy. Instead, they typically list different kinds of energy (e. g., nuclear, mechanical, electrical), as though a collection of familiar terms can substitute for a cogent definition.[i] How could it possibly be true that there are distinct types of energy? None of the familiar equations for energy account for separate types. Surely, all energy is equivalent and interchangeable. To make matters worse, we pretend that energy can be classified as being kinetic or potential. What in the world is potential energy? What, exactly, does potential energy do? Does it merely wait around until it has the chance of converting into kinetic energy? Nonsense!

To understand the meaning of energy, we need to return to the symmetry that establishes the law of energy conservation; namely, time-shift invariance. When we perform an experiment at any moment in time, we can be certain that if we return at a later time and perform the same experiment under the same conditions, we can expect to get the same results. Therefore, the total energy available to the experiment is conserved. The energy can be functionally defined as that thing that is being conserved by time-shift symmetry, but it is more practical to think of energy as that which makes the experiment happen. Insofar as all experiments involve motion of a sort (e. g., particles moving, molecules vibrating, electromagnetic waves propagating through space), we can think of energy as whatever makes things move.

Another way of thinking about the relation between energy conservation and time-shift invariance is to imagine the universe at any moment in time. If we allow a small amount of time to pass (let us say 1 millisecond), and we observe the universe again, we find that things are much the same, but the contents of the universe have moved a little bit. Planets have progressed a bit in their orbits. Light waves have traveled 186 miles. Some electrons have shifted a quantum up or down in their orbits. The difference between one moment in time and the other is observed as motion. Therefore, time-shift invariance tells us that the conserved quantity must be whatever is responsible for the motions that occur during the time shift. Because energy is conserved over time, we know that the totality of energy does not change over time (by definition), and that whatever energy is responsible for motions occurring over time must be exactly balanced by energy that could potentially restore the universe to its state prior to the time shift. The dual concepts of the energy associated with movement (i. e., kinetic energy) and the restorative energy associated with a return to some original state (i. e., potential energy) are together accountable for all the equations of motion developed for classical and quantum physics. We will return to the concepts of kinetic energy and potential energy when we introduce the Lagrangian and least action later in this chapter.

---

i A fairly comprehensive list includes: nuclear, electrical, thermal, chemical, gravitational, radiant, acoustic, mechanical, elastic, and energy of motion.

Keeping the relationship between time and energy in mind, let's briefly consider a moment 14 billion years ago, when the universe as we know it was born in a Big Bang. To the best knowledge of physicists, none of whom were witness to the event, the Big Bang produced everything in our universe, emanating from a point in spacetime.

How much energy did the Big Bang produce? In accordance with time-shift invariance, the Big Bang produced a net zero of energy. There was no energy at the moment preceding the Big Bang; therefore there must have been no net energy at the moment after the Big Bang. But the universe is full of energy! Where did it all come from? The easiest explanation is that the Big Bang produced equal amounts of positive energy and negative energy. We have become accustomed to observing positive energy, but the concept of negative energy evades description.[ii] Time-shift invariance tells us that a net positive energy cannot arise from a net zero energy, indicating that our universe began with nothing and we have been running on empty ever since.[117]

Einstein's special theory of relativity was developed from two fundamental assertions about the universe [40]:
1. Physics works in the same way in all uniformly moving frames of reference (an extension of Galileo's principle of relativity).
2. The speed of light is the same in all frames of reference (as demonstrated by the Michelson–Morley experiment [179]).

When we think about it, Einstein only needed the first assertion to develop his remarkable theory. The second assertion follows as a consequence of the first. When we say that the observers in any frame of reference observe the same physical laws and the same outcomes of those laws, then we can infer that the observers must observe the same measurement of the speed of light in every frame of reference. Therefore, the speed of light must be the same in every frame of reference. Michelson and Morley's experiment merely confirmed a foregone conclusion.

We can also interpret the constancy of the speed of light as a conservation law. Time-shift invariance tells us that experiments performed at a certain time will produce the same results if they are performed under the same conditions in another time. Hence, whether we measure the speed of light today, or tomorrow, within our frame of reference, the speed of light remains the same, never slowing down or speeding up. We also know that the same constancy of light will be true in every frame of reference. Hence, the speed of light is constant in every frame of reference, at all times (i. e., the speed of light never changes); and this is a conservation law.[iii] [118]

---

ii It has been speculated that negative energy is what holds protons together, in a dense nucleus, against the repulsive forces that would otherwise tear apart particles having the same charge (i. e., positive-positive repulsions among protons). Gravity is also considered a possible source of "negative energy." We will not pursue either hypothesis.

iii Scientists could have deduced, from casual observation, that the speed of light must be constant, in all reference frames. In point of fact, the constancy of the speed of light is something that could have

Einstein went on to demonstrate that so-called normal matter cannot travel at the speed of light, no matter how much energy is devoted to providing an accelerating force. At the same time, no added energy is required for a beam of light to maintain its constant speed forever, as it travels endlessly through the universe. This profound difference between matter and light suggests that these two objects belong to two fundamentally different classes. As it happens, fermions are responsible for atoms and light is mediated by bosons, each having its own representations of a particular symmetry group. More about this later.

In 1904, Hendric Antoon Lorentz introduced a set of mathematical variables that preserves the symmetry of transformations in space and time.[iv]

$$c^2 t_1 t_2 - x_1 x_2 - y_1 y_2 - z_1 z_2 = c^2 t_1' t_2' - x_1' x_2' - y_1' y_2' - z_1' z_2'$$

Time is denoted by "t", and the spatial coordinates are x, y, and z. The primed variables indicate coordinates in an alternate frame of reference. The unit of time is converted to distance by multiplying by the lightspeed constant, c, measured by distance per second, times "t", measured in seconds. Solutions to translations in time and space that are invariant with frame of reference must satisfy the Lorentz transformation.

At the time, Lorentz did not fully appreciate the physical significance of his own work [40]. Soon afterward, Poincaré demonstrated how the Lorentz transformations could be extended to represent all the invariances described in Maxwell's equations.[v] Einstein's theory of relativity was soon re-interpreted as the physical expression of Poincaré's extension of the Lorentz transformations. The invariances in these great works stemmed from the observation that electromagnetic waves move at a constant speed in all frames of reference. At last, we saw the light!

Suppose there is a hammer on the left side of your desk, and you choose to move it over to the right side of your desk. This little maneuver, like the hammer itself, is totally within your grasp. As the hammer is horizontally translated across the desk, all of the molecules of the hammer move in perfect unison, maintaining their locations with respect to the other trillions of molecules of the hammer. The hammer is unchanged, whether it is on the left of your desk, or the right. We owe this amazing outcome to the homogeneity of space and the molecular organization of the hammer. The individual molecules of the hammer can take up residence in some new location, without forfeiting their positions in the hammer, because the relationships among the molecules of the hammer are determined by atomic forces that act independently of the hammer's spacetime coordinates. One point in spacetime is much like another, and any point will do.

---

been acknowledged centuries before the Michelson–Morley experiment. An extended discussion of this topic is provided as note 118 in Chapter 11.

iv A symmetry is an equivalence after transformation.

v Maxwell's equations describe the relationships between electric and magnetic fields, and the changes that occur in these fields due to electric charges and currents.

Applying the homogeneity of empty space, we can derive the inverse square law, with a little help from the law of conservation of energy. When energy emanates through empty space from a point source it spreads through time in all spacetime coordinates equally. This tells us that a point source of light will spread out as a sphere that expands over time. At any moment, any point on the surface of the sphere will have the same light exposure as any other point on the surface of the sphere. Due to conservation of energy, the total energy (from the point source) that moves through a spherical shell at radius "r" is equal the total energy moving through a spherical shell of radius "$2r$" and of radius "$4r$" and so on.

The area of a 3-dimensional sphere is $4\pi r^2$. This means that the total energy moving through a sphere of radius r will be equally distributed over the area $4\pi r^2$. As the energy expands outward, the total energy (a constant) moving through a sphere of radius $2r$ will be distributed over an area of $4\pi(2r)^2$, or 4 times the total area of the smaller radius. This means that the passing energy measured per unit surface area is reduced by 1/4, when we double the distance from the original point source.

Generalizing, the expanding energy received per unit area reduces as the inverse square of the distance from the emanating point source. This is known as the inverse square law. Physicists have shown this law to be true, by direct measurements. We can say that our conclusion was made independently of the experimental evidence, and solely on the basis of the three conditions: the conservation of energy, the invariance of spacetime, and the Euclidean geometry of our universe.[vi]

As an aside, we can use the homogeneity of space to establish the conservation of momentum law. Because no location in empty space is any different from any other location, the momentum of an object is conserved when the object is moved from one point to another. Hence, the homogeneity of space accounts for the law of conservation of momentum. Likewise, the conservation of momentum laws under rotational translation leads to the conservation of angular momentum. An undisturbed sphere that is set spinning in empty space will continue to spin at a constant angular speed, forever.

A spectacular example of the law of conservation of angular momentum is observed during the creation of a neutron star. When a star collapses, it shrinks in radius, but maintains its mass and it's angular momentum. Because angular momentum is proportional to the mass of the rotating object multiplied by rotational speed and by its radius, we can infer that as the star shrinks, its angular velocity must increase (to preserve a constant angular momentum). When the radius of the star shrinks by a factor of 100,000

---

vi Employing the same reasoning, the inverse square law for n-dimensional space, can be generalized as:

$$I \propto \frac{1}{r^{n-1}}.$$

In the general inverse power formula for n-dimensional Euclidean space. The intensity of falls off as the inverse $(n-1)$th power as the radius (distance from light source) grows.

(as it might during the creation of a compact neutron star), then the angular velocity increases by a factor of 100,000. For this reason, we observe that neutron stars may revolve hundreds of times per second.

For a system (e. g., the contents of a box, or of a universe), the condition of invariance implies that whatever quantity comes into the system must result in an equivalent quantity exiting the system. Physicists have a simple equation that formally expresses system invariances: the general continuity equation.

$$\frac{\partial \rho}{\partial t} = -\nabla \cdot \mathbf{J} \quad \text{general continuity equation.}$$

Here, $\nabla\cdot$ is divergence, ρ is the amount of the quantity q per unit volume, j is the flux of quantity q, and t is time,

Conservation laws, when expressed mathematically, have a form that closely resembles the continuity equation, allowing for a different set of variables depending on which quantity happens to be conserved. In fact, the continuity equation is identical to Maxwell’s equation for charge conservation.

$$\frac{\partial \rho}{\partial t} + \nabla \cdot \mathbf{J} = 0.$$

In the equation for charge conservation, $\mathbf{J}$ is the current density, measured in amperes per square meter, and ρ is the charge density measured in coulombs per cubic meter.

9.2 The “symmetry first” universe

It is only slightly overstating the case to say that physics is the study of symmetry

– Philip Anderson

Up until now, we have stressed invariance as a key condition accounting for much of what we know about the operational laws of the universe. Modern textbooks of physics have largely abandoned the term “invariance,” and have emphasized a related term that closely aligns physics with mathematics. This new term is “symmetry”. The term “symmetry” has been used in the field of geometry to indicate that a geometric object can be moved in space (i. e., rotated, flipped, reflected, or translated laterally) without changing the way it looks. For example, when a circle is bisected along its diameter, the left half of the circle is identical to the right half of the circle, and the bisection of the circle provides an axis of rotational symmetry. If we have an equilateral triangle, we can flip it any way we like, and it will fit perfectly into its original silhouette. Modern physicists have extended the notion of symmetry to operations upon anything (not just geometric objects) in which the result of the operation is invariant. Thus, we can start to think of the symmetries that might involve time, matter, energy, gravity, and all of the fundamental components of our universe. Furthermore, the invariances that we have described earlier in this chapter all qualify as symmetry operations inasmuch as each of the invariances involves an operation in spacetime (movement at a constant velocity, angular rotation of an object, flipping one object and another of the same type). As an example, let us return to our example of the speed of light, which is invariant (the speed does not slow down or speed up, and all beams of light travel at the same speed), and a conserved quantity (light does not involve the addition of energy to maintain its speed, and light is not lost as it travels in empty space). Does the speed of light have symmetry? Yes. Light exhibits translational symmetry in that it is constant in all frames of reference!

The fundamental relationship between invariances, conservation laws, and symmetries is described mathematically as Noether's theorem, published by Emmy Noether, in 1918 [187]. Paraphrasing Noether's theorem: Wherever we have an invariance, we have a symmetry, and wherever we have a symmetry, there is a conservation law. A proof of Noether's theorem is provided as a note in Chapter 11.[119]

As a simple example how Noether's theorem may be applied, let us return to our example of time-shift invariance, an observable feature of our universe that tells us than an experiment done on Tuesday yields the same results as the same experiment performed on Wednesday. Time-shift invariance is a type of symmetry. Just as geometric translation symmetries produce no change in objects, time-shift is another type of translation symmetry because the end result is an invariance (i. e., Tuesday's result is the same as Wednesday's result).

Under Noether's theorem, all conservation laws can be represented as symmetry transformations. For example, we know that the energy of a system in space is independent of the direction of our observation. Specifically, the energy levels of an isolated atom do not depend on the direction from which the atom is observed [40]. This tells us that atoms have rotational symmetry (i. e., we can rotate an atom in empty space and the energy does not change). According to Noether's theorem, we can infer that the energy associated with rotation is conserved. We shall soon see that transformations of a kind can be represented as a mathematical group (i. e., a symmetry group).

In one sweeping contribution to physics, Noether connected the classical laws of conservation (e. g., conservation of energy, conservation of momentum, conservation of angular momentum, uniform center of mass motion, time and space translations and rotations, and Galileo's and Lorentz's frame of reference transformations), with symmetry transformations. In fact, Maxwell's equations, which are basically conservation laws, can be interpreted as examples of Noether's theorem.

Before we move more deeply into the subject of symmetries, let's take a look at one very simple example that demonstrates all of the concepts that we have recently introduced: phase, phase invariance, symmetry, conserved quantity, and wave quanta. Consider a clock face with a minute hand but no hour hand and no sweep (second) hand. What conclusions can we draw from this initial set of conditions?

1. The movement of the minute hand is harmonic, repeating its movements as it moves around and around the clock face.

2. The movement of the minute hand has a period (of duration one clock face, or every 60 minutes, or every 2π rotation).
3. The movement of the minute hand can be described by a wave equation ($\psi = e^{i\theta}$), where θ is the angle of the minute hand from some arbitrary position on the clock.
4. The minute hand has a probability of 1 of being located somewhere on the clock. Interpreted as a wave, the probability of finding the minute hand somewhere on the clock is equal to the wave function times its complex conjugate, or ($e^{i\theta} \times e^{-i\theta} = e^0 = 1$).
5. The location of the minute hand within its period is phase invariant. You can determine the minute, but you cannot determine when the wave (harmonic) began. The beginning of the wave is an arbitrary time. Put differently, a wave has no absolute phase.
6. The minute hand has measurable phase difference compared with other clocks of the same type. If one clock has the minute hand pointing to 3 and another clock at the same moment has the minute hand pointing to 4, then the two clocks are 5 minutes out of phase with one another.
7. No two clocks can be more than one hour out of phase with one another. One hour is the length of a single period or rotation around the clock.
8. If we compare one clock with another clock that is running faster, we can find a transformation formula between clocks. This transformation formula would be analogous to the Lorentz transformation for spacetime measurements for observers in different frames of reference.
9. There is a symmetry associated with the phase invariance. Specifically, a clock's minute hand can be rotated by an hour (i. e., 2π radians)and return to the identical position.
10. The system has its own quantum (i. e., the minute). The hand of the clock advances by small jumps from one minute to the next. There is no observable way of dividing the quantum of time (i. e., a minute) into smaller units. Remember, our clock does not have a sweep (seconds) hand.
11. There is a conserved quantity corresponding to the quantum of the clock's time. The clock's quantum (i. e., the minute) is conserved. You cannot simply drop or erase any of the minutes of the clock's face without creating chaos. In this example, the conserved quantity of time expresses Noether's theorem quite nicely. Noether's theorem tells us that a conserved quantity can always be found when the system has a symmetry. In this case, the symmetry is the invariance of the position of the minute hand when the hand has been rotated by 2π. The conserved quantity is the quantum of time (the minute).

Invariances and symmetries are alternate forms of the same concept. We can see a conserved quantity as an invariance, or we can see a conserved quantity as a symmetry (because a symmetry transformation conserves the quantity). The advantage of the symmetry concept is that invariant transformations can be belong to a particular mathematical abstraction known as a group (specifically, a symmetry group). Symmetry groups have an abundance of interesting properties that permit us to draw scientific conclusions that would have been extremely difficult to attain by other analytic approaches. Bluntly put, symmetries have groups, invariances do not.

At this point, we need to make a slight diversion to discuss the concepts of groups and, in particular, symmetry groups.

Let us begin by defining a group. A group is a set having the properties of closure and associativity under an operation; having an identity element; and having an inverse element for every member of the set.

Specifically, for some operation (which we will call "x" in this example), Closure tells us that for all a, b in G, the result of the operation, $a \times b$, is also in G.

Associativity tell us that for all a, b and c in G, $(a \times b) \times c = a \times (b \times c)$.

To have an identity element, there must exist an element e in G such that, for every element a in G, the equation $e \times a = a \times e = a$ holds. The identity element is unique within a group, and we speak of "the" identity element.

To have an inverse element, there must exist for each a in G, an element b in G, such that $a \times b = b \times a = e$, where e is the identity element. No matter what a group's operator happens to be, we know that we can always apply operations on group elements to return the original element. Specifically, there will always be a b such that $a \times b \times a = a$. Hence, every group has an operational symmetry that is built into its definition.

A simple group is a group G that does not have any normal subgroups except for the trivial group and G itself. A finite simple group is a simple group having a finite number of group elements.

Before we continue, we need to discuss the term "symmetry group". A symmetry group, more accurately known as a symmetry transformation group, is a group consisting of transformations on objects that preserve certain defined properties of the objects. The elements of the group are the transformations. Not to put too fine a point on this concept, but when mathematicians refer to the symmetry of an object, they are most often referring to the group of symmetry transformations that preserve the object (i. e., the symmetry group of the preserved object). When physicists refer to "groups" or to "symmetries", they are most likely referring to symmetry transformation groups.

One of the greatest conceptual hurdles that non-mathematicians have, when approaching the topic of symmetry groups, involves understanding why group theory has been conflated with the study of symmetry. What does one thing have to do with the other? The reason for the confusion lies in the subtle way in which mathematicians use abstract transformations (not numbers or variables or the objects upon which the transformations act) as group elements.

For example, the very first group that is commonly described in introductory group theory textbooks is the group of transformations that can be performed on an equilateral triangle that produce a new triangle that is the same size and shape as the original triangle. The group consists of six transformations (three rotations including the identity rotation) and three reflections. This group (i. e., the symmetric transformations on an equilateral triangle) happens to be equivalent to S3, the group of all possible permutations of 3 objects. In either of these cases, the group consists of a set of transformations (i. e., transformations on a triangle or transformations (permutations) on 3 objects). The triangle is not an element of the triangle symmetry group. The triangle is just something that the group of transformations acts upon. In the case of the permutation group, the set of 3 objects is not included in the group. The group is the set of permutations (another type of symmetry transformation) that act upon three objects. What we call a symmetry group is any group in which the group elements are symmetry transforms. It just so happens that most of the interesting applications of group theory have involved symmetry transformation groups. Hence, physicists have little or no interest in groups other than symmetry transformation groups. Whenever we read any scientific literature that deals with symmetry transformation groups, we might find that the authors truncate the term to just “groups” or even just “symmetries.” In this matter, we must indulge the mathematicians and physicists.

Felix Klein summarized the situation in 1872, when he wrote (translated from German) that “what is important are not geometrical figures, but the groups of transformations under which they remain invariant” [40].

Symmetry groups are commonly named for the types of structures that their transformation elements preserve. For example:
- A crystal group preserves a crystal lattice;
- A fixed point group preserves one particular element;
- A homeomorphism preserves a topology;
- A linear group preserves a linear structure;
- An isometry preserves a metric;
- An isomorphism preserves an algebraic structure;
- An orthogonal or unitary group preserves distance from a fixed point.

Let us look at one example where physicists have used the concept of geometric symmetry groups to establish physical law. The group of all rotations about the origin of 3-D Euclidean space is known as SO(3), an abbreviation for special orthogonal group for 3 dimensions. In SO(3), a rotation about the origin is a transformation that preserves the origin, Euclidean distance, and orientation. Every rotation is described by an orthogonal 3×3 matrix which, when multiplied by its transpose, results in the identify matrix. All of the laws of physics are SO(3) invariant because they apply equally when observed from any direction in space. Specifically, the rotational symmetry between time and space coordinate axes gives us the Lorentz transformations which account for the theory of special relativity. SO(3) can be represented by equivalent groups whose elements are matrices and whose group operations (i. e., the symmetry transformations) are matrix operators. The matrix representations of SO(3) led to the full elucidation of the Pauli’s exclusion principle. The modern atomic model, represented in the periodic table of the elements, is in turn based on the principles of rotational symmetry and on the Pauli's exclusion principle.

A story is told that when Einstein published his special theory of relativity, there were only a handful of physicists capable of understanding his contributions. A bit of hyperbole may be at play here, but it is certainly true that most classical physicists living at that time were puzzled by Einstein's theory. The scientists who were most definitely not challenged by Einstein's special relativity were the group theorists, who saw special relativity as a straightforward application of their own abstract work, into the physical realm [40]. For example, Felix Klein accepted special relativity as the invariant theory of the Lorentz transformations [40]. Similarly, Hermann Minkowski equated all of special relativity to the observation that Maxwell's equations happen to be invariant under the group of Lorentz transformations [40]. Poincaré had observed that the Lorentz transformations form a group that could be characterized by those linear substitutions of the four variables x, y, z, t, which leave the quadratic form $x^2 + y^2 + z^2 - c^2t^2$ invariant. This group is known eponymously as the Lorentz group or, using formal group nomenclature, is known as the SO(3, 1) Lie group. The transformations of the group correspond to all rotations about an origin in 3-D space that preserve the origin, the Euclidean distance, and the orientation.[vii] [120]

To summarize, some very clever physicists have seized upon an abstract mathematical group to encapsulate our 3-D reality, which preserves the invariance of the speed of light in all frames of reference. How did they accomplish this feat? Did they pull SO(3, 1) out of a bag of mathematical abstractions? Not quite, but almost. As it happens, SO(3, 1) happens to have properties that fit closely with what we know about the basic rotations of object in space and preserves invariance under the Lorentz transformation. Because it is a Lie Group, it supports continuous rotations in space (i. e., smooth rotations through all angles).[viii] SO(3, 1) is a good fit for special relativity. There are, as always, a few nagging problems. Most importantly, SO(3, 1) doesn't serve us when we explore quantum phenomenon. For example, when we consider spin, we rely upon a different group (SU(2)), with a different set of properties, to accommodate the experimentally demonstrated fact that electrons have one-half spin [266]. Then there is the group U(1) for electromagnetism, and the group SU(3) for the strong force (that confines quarks and gluons) [266].

Because physicists use different groups to model quantum phenomena, we are left with no single way to encompass quantum reality. Accordingly, physicists searched for longer than a half century for some unifying group that may account for special relativity, quantum phenomena, and general relativity. All to no avail. For the most part,

---

vii A useful feature of the SO(3) group is that every rotation can be described by an orthogonal 3×3 matrix which, when multiplied by its transpose, results in the identity matrix.

viii Smoothness is a mathematical concept. Functions are smooth if they are differentiable everywhere.

they have turned to large, high-dimensional Lie groups (groups with continuous transformations). When doing so, their models have yielded predictions that have failed to find experimental confirmation. Perhaps they should look elsewhere. The finite simple groups support multi-dimensionality and are particularly appropriate for modeling interactions among a finite number of fundamental particle types [266].[121] Perhaps we are just fooling ourselves when we insist that reality has a unifying mathematical principle. It is all very frustrating because until we can get a better grip on reality, we can't really understand how our reality began.

Physicists interpret Noether's theorem as an expression of the symmetries of the Lagrangian. For us to appreciate how Noether's theorem is employed by physicists, we need to discuss the Lagrangian, and its action integral. The Lagrangian is a mathematical function that encapsulates the forces operating within a space. It is suited for solving sets of equations where the forces are conserved.

In mathematical terms, the Lagrangian is the difference between the kinetic energy of the system (i. e., the sum of the kinetic energy of all the parts of a system) and the potential energy of the system. In effect, the Lagrangian describes the physical system[122]

$L = T - V$ Lagrangian is the kinetic energy (T) minus the potential energy (V).

When we speak of symmetries under Noether's theorem, we are referring to symmetries of the Lagrangian. That is to say that we are dealing with those symmetries that leave the Lagrangian unchanged. The "Action" of the Lagrangian is the Lagrangian's integral.

$$S = \int_{t_1}^{t_2} L\,\mathrm{d}t \quad S = \text{Action}$$

The mathematical evidence of an unchanged transformation of the Lagrangian (e. g., unchanged over time, unchanged over translation by distance) is determined by the so-called *least action* of the Lagrangian.

$$\delta S = \delta \int_{t_1}^{t_2} L(q, \dot{q}, t)dt = 0 \quad \text{General principle of least action.}$$

The equation for least action. When the action does not change, its derivative (the least action) is zero (or at least, minimized). All of Newton's laws can be derived from the least action, as applied to objects in motion.[ix]

---

ix There is just one more equation that we need, if we want to be on an equal footing with physicists; the Euler–Lagrange formula:

$$\frac{\mathrm{d}}{\mathrm{d}t}\left(\frac{\partial L}{\partial \dot{q}_j}\right) = \frac{\partial L}{\partial q_j} \quad \text{Euler–Lagrange formula}$$

The principle of least action holds that physical systems do not change the value of the action. The symmetry transformation groups that obey the principle of least action (i. e., whose transformations leave the Lagrangian stationary) are the symmetry groups that represent the physical laws of the universe. The principle of least action is a mathematical representation of the conservation laws that account for most of the laws of physics. The process for developing a conservation law can be summarized in 3 steps:

Step 1. Compose a symmetry in terms of the Lagrangian formula (i. e., compose a Lagrangian that preserves an invariance under transformations of the Poincaré group).

Step 2. Derive the Euler–Lagrange equation for the Lagrangian.

Step 3. Solve the Euler–Lagrange formula, thus deriving a conservation law.

When we organize our thoughts around symmetries as the origin of fields, motion, matter, forces, and physical laws all fall into place.

- Symmetry in the universe tells us that physical phenomena can be represented by mathematically defined groups, and analyzed with the methods of group theory. Group theory is a field of mathematics that was well-established by the middle of the nineteenth century.
- Noether's theorem tells us that where there is a symmetry in a physical system, there is an associated conservation law.
- Conservation principles determine many of the fundamental laws of energy and matter, including all of Newtonian physics, Maxwell's equations, and special relativity.
- Much of modern physics involves relating invariances and conserved quantities to symmetry groups. How do we identify the symmetry group that represents a physical reality (e. g., motion, rotation, force carrier (i. e., particle), energy, field)? The symmetries we find are symmetries of the Lagrangian. That is to say that the laws of physics are based upon the various transformations that leave the Lagrangian unchanged; the symmetry of the Lagrangian being the mathematical expression of Noether's law.

  Once we find the group that corresponds to a physical symmetry, we can apply the known features of the group (or, more often, of the matrix representation of the group), to derive physical laws. When we impose symmetries into our theories (e. g., spatial homogeneity, spatial isotropy, time-shift invariance), the conservation laws of physics pop out of our equations.
- Once we have identified the particular groups that represent the fundamental particles of matter and the carriers of energy, we can use the mathematical properties of the group to predict the properties of matter and energy.

When a Lagrangian is shown to be a symmetry, physicists apply the Euler–Lagrange formula to describe the behavior of the system.

- The predictions based on group-theoretical analyses of matter and energy have been validated by experimental studies, and have attained the status of physical law.

We seem to be following this somewhat sketchy path to enlightenment:
1. A symmetry leads to,
2. Transformations in a field, which lead us to,
3. Conservation laws conferring properties to the field, which leads us to,
4. Constituents of the field (e. g., conserved field quantities, energy, field forces, and force-carriers, waves, wave packet particles), which leads us to,
5. Interactions among many different constituents of several different field, which leads us to,
6. Emergent systems that constitute observable reality.[x]

9.3 Does mathematics fully describe reality?

> Reality is nothing but a collective hunch.
>
> – Lily Tomlin

Let us summarize what we know about energy and matter. It would seem that a photon is not really a particle (it is massless and has no measurable size), and it does not move through time (it persists in the eternal now), and it really does not move through space (only reference frames move). Furthermore, light waves are not really waves (they are coordinated oscillations of the electromagnetic fields). Grasping for sensibility, we resign ourselves to accept that solid matter is composed of non-material units best described by a wave function (even though solid matter is obviously not a wave). Energy is not the ability to move matter against a force. Energy is what is conserved by time-shift symmetry. In point of fact, all of our physical laws seem determined by symmetries (either geometric symmetries or gauge symmetries), and these symmetries are intangible relationships, supposedly represented by specific groups that are themselves mathematical abstractions. Hence, a set of mathematical abstractions are responsible for everything we observe or indirectly measure (fields and their consequent forces, waves and particles and their respective laws).[xi]

---

x Summarized even more succinctly, physical reality has symmetries, and symmetries have laws that apply to mathematical groups.

xi The relationship between mathematics and reality can be formulated as an origin paradox, as follows. It would seem that our reality owes its existence to mathematics — specifically, the symmetry relationships that form the basis of the physical universe. At the same time, the relationships of mathematics owe their existence to the reality that they represent. Therefore, both reality and mathematics could not

Did the preceding paragraph clarify anything? Probably not. Wolfgang Pauli remarked:

> The best that most of us can hope to achieve in physics is simply to misunderstand at a deeper level.

As we move deeply into the fundamentals of physics, in search for the beginning of reality, we find ourselves mired in a universe that makes no sense to us. Nonetheless, we know that the universe must make sense because it is non-chaotic. It operates on a set of rules that produces galaxies, stars, planets, and solid matter all operating in the same fashion, everywhere we see. If we fail to make sense of the world, then fault must lie within ourselves.

We have a tendency to think metaphorically, always finding familiar concepts to compare to unfamiliar concepts. A rigid fundamental particle is likened to a golf ball. A light wave is likened to a water wave. A permutation group is likened to a collection of colored marbles. Reasoning by metaphor is characteristic of human intelligence; we cannot do otherwise. As we move deeper in our understanding of nature, our metaphors become inadequate. We become frustrated because we learn that we can no longer depend on something that we, as humans, have come to depend upon. This being the case, what are our intellectual options? For the mathematician, the choice seems dour. We must abandon any hope of explaining reality based on what we see in the observable world, and we must switch to a wholly mathematical explanation of everything.

Because we seem unable to understand modern physics with words or metaphors, we have allowed mathematics to intrude where imagination has failed. Most of the consequences of this shift toward mathematical analyses have been immensely beneficial for science. It should be no surprise that for more than a century, seminal advances in physics and chemistry have required mathematics. As our understanding of mathematics advances, so does our understanding of our physical reality.

Of course, there is a catch. As we come to rely upon mathematical abstractions, we find ourselves burdened by all manner of theories that have a sound mathematics basis, but for which there is no physical confirmation. Many of the wondrous possibilities described in the popular literature (e. g., the many worlds hypothesis, supersymmetry, string theory, and a large assortment of hypothetical subatomic particles) have never been substantiated and cannot be tested. As the energy of particle collisions has increased, at great expense to funding agencies, the novel particles predicted by theory have not materialized. With the exception of the Higgs particle, the Large Hadron Collider has failed to validate the vast majority of mathematical predictions in the field of particle physics. Mathematics has led us to a place where reality has not followed.

---

exist without the pre-existence of the other; and this is a paradox. Like most other "origin" paradoxes, it is solved when we accept that mathematics and reality are equivalent to one another.

9.4 Limitations on what we may prove

There is a crack, a crack in everything. That's how the light gets in.

– Leonard Cohen, from his song *Anthem*, released in 1992, in the album, *The Future*

Mathematics sets us up with conditions that we cannot violate. For example, it happens that there are exactly 5 regular polyhedra in three-dimensional space, tetrahedron, cube, octahedron, icosahedron, dodecahedron. The regular polyhedra are constructed with repeating congruent (identical) and regular (all sides equal) polygonal faces, with every vertex of the solid formed by joining the same number of polygon faces. The ancient Greeks wondered why there were exactly 5 such polyhedra. Theaetetus, a mathematician living at the same time as Plato (circa 350 BCE), proved that no other regular polyhedra could possibly exist. As previously discussed, Plato famously (but incorrectly) asserted that each of the 5 classical elements known at that time was composed of its own polyhedron. As a result, they are known today as the Platonic solids. Few of us remember Theaetetus' greater contribution, but it was Theaetetus who taught us that mathematical abstractions assert absolute limitations on our solid reality. No power in the universe can create a 6th Platonic solid. The mathematics has been proven.

Mathematics tells us not only the limitations of reality, it can also prove to us its own limitations! Let us take a look at Gödel's Incompleteness Theorem. It would seem obvious that all true statements must be provable, otherwise we lose our ability to distinguish reality from unreality. In fact, all of science is built on the premise that assertions of truth are testable. It may come as a shock to know that not everything that is true is provably true, and not everything that is false is provably false.

The mathematician and philosopher Kurt Gödel published in 1931 his "incompleteness theorems," proving that there are limits to what we can say, with any confidence, about our universe. His first incompleteness theorem stated that "any consistent formal system F within which a certain amount of elementary arithmetic can be carried out is incomplete." Simply put, there are assertions that can neither be proven or disproven. Gödel devised a clever, but somewhat convoluted way of proving his theorem that we won't try to summarize here. It suffices to say that the key step of his proof involves a self-referential statement that poses a logical contradiction. Basically, the statement "All true assertions can be proven to be true," establishes a contradiction that nullifies the proof of the statement.

Let us dawdle a bit and imagine to see where Gödel is leading us. Let us agree that all statements must be either true or false, and we will begin with a statement that we shall call "Q" that asserts that Q cannot be proven.

1. Q asserts that Q cannot be proven.
2. We assume that Gödel was just plain wrong and that all true statements can be proven to be true.
3. If we assume that Gödel was wrong, then Q can be proven to be true whenever Q is true.

4. But the statement Q asserts that Q cannot be proven.
5. Therefore, when Q is true, Q cannot be proven to be true.
7. Therefore, when we assume that Gödel's incompleteness theorem is wrong, we run into a contradiction, thus proving that Gödel's incompleteness theorem is correct.

Basically, if all true statements were provable, we would have a logically untenable situation. Therefore, the assertion "All true assertions can be proven to be true," must be false. In one stunning manuscript, Gödel managed to create an abstract limit on what we may know about our real universe [106].

Mathematics is just one of those things that is there if we say it is there. The relationships that mathematicians have discovered are timeless and can arise out of nothingness. Real, irrational, transcendental numbers, matrices, functions, equations, and transformations are all waiting to be applied to anything or nothing.

Setting aside our awe, let us stop a moment to discuss the unique property at the heart of all mathematics. Mathematics is a formal way of establishing relationships among abstractly defined objects. Such objects are created and defined by mathematicians, but the real beauty of the mathematical approach centers on the relationships among types of one class of object with other classes of objects. For example, we can define the concept of a group, but the group definition only becomes interesting when its properties can be related to other groups and to other forms of defined objects. For example, when a group is related to a geometric symmetry, and the geometric symmetry can be represented by a Lagrangian function and transformed to a matrix domain or represented by some physical law.

Students of mathematics flex their muscles whenever they take one or more equations and develop some new equation (i. e., a new mathematical relationship) after applying a few clever operations (e. g., functions, transformations, substitutions). Mathematicians have formalized these pursuits with a symbolic language, but it all comes down to the same sort of reasoning employed by scientists working in any discipline. The approach can be described as follows:

1. The objects of the discipline (e. g., numbers, forces, fundamental particles, atoms, chemicals, bacteria, college professors) have observable relationships.
2. Objects can be placed into classes according to their particular relationships, shared by all the members of a class.
3. Every class has a set of relationships particular to the class that establish a set of class properties. The different classes can be related to one another based on these class-specific properties.
4. The relations between the objects of any particular class and the relations between different classes can be examined to attain a deep understanding of the properties that emerge from the collected classes.

Mathematicians study groups and their defined symmetry operations from which they determine relationships, often taking the form of isomorphisms and transformations, that support new proofs of the fundamental characteristics of classes of reality. Physicists study the relationships among fields, particles, energy, and spacetime geometry to better understand the universe, which is fundamentally a classification of its contained objects. Biologists use the relationships between living organisms of a class, and between different classes of organisms, to better understand the classification of all living organisms. In summary, all of science is concerned with generalizing objects into classes, examining the relationships established by the classes, and making predictions based on the classification. The relationships of mathematics happen to reside in an abstract realm. The study of classes of relationships, not of particular objects within classes, is the essence of science and of mathematics. Any emergent universe must have the properties found in the collection of available mathematical relationships.

On a cautionary note, the very feature that makes mathematics powerful (its ability to find relationships using analytic methods), stops us from fully understanding what we have done. We believe our final conclusions because we know that all the steps leading to the conclusion are mathematically valid (without our being smart enough to keep all those steps in our mind at any one instant). Therefore, if there is any error in any step, we can reach an absurd conclusion and never know that it was wrong. Most mathematical derivations depend upon representations of reality that we cannot rigorously prove. Consequently, we tend to endow equations with more credibility than they merit. It is not until we find inconsistencies and demonstrably false conclusions that we are forced to look back and find the follies of our way.

9.5 Summary remarks concerning creation

> Biologists think they are Biochemists,
> Biochemists think they are Physical Chemists,
> Physical Chemists think they are Physicists,
> Physicists think they are God,
> And God thinks she is a Mathematician.
>
> – Anonymous and widely repeated quip

> Mathematics is the part of science you could continue to do if you woke up tomorrow and discovered the universe was gone.
>
> – Dave Rusin

They say that reality is the thing that can kill you, whether you believe in it or not. A less brutal way of saying the same thing is that reality is the thing that does not change, regardless of how you interpret it. But the universe is constantly changing; getting larger, losing information (via black holes), decreasing energy density (as the cosmic background radiation stretches its wavelength), dropping its temperature steadily in all the years following the Big Bang, and increasing its entropy in accordance with the second law of thermodynamics. This being the case, it seems that mathematics is the only true reality insofar as it never changes and is unswayed by our opinions. If mathematics is the true reality, then is our observed universe just a cheap imitator posing as the real thing?

Scientists distinguish mathematical descriptions of observed phenomena (such as we find in Newton's Principia), from mathematical rules by which physical reality must abide (such as the Pauli's exclusion principle). We never assert that Newton's laws of motion propel objects through the air, but we insist that the Pauli's exclusion principle prohibits two identical objects in the same orbital shell from occupying the same quantum state. In the latter case, mathematics creates a reality in which we cannot walk through walls, and we cannot fit all the atoms of the universe onto the head of a pin. When we approach the fundamental properties of the universe, we may be entering a realm wherein reality becomes purely mathematical, and what we experience in the physical realm is nothing other than a mathematical simulation. In this were the case, then fields and their group representations are the abstractions that describe so-called reality, and there simply is nothing else to know.

Before we surrender ourselves to a deterministic universe created by mathematics, let us take a look at an abstraction that seems to intrude itself upon reality at every moment of our petty lives: chance. Imagine a random number generator that provides output upon request. We ask for a number, and it gives us "526943". This randomly selected number just happens to be a prime number, and mathematicians have provided a variety of theorems related to the mathematics of prime numbers. The prime number generator has inserted into our reality a random number that just happens to be a member of a group of numbers that all conform to a collection of theorems and that are related to other numbers by their common properties. Has not the random number generator effectively created a new universe whose reality began with a number and all of the imposed properties that arise therefrom? The "Big Bang" of this new universe began with the appearance of "526943," and developed from the properties that happened to be associated with that particular number. The "526943" universe begs the question of who/what may have created the random number generator. Because "526943" is a random number, it could not have been predicted by a mathematical formula.

Let us return to another example, but this time, we will dispense with the random number generator and focus only on the existence of disembodied chance. Imagine that out of nothing but abstract chance, there emerged mathematical abstractions (such as fields and groups and symmetries and dimensionality, and nullity) that determine an effective operating system for the universe that we inhabit. In this case, what we recognize as particles are simply focal intensities in fields. What we recognize as energy is simply the mathematical flux of field intensities. Spacetime becomes a dimensional property of symmetry groups. In this weird setting, chance alone provided the fundamental mathematical relationships that we observe as our physical reality.

Let us remember our earlier discussions of fundamental properties of particles and of energies. Whenever we reduce matter to its fundamental particles, we always arrive at objects composed of nothing but properties. There is no material "stuff" that makes a quark or a gluon or even an electron. The fundamental particles are "points" without size. Likewise, our forces are abstract fields following mathematical equations. Everything we see as "happening" in our universe, comes to us courtesy of the Schrödinger's wave function, which itself is a probabilistic mathematical model that can be applied to any and all fundamental particles. When we dive deeply into reality, we find that there is nothing there but equations, and that the status of our current time has developed according to a wave equation, from initial conditions. The initial conditions, if we wish, can be imagined as the result of chance favoring one set of equations over another.

It is tempting to summarily reject the idea that mathematics created and maintains the universe. A foundational principle of science holds that reality is observable and our understanding of reality is based on observation. For over a century, we have known that the act of observation distorts our observations. We refer here to the collapse of the Schrödinger's wave function (discussed in Chapter 6, "Fundamental Constants"), a phenomenon that has never been resolved.

We have also learned that there are strict limits to observation. For example, we cannot observe objects smaller than the Planck length and we cannot observe electromagnetic waves that oscillate higher than the Planck frequency. The reason being that observation requires energy. When the object is small, we must concentrate our energy to a very small area, requiring an energy density sufficiently large to create a black hole. Consequently, the universe is something of a blur, with no precise edges. We cannot directly observe point particles, because they have no size, but point particles account for all of the matter in the universe. Ultimately, we cannot objectively observe much of anything. It seems unreasonable to encumber our notion of science with the impossible requirement of observability. What we seem to be finding is that the laws of our universe are nothing other than abstract mathematical relationships that account for the existence and behavior of the universe and its contents.

When we play with the idea that mathematics creates reality, we find that several of our deepest questions regarding the beginning of everything are immediately answered:

- Mathematics is timeless. It has no beginning and no ending, and this fits with our intuitive sense that existence always exists.
- Mathematics is fully descriptive of reality. We think of reality as spacetime and the forces and particles operating in spacetime. All of this is described mathematically, and there does not seem to be anything we can observe that cannot be described (or determined) mathematically.[xii]

---

xii Max Tegmark, a proponent of the mathematical theory of everything, has made the argument that mathematics is isomorphic to reality, meaning that there is a 1 to 1 mapping between mathematical description and reality [236]. If two structures are isomorphic, he reasons, then there is no meaningful distinction between them. Therefore, the isomorphism between the mathematical structure and reality establishes their equality. Hence, reality is a mathematical structure.

- Other than mathematics, there seems to be no alternative way to explain what we think we know about reality. Let us review a few assertions about the nature of reality that seem to require mathematical explanation.
  1. Matter is composed of point particles (better known as fundamental particles) that have no size and no precisely discernible location. The point particles are not composed of "stuff". They are made of nothing but their properties. When we encounter large, non-fundamental particles, such as protons and neutrons, we find that they are composed of point particles and the forces that bind them. The size of a proton or a neutron corresponds to the size of the containment field (that holds their constituent quarks), and the mass of the proton or neutron is basically the mass-equivalent of the containment field's energy.
  2. We can only understand matter, force, and energy in terms of their properties. That is to say, everything that is responsible for movement and for reality to "happen" consists of its own description.[xiii]
  3. The universal determinates of the relations that govern the universe are dimensionless constants. It is difficult to imagine how a small coterie of numbers could have such sway in anything other than in a purely mathematical reality.
  4. Symmetries account for the laws of the universe, but the symmetries have only mathematical meaning. Arguably, in the case of the geometric symmetries, we can fit the symmetries into our understanding of how things work in the "real" world (e. g., flips, transverse movements, rotations). In the case of the gauge theories, though, the symmetries have no correlates in anything tangible.
- Mathematics is inviolate. Gödel's incompleteness theorem tells us that there will always be true statements that cannot be proven. Does Gödel's incompleteness theorem apply to all of reality? Yes. Suppose there existed an all-powerful being who had created the universe and all of the laws that govern reality. We might believe that it would be within the power of an all-powerful being to prove any true statement that happens to be true. Not so. No power can violate Gödel's incompleteness theorem. Mathematics will not budge.

Perhaps our reluctance to embrace the notion of a mathematical universe extends from the popular belief that reality is just a bunch of stuff prepared for us by a power beyond our comprehension. In consequence, many of our observations have an anthropic bias. We say that the sun rises in the morning, when we know that the sun does not rise and the concept of morning has relevance only to us Earthlings. We say that the moon waxes and wanes, when the moon does neither. Despite evidence to the contrary, we say that an electron follows an orbit around the nucleus, much the same way as the Earth orbits the sun. We know that this is not true, but we persist in our use of scientifically

---

xiii Paraphrasing the mathematician and philosopher Bertrand Russell, we do not really know what matter is. We only know the mathematical description of the properties of matter [212].

discredited metaphor because we like to think that everything can be interpreted in terms of human experience. If instead we had been introduced, early in our education, to the notion that the universe is determined by group symmetries, then we might all be in a better position to understand the mathematical nature of our world.

Back in Chapter 2, "Beginning with Life", we asked ourselves whether viruses qualified as living organisms. If you remember, we concluded that life has no certain definition. If we chose to say that viruses are alive (and not just packets of nucleic acid), then we were free to do so. If we maintained that viruses are nonliving, then this opposing opinion is also valid. Life, we conceded, is what we choose to call it. Maybe the same can be said about reality — it is whatever we choose to call it. If we choose to say that mathematics is our only reality, then who can say that we are wrong. Contrariwise, if we choose to say that mathematics is abstract and that abstractions cannot account for reality, then we would be hard pressed to prove otherwise. It is all good.

10 Recap

Let's wrap up with a recapitulation of the points listed in Chapter 1, "A preview of book contents", adding short summaries of the explanations discussed in the book chapters.

1. It all begins with a particular kind of nothingness; a nothingness that accommodates wave motion.
   We can easily show that nothingness has properties.
   (a) If nothingness lacked all properties,
   (b) Then, it would have "absence of properties",
   (c) But "Absence of properties" is itself a property,
   (d) Therefore, nothingness must have properties.
   If nothingness must have properties, then the bare-bones minimum number of properties it must have is "1". As its one and only property, the ability to carry waves (i. e., harmonic oscillations) seems to be a modest condition.
   An alternate hypothesis (championed in the Old Testament) is that the universe began as chaos. Maybe so, but chaotic systems are, by definition, unpredictable and unstable. Eventually, chaotic systems yield non-chaotic systems that are predictable (i. e., operate according to immutable physical laws)[i] Accordingly, we might suggest that had the universe began in chaos, it may well have developed into a non-chaotic universe composed of nothingness (i. e., vacuum) and waves.
2. Wave propagation implies the presence of energy (to move the waves), plus spacetime (in which the waves move), plus momentum.
   Everyone, even physicists, have difficulty pinning down the definition of energy, but we can safely say that every kind of energy (heat energy, nuclear energy, wave energy) is associated with movement. Movement, in turn, occurs in spacetime (i. e., movement travels through the coordinates x, y, z, and t).
   Where there is energy needed for movement, there must be momentum (i. e., the resistance to changes in movement). If there were no such thing as momentum, then changes in movement would not require energy. But we have accepted that movement requires energy. Therefore, energy and movement require momentum.
3. Because the bare-bones universe consists of a vacuum through which waves can propagate, the evolution of the universe over time can be described by a wave equation.
   The Schrodinger's wave equation was derived in Chapter 3, Section 3.2, "Nothing but waves."
4. A universe of waves moving through vacuum places limits upon observability (and hence, limits on the precision of observations).

---

i The transformation of chaotic systems into non-chaotic systems has been explored by the Nobelist Ilya Prigogine. Emerging non-chaotic systems become increasingly structured (i. e., increasingly non-chaotic) over time [204].

https://doi.org/10.1515/9783111498690-010

All waves come with two properties: frequency (number of oscillations per unit time) and wavelength (the length traveled by the wave in one oscillation). The wave frequency multiplied by the wave length yields the wave speed.
Knowing nothing at all about the nature of the wave (i. e., whether it is an electromagnetic wave, or a water wave, or a gravity wave), we can deduce that there must be some physical limits put upon its wavelength. If its wavelength were infinitely small (or, equivalently, its frequency infinitely high), then the wave's oscillating position would be unmeasurable. The reason being that observations of a wave must be made with a detector whose scanning wavelength is smaller than the wavelength being observed. Otherwise, if the detector's wavelength were longer than the wavelength to be observed, then the detector could not distinguish exactly where the wave cycle began and ended. At best, the detector would register that the wave's oscillating position was something smaller than the wavelength of the detector.[ii]
As a wave's length approaches zero, its frequency approaches infinity, along with its associated energy (which is directly proportional to frequency). Prior to attaining a zero wavelength size, the energy density within the space containing the wave exceeds singularity, and a black hole suddenly develops. Hence, at some point, as the wavelength of light becomes smaller and smaller (and its energy grows larger and larger), the ability to observe and measure the wave's oscillating position becomes impossible.

5. Heisenberg's uncertainty principle sets limits on the precision by which we can simultaneously measure both position and momentum or both energy and time.
Knowing that there are limits placed upon wave oscillations, and upon the observability of waves, it should come as no surprise that we can determine what those limits might be for various types of waves, if we know the mathematical formulations of their properties. Because wave-forms account for everything in the universe, the limits of waves are the limits of reality.
Heisenberg's uncertainty principle asserts that we can never measure a change of momentum $\triangle p$ and a change of position $\triangle x$ to within $\frac{\hbar}{2}$, where $\hbar$ is the reduced Planck constant.
Expressed as an equation, we have:

$$\triangle x \; \triangle p \geq \frac{\hbar}{2} \quad \text{Heisenberg's uncertainty for momentum and position.}$$

The uncertainty principle also sets limits to the precision with which we can measure energy and time.

---

ii Similarly, microscopes that use visible light optics cannot help us visualize cellular organelles whose size is smaller than the wavelength of visible light. To detect very small cellular organelles, scientists use electron microscopes that employ a beam of very small wavelength and very high energy.

$$\triangle E \triangle t \geq \frac{h}{4\pi} \quad \text{Heisenberg's uncertainty for energy and time.}$$

In Chapter 11, as a note titled "Derivation of the Heisenberg's uncertainty principle," we show that when we have variables that are Fourier conjugates of one another, Heisenberg's uncertainty principle applies.[58] Put another way, the uncertainty relationship holds for any pair of variables (e. g., momentum and position, energy and time) where each of the two variables can be expressed as a Fourier sum of waves of the other. Living as we do in a universe consisting of waves and vacuum, we must accommodate ourselves to the limitations imposed by Heisenberg's uncertainty principle.

6. The uncertainty principle tells us that there can never be a measurably zero energy in space. Hence, the energy at every point in space is in constant flux.
   The unpredictable but omnipresent energy flux in our universe was predicted as Casimir's effect and has been experimentally validated. A discussion of Casimir's effect is provided in Chapter 3, Section 3.7, titled "Nothing, the key to everything."
7. Conserved physical quantities, and the laws that describe their conservation, arise as a consequence of symmetries (Noether's law).
   A symmetry is an equivalence that holds after a transformation (e. g., spatial translation, spatial rotation, mirror image flips, time shifts). Physical symmetries lead to all of the conservation laws. For example, time-shift symmetry accounts for the conservation of energy. Here is why:
   (a) If we have time-shift symmetry, then an experiment performed today will yield the same results when repeated a billion years from today.
   (b) If the results of the experiment are the same, after a time-shift, then the energy required to produce the results must also be the same, before and after the time-shift.
   (c) If the energy stays the same over time, then the energy is conserved.
   (d) Therefore, the symmetry of a time shift accounts for the law of energy conservation.

   The relationship of symmetry and conserved quantities is described mathematically in Noether's theorem, and a proof is provided as a note in Chapter 11.[119]
8. Symmetries of energy and motion are preserved in different frames of reference. Reference frame symmetries impose a limit upon the speed of light and yield Einstein's laws of special relativity.
   The time-dilation equation of special relativity is discussed in Section 4.2, titled "Limits due to special relativity", and derived in Note 62 in Chapter 11.
9. The laws of special relativity account for the concept of mass, representing that which resists acceleration by energy.
   Everyone believes that they have a good understanding of the concept of mass. Actually, "mass" does not fit into anything that we might describe as a constituent of the universe (i. e., it is neither energy, nor particle, nor wave). Its meaning is basically operative, being a resistance to the acceleration exerted by energy.

If we wished, we could take Einstein's famous equation ($E = mc^2$) and refer to mass only by its relationship to energy and the speed of light (i. e., $m = E/c^2$).

10. Symmetries can "break" to form additional symmetries. A symmetry that arises from another symmetry has its own properties while also preserving the properties of the symmetry from which it arose.
    Casimir's effect consists in the fluctuations in energy, everywhere in space, that result from the Heisenberg's uncertainty principle. Such fluctuations can be construed as perturbations in symmetries. Because the magnitude of such perturbations are unpredictable, we must accept that an occasional perturbation may arise with sufficient energy to "break" the symmetry of its field, essentially creating, if only for a moment, a new field operating under an additional complement of physical law.[iii] It is through the phenomenon of symmetry breaking that new fields, capable of interacting with the field from which they arose, are thought to form.
11. Symmetries account for the fields described by the Standard Model of quantum physics.
    Our hypothetical primordial symmetry had the properties of vacuum and waves. We would expect to find these two properties to carry into every descendant symmetry. This appears to be the case. Schrödinger's wave equation serves as a reasonable way of describing how our universe (which preserves the different symmetries of the fields described in the Standard Model) evolves over time. Schrödinger's wave equation was discussed in Section 3.2 titled "Nothing but waves", and derived in Note 52 of Chapter 11.
12. Each of the fields described by the Standard Model of quantum physics accounts for a physical force (i. e., the force distributed through the coordinate space of the field). The Standard Model forces, plus the force of gravity are:
    - The strong force (binding the nuclear subatomic constituents);
    - The weak force (accounting for nuclear decay processes);
    - Electromagnetic force;
    - Gravity.
13. The interactions between fields described in the Standard Model require fundamental constants.
    That is to say that dimensionless fundamental constants "pop" out of the equations that describe the Standard Model.
14. Each field of the Standard Model is characterized by its own particular quantum particle.[123]
15. The identicality of particles of a kind leads to Pauli's exclusion principle.
    The exclusion principle asserts that for the class of particles known as fermions (including electrons and neutrinos that move more slowly than light waves) super-

iii As a reminder, a field is the infinite spatial realm in which a symmetry operates. For the purpose of mathematical analysis, a field is a quantity defined at every point in space and time.

position is impossible. That is to say, two identical fermions, in the same quantum state, cannot occupy the same location at the same time.
Superposition can occur for the second category of quantum particles, bosons. Photons are a type of boson, and light beams can pass through one another.
A proof of the Pauli Exclusion Principle is provided in Section 6.10 titled "Identicality and Pauli's exclusion principle".

16. Pauli's exclusion principle accounts for much of what we recognized as manifestations of the material universe. These include:
    - The two general forms of quanta (fermions and bosons).
    - Limitations on the condensation of fermionic matter.
      For example, the atoms of a chair cannot compress down to a single point in space, and the matter in the chair cannot merge with the matter in the floor.
    - Atomic valence.
      Pauli's exclusion principle requires that no two electrons may occupy the same set of quantum states as any other. In atoms, electrons distribute themselves about a nucleus in a pattern determined by their quantum states.
    - Chemical reactions.
      The reactivity of elements is largely predictable based upon the valence of the outer shell of atoms. Two atoms are considered to be stably "bonded" together when the energy required to separate the atoms (i. e., break the bond) is unavailable under the prevailing conditions.
    - Periodic table of the elements.
      The arrangement of the elements in the periodic table is largely determined by the total number of electrons in the neutral elemental atom and the valence of its outer shell of electrons.
17. The first atoms to form after the Big Bang are the two smallest atoms in the periodic table, hydrogen and helium, with hydrogen predominating.
    In the early moments after the Big Bang, the universe was small and hot - too hot for electrons to maintain orbits around protons. Over time, the universe expanded and cooled. About 300,000 years after the Big Bang, the universe was cool enough to permit electrons to associate with nuclei (protons and neutrons), to form atoms. At the moment that atoms (hydrogen and helium) formed, photons that had been locked into dense plasma of quantum particles, escaped to move freely in space (without immediately been absorbed by plasma particles). The escape of light moving through the universe resulted in the cosmic background radiation.
    To this day, the cosmic background radiation pervades space, and its radiation accounts for the majority of observable energy in the universe. Also, the vast majority of atoms in the universe are hydrogen and helium atoms, and the majority of those atoms are the very same atoms that formed nearly 14 billion years ago (i. e., a few hundred thousand years after the Big Bang).

18. Gravitational attraction produces concentrated collections of hydrogen and helium. At a critical point, dense collections of atoms commence a sustained nuclear fusion. Stars are born and heavy elements are produced.
    It is thought that nucleosynthesis within stars is responsible for the generation of all elements of atomic number up to an including that of Iron. Synthesis of atoms heavier than iron seems to require higher energies than are generated in gaseous stars and is thought to arise from colliding neutron stars.
19. Molecules form from elemental atoms.
    Bonded assemblies of different elemental atoms are called molecules. A molecule is considered stable if prevailing conditions do not provide sufficient energy to break the bonds holding the molecule together.
20. Additional types of molecules form from reactions between existing molecules.
    The particular structural properties of molecules leads to the acceleration of reactions in which new molecular species are produced when one molecule bonds to another.
21. Catalysts facilitate chemical reactions.
    Catalysts are materials whose physical conformation allows substrates to fit in a position that is favorable for chemical interactions to proceed between a substrate and a reactant. The first catalysts on earth were widely available substances such as rocks and minerals.[iv] Catalysts made of proteins are called enzymes. In the case of enzymes, the product of the reaction is immediately released, leaving the enzyme with a conformational "slot" to be filled with another substrate molecule. The enzyme itself is not consumed by the reaction.
22. Chemical reactions lead to chemical pathways.
    A chain of chemical reactions (wherein the product of one reaction serves as the substrate for the next reaction) may lead to a product that would not have been produced by any single chemical reaction.
23. Reactions emerge that create new enzymes.
    The first catalysts were inorganic materials found on rock surfaces, and chemical reactions could only proceed in the vicinity of these immobile catalysts. As a growing variety of chemical reactions produced novel types of molecules, it was inevitable that some of these new molecules had catalytic activity. These newly created biological catalysts, called enzymes, permitted chemical reactions to occur almost anywhere (i. e., away from inorganic catalysts confined to rock surfaces). The availability of synthesized enzymes vastly accelerated the rate at which new pathways, new chemicals, and new enzymes were created.
24. Chemical pathways yield reproduction of molecular species

---

iv The preferred term for an inorganic enzymes is "catalyst". We use the term "enzyme" liberally to refer to any organic substance with catalytic activity. To this day, industry uses a variety of catalysts that are composed of metals or amalgams.

The variety of molecules produced by chemical reactions eventually leads to the emergence of a species of molecule that can participate in its own reproduction through a devoted chemical pathway.

25. Circumstances arise in which self-reproduction proceeds without limit.
Pathways that yield energy for self-reproduction, and which have an unlimited supply of substrates, may reproduce without limit.
26. Self-sustaining chemical systems established within small, water-filled, lacunae and recesses inside volcanic rock (e. g., pumice). These chemically active containments can be described as non-living cells.
A chemical manufacturing assembly requires some degree of isolation, otherwise its products would diffuse into the environment and could not participate in a complex set of coordinated reactions. Some types of volcanic rock (such as pumice) are pitted with small holes during their formation. These hole (referred to here as lacunae) can be of micron scale, roughly the size of many unicellular organisms existing today. Water may permeate these porous rocks, yielding crucibles in which sustainable chemical reactions might proceed.
27. The chemical machinery within rock cells (i. e. water-filled spaces) floated between rock lacunae, permitting a crude form of cellular reproduction.
When the reactions within one lacunae can slide into the adjacent lacuna, we can begin to imagine that the original lacuna has reproduced itself.
28. Specialized enzymes emerge containing a reaction template that could be exactly reproduced.
At this point molecules emerge that contain all the information for consistent enzyme self-reproduction. The method of self-reproduction from a template may involve a sequence of complex chemical pathways. As a crude example, we should try to imagine a very early form of messenger RNA that served as a template for the accumulation of a string of connected amino acides that in turn served as a template for the synthesis of a copy of the original messenger RNA molecule.
29. Early forms of virus emerge.
Self-coding, self-reproducing template molecules moved from lacunae to lacunae, merging with or replacing the chemical contents of other rock-encased cells. Such molecules can be considered the very earliest form of viruses.
30. Early self-coding templates eventually lead to templates that could code for multiple reactions in addition to coding for its own self-replication.
Here we are referring to an early genome, which codes for the machinery to replicate itself and which also contains genes that code for proteins.
31. New chemical pathways synthesize lipids, and these synthesized lipids self-assemble as an outer membrane enclosing the watery chemical machinery housed within rock lacunae.
Many forms of lipid have been shown to self-assemble into membranes that cover a water surface. One side of the lipid layer is hydrophobic and the other side is hydrophilic.

32. Lacunae spew their contents into the primordial ocean, in lipid-lined bubbles.
    Eventually, lipid-lined lacunae will squeeze out of their rock caves to become free-floating, fluid-filled enclosures.
33. Living cellular life begins.
    Eventually, the reactions within some of these free-floating bubbles will yield an expansion of membrane and an increase in cellular contents to the point where two bubbles are formed from one bubble. The self-replication of cells marks the beginning of life on earth. In this scenario, life arises from non-living matter, a phenomenon we know as abiogenesis.
34. DNA emerges as the template molecule of the genome of all cellular life.
    Genome reproduction requires the faithful replication of the template molecule that reproduces itself and that encodes the instructions for cellular machinery (e. g., enzymes, structural proteins, regulatory molecules). RNA molecules could not provide the molecular strength and stability needed to sustain generations of complex cellular reproduction. A closely related molecule, DNA, emerged as a template molecule able to support multiple rounds of replication, and capable of accruing other DNA fragments, thus attaining enormous size and complexity. DNA became the universal template for all cellular life on earth.
35. Gradual alterations in the genetic machinery of cells lead to heritable alterations in cellular organisms (i. e., evolution).
    Small alterations to DNA inevitably occur, due to disruptions by chemical or physical agents (e. g., alkylating agents, intercalating agents, ultra-violet radiation) or through errors of replication. Because DNA is the instrument of cellular heredity, alterations in the structure of DNA may lead to heritable changes in cells. Natural selection occurs when heritable variations in cells confer a survival advantage.
36. Fusions of different cells or the capture of parts of different cells lead to an increasing variety of cellular species, with heritable properties obtained directly from other cells of the same or of different cell species.
    Examples of the direct transfer of heritable material from cell to cell or organism to organism are found to occur within all of the kingdoms of life on earth (i. e., Class Eubacteria, Class Archaea, Class Eukaryota, and Class Viridae).
37. The first eukaryotes emerge from prokaryotic forms of life.
    The first eukaryotic cells (cellular organisms containing a membrane-enclosed nucleus and one or more membrane-enclosed mitochrondrion) may have developed from the fusion of pre-existing cell types.[v] The fusing cells might have included members of either or both of the prokaryotic kingdoms (i. e., species of Class Bac-

v Some species of single-celled eukaryotic organisms do not have mitochondria. Such amitochondriate species are presumed to have descended from species having mitochondria. Amitochondriate species often retain a vestigial remnant of mitochondria, such as peroxisomes.

teria, or of Class Archaea, or both), and with possible contributions from the viral kingdom (such as members of Class Megaviridae (giant viruses)).[vi]

38. The mitotic pathway of cellular replication evolves in eukaryotic organisms.
Mitosis involves the replication of both of the paired chromosomes of the nucleus followed by a split of the cell body, with a paired set of chromosomes migrating to each of the two halves of the split cell. The process results in two so-called daughter cells, replacing the parent cell. The two daughter cells each have the same complement of genetic material as the parent cell and each contain a similar amount of the parent's cytoplasmic material.
Mitosis emerges as the most common form of cellular replication among the somatic cells of eukaryotic multicellular organisms.[vii]
39. Meiosis, the process by which germinal cells replicate, evolves from the biochemical machinery of mitosis.
In meiosis, cell division of a germinal cell leads to the production of four cells, each having a haploid set of chromosomes. Many of the same enzymes, genetic materials, and metabolic pathways involved in mitosis are also found in meiosis. Because simple, single-celled eukaryotic organisms have mitosis, and because meiosis is a feature seen in complex multicellular organisms, it is presumed that meiosis evolved from mitosis.
40. Collections of cellular organisms working as a unit led to multicellular organisms.
The first eukaryotes were single-cell organisms, and such single cell organisms account for the predominant class of eukaryotic organisms living on earth today (in terms of numbers of organisms). Under particular environmental conditions, single cell organisms will group together as tight clusters. It is presumed that the multicellular arrangement became vital for the survival of some organisms.
41. The earliest multicellular organisms arise as clonal clusters of eukaryotic cells dependent upon the multicellular lifestyle for their continued survival.
A clone is a collection of cells arising from generations of cell division stemming from a single ancestral cell. Clonal identity is a property that helps us to distinguish one organism from several different organisms that happen to congregate together.[viii]

---

vi The nucleus is synonymous with the karyon (from the Greek word meaning "nut"). The prokaryotes (bacteria and Archaeans) have no nucleus. The eukaryotes are cells having a nucleus (i. e., a concentration genetic material and enzymes wrapped in a membrane and floating as in organelle in cell cytoplasm). All animal, plants and fungi are eukaryotic organisms.

vii Somatic cells are cells of the organism other than the germ cells (i. e., oocytes and spermatocytes). The word "soma" comes from the Greek word meaning body.

viii A rare example of an organism that is not clonal is the lichen. A lichen is a composite organism that contains up to three species of fungus plus up to two species of algae or cyanobacteria. Although the lichen is currently considered to be an organism, it would be more accurate to think of the lichen as a community of organisms.

42. Subclones of cells within the clonal organism demonstrate specialized properties, through a process known as differentiation.
    Such cells have the same genome as the other cells of the organism, but they look and behave different from the others.
    For example a cell on the underside of the clone may be specialized for absorbing food. A cell on the edge of the clone may be specialized for sensing the presence of food. Another cell, in the same general location may have locomotor skills. Specialization of particular subsets of cells within a clonal organism is known as cellular differentiation. As a general rule, differentiated cells that divide will produce other cells that look and behave like themselves (e. g., a dividing liver cell produces two liver cells)
43. Evolving gene pools result in speciation.
    A species is collection of individuals with reproductive access to one pool of genes that is unavailable to other species (i. e., each species has its own gene pool).
44. Meiosis leads to sexual reproduction.
    The haploid progeny of the germinal cell can fuse with the haploid progeny of germ cells of another individual of the same species. When successful, this process yields a diploid cell (a cell with double-haploid sets of chromosomes); one haploid set from each of the parent cells. This is what sex is all about. Because meiosis first appeared early in eukaryotic evolution, we observe that sexual reproduction is a reproductive option available to most eukaryotic species.
45. An alternating life cycle of haploid and diploid organismal stages develops in the animal, plant, and fungal kingdoms.
    Meiosis occurring in a diploid organism may lead to haploid cells, which can grow as organisms, and the fusion of two haploid cells from the same or from different haploid organisms may produce diploid cells that can grow as diploid organisms. In fungi, haploid organisms are the dominant form. In plants, the haploid stage is the gametophyte, which can grow as a gametophytic plant. Gametophytic single cell organisms are the pollen grain (male haploid organism) and the egg cell (female haploid organism). In animals, the haploid organisms are the male spermatocytes and the female oocytes.
46. Two major kingdoms of cellular life (animals and plants) evolve a method of organismal development through progressive stages, wherein each stage is itself a distinct organism.
    Each stage of plant and animal development contains a genetic set of instructions to develop itself, as well as the instructions to build the next stage of development. Some insects undergo a complete metemorphosis with each stage: egg, larva, pupa, and adult. Many animals and all seed plants (i. e., plants of Class Spermatophyta) produce an embryonic organism, out of which the adult stage of the organism may develop.
47. Evolution of the various cell types in animal organisms eventually yields the neuron.

The ability to use ion flux to carry an electrical action potential along a cell membrane had evolved early, in single-celled eukaryotic organisms. Single-celled organisms were also capable of synthesizing and secreting chemicals that could alter the behavior of nearby cells. The neuron evolved in multicellular animals to coordinate both of these early functions. We don't know when the first functional neurons appeared, but coordinated neuronal activity was certainly present before the Cambrian explosion (roughly 600 million years ago).
In a fully functional animal neuron, electrical impulses propagate along the full length of a cell body extension (i. e., axons and dendrites). At the terminus, the electrical pulse prompts the release of molecules (at the intercellular synapse) that triggers an immediate response in the adjacent cell.

48. Organization of small sets of sensory cells and effector cells yield bi-directional signal pathways that can be described as reflex arcs.
    Neurons became the principle instruments whereby sensory information is registered and cellular responses are initiated.
49. Primitive subconscious thought emerges.
    Collections of neuronal cells evolve to coordinate senses (e. g., pain) and choose actions (e. g., muscle contractions).[ix]
50. Evolutionary enhancements of the subconscious produce improved thought processes and inherited instincts.
    Organisms with the best thought processes will make the choices that improve their survival chances. These improved thought processes are selected through genes.[x] Instincts are inherited behaviors (i. e., genetically programmed subconscious thoughts and drives).
51. The conscious mind emerges from the subconscious mind.
    Eventually, neuronal collections within organisms hit upon the technique of purposeful action (action with intent). After an organism has acquired the capacity of intentional action, it is a short evolutionary hop to reach the point when conscious decision-making occurs.
52. The conscious and subconscious mind produce intellect.
    Something akin to rational thought emerges from the interplay of subconscious drives and conscious deliberations.
53. A trait emerges in Homo sapiens wherein thought is conducted in words, leading to the development of complex languages.

---

ix It is impossible to adequately define the earliest form of "thought". For the sake of argument, we can say that the first thought occurred when an organism employed a collection of its neurons to choose a particular course of action from among several possible choices.

x The only way by which subconscious thoughts or behaviors can be passed on from generation to generation is through the genes.

Before there was language, we can presume that thought was conducted by a combination of instinctual and learned behaviors, aided by an imagined set of outcomes. When language emerged in humans, words served as pointers that substitute for complex imagined scenarios. Language permits us to express a sequence of things, actions, and concepts, using a small number of chosen words. All humans use language to conduct a continuous inner monolog of the conscious mind. To the best of our knowledge, no other organism on earth uses language to conduct thought.

54. An important intellectual achievement, occurring as an accessory to language, is reading and writing.
    Not every language has a written form, and not every language with a written form has an alphabet designed to reproduce the component sounds of the words of the language. Some written scripts are collections of metaphoric symbols. Others are pictograms, expressing a story in pictures. All written languages can be understood by at least some subset of the speakers of the language.
    Reading and writing are examples of "invented" products of Homo sapiens. Evolution played no role.
55. Additions to the body of written language (i. e., books) lead to advances in science, medicine, engineering, and the arts.
    With the advent of the written word, humans were no longer limited by their personal memories and experiences. They had access to the intellectual advancements of humans who lived in past generations, and they could contribute their own discoveries to the growing historical record.
56. The curiosity of scientists leads to an tentative understanding of the origins of reality.
    It was only natural for humans to wonder about the history of the universe, and its destiny.
57. The culmination of all the preceding steps leads us to the publication of *Fundamentals of Reality*.
    Res ipsa loquitor.

We began this book with a quotation from Groucho Marx, and we will end with a quotation from Karl Marx.

> Last words are for fools who haven't said enough.
>
> – Karl Marx's own last words

11 Notes

[1]**Amateurs.** We often assume that science is a pursuit best left to the professionals. This assumption is not true. Many of our most important breakthroughs in science have come from rank amateurs. For example, Antonie van Leeuwenhoek (1632–1723), one of the earliest developers of the compound microscope, who is sometimes credited as the father of microbiology, was a janitor. Augustin-Jean Fresnel (1788–1827) was a civil engineer who found time to make significant and fundamental contributions to the theory of wave optics. Johann Jakob Balmer (1825–1898) earned his living as a teacher in a school for girls, while formulating the mathematical equation describing the spectral emission lines of hydrogen. His work, published in 1885, led others, over the next four decades, to develop the new field of quantum mechanics. Most famously, Albert Einstein was a patent clerk who found time, in 1905, to publish three papers that forever changed the landscape of physics. Today, many of our most successful amateurs are autodidacts with access to public data resources [48, 177, 60, 61, 64, 181]. In the field of medicine, some of the most impressive feats of logical deduction have come from laypersons who have availed themselves of publicly available resources to research their own conditions [82, 87, 237].

The term "amateurish" is sometimes used to describe products that are unprofessional and second-rate. These insults should stop. We are living in a post-information age, where members of the public often have access to the same kinds of data and analytic tools that were formerly limited to academics [25]. At this point, all the professional scientists who are living today could not possibly exhaust the information available to anyone with Internet access. Amateurs are picking up the slack. In the logical analysis of widely available scientific data, amateurs can compete on a level playing field with professionals [24].

[2]**Loose ends.** Laymen may hold the opinion that the top researchers always work at the cutting edge of science. In truth, the "cutting edge" metaphor does not always apply. In many instances, the best science research is focused on the so-called "loose ends", these being the nagging inconsistencies and valid objections raised by yesterday's "advances". Historians will confirm that nearly all of modern physics was developed in response to age-old questions that classical physics left unanswered.

Science is forever frayed at its edges. The Standard Model, representing today's central dogma of quantum physics does not unify all the forces of nature (i. e., it excludes gravity), and does not tell us what happened in the first 10^{-43} seconds after the Big Bang.

The following questions tug at the loose ends of science:

- Can we unify all of the forces of nature under one model?
- Where did the fundamental constants of nature come from?
- How did spacetime come into existence?
- How can we account for the rate the universe's expansion?

https://doi.org/10.1515/9783111498690-011

- How can we account for the clustering and the rate of spin of galaxies?
- What does the collapse of the Schrödinger wave function tell us about our universe?
- Does our classical conception of causality, wherein the cause always precedes the effect, always apply?
- Is our universe deterministic (and do we have free will)?
- Is all information in the universe conserved?
- What is consciousness?
- What is the ultimate fate of the universe?

[3]**Proof in the natural sciences.** Proof in biology, chemistry, physics, and their associated sub-disciplines is fundamentally different from proof in mathematics. In mathematics, proof is absolute, given an initial set of assumptions. In the natural sciences, absolutism is never attained. Instead, proof of an assertion (which can be in the form of a law, a rule or even a theorem), exists when the following three conditions are established:

1. The assertion is consistent with an observation (upon which a credible hypothesis is based).
2. The assertion is consistent with every observation to which it applies. This means that the assertion is generalizable and has not be disproven by additional observations.
3. The assertion is consistently predictive. This means the assertion lets us predict what will happen when the criteria of the assertion are met, and that all such predictions fit subsequent observations.

These three conditions reveal that proof in the natural sciences is tentative. We can never be entirely certain that new or otherwise ignored observations might invalidate our assertion. We can never be entirely certain that the assertion will hold true for eternity. Nonetheless, our understanding of reality is rational and is based upon logical inferences drawn from observations.

There is one additional limitation to proofs in the natural sciences (that does not apply to mathematical proofs). Here I am referring to the issue of the interpretation of observations. We have no adequate way to prove the legitimacy of interpretations (which are themselves products of the human imagination). You may notice, scattered through this book, qualifying phrases such as "We seem to find," or "It seems that". The reason for doing so is simple. Our interpretations of reality arise from the human imagination and must be approached with caution.

[4]**Gene regulation.** Gene expression is influenced by many different regulatory systems, including the epigenome (e.g. chromatin packing, histone modification, base methylation), transcription and post-transcription modifiers (e.g., transcription factors, DNA promoter sites, DNA enhancer sites, cis- and trans-acting factors, alternative RNA splicing, miRNA and competitive endogenous RNAs, additional forms of RNA silenc-

ing, RNA polyadenylation, mRNA stabilizers), translational modifiers (e. g., translation initiation factors, ribosomal processing) and post-translational protein modifications.

Disruptions of any of these regulatory processes produce disease in humans and other metazoans [86, 234, 122, 203, 224, 3, 17, 63, 169, 175]. Moreover, anything that modifies any regulatory process (e. g., environmental toxins, substrate availability, epistatic genes) can influence gene regulation; potentially causing disease.

[5]**RNA and DNA.** The two types of nucleic acids found in living cells. Both are composed of combinations of four nucleotides. We find differences in one of their nucleotides (Uracil in RNA and Thymine in DNA); in strandedness (DNA most often begin double-stranded and RNA most often being single-stranded); in chemical structures (DNA having a deoxyribose structural component and RNA having a ribose structural component); in function (DNA generally contains the genes that are translated into RNA copies); and in form (there are several forms of RNA molecules, all serving different functions). All cells, including viruses, contain RNA or DNA or both.

[6]**Abiogenesis.** In 1668, the world was modernized, in many ways. We had the fundamentals of cryptography (Viete, 1589), Pi calculated to 20 decimal places (Ludolf, 1596), logarithms (Napier, 1624), Fermat's last theorem (1637), the fundamentals of probability (Pascal, 1654), and the ability to diagnose cancers by pathologic examination (Malpighi, 1659). Differential equations were understood (1662), and the last details of integral calculus were being written, independently, by Isaac Newton and Gottfried Wilhelm Leibniz. Despite all of these scientific advances, the world believed that the life of very small organisms arose spontaneously, from thin air, or possibly from inanimate particles of dirt. An act of God, it was presumed, brought life into the lifeless.

Francesco Redi, in 1668, designed an experiment to test whether maggots arose through spontaneous generation. He incubated meat in flasks; some flasks were covered to stop the entry of flies, and others were left uncovered, permitting flies to enter. After some time passed, Redi examined the meat in both sets of flasks. Only the meat from the open flasks contained maggots. Redi correctly concluded that maggots do not generate spontaneously from dead meat. We now know that maggots generate from tiny eggs laid by flies.

Redi's experiment had very little influence on the prevailing belief systems. Seventy-two years later, John Turberville Needham conducted his own meat-related test for spontaneous generation. He heated mutton broth in a closed container, and examined the contents a few days later. The container swarmed with micro-organisms, proving, to the satisfaction of many, that micro-organisms arise by spontaneous generation. In retrospect, we can assume that the broth was not heated sufficiently to kill all of the organisms initially present in the container, or that the container was not closed tightly. For some time, though, Needham's experiment vanquished lingering doubts regarding the validity of spontaneous generation.

In 1768, a full century after Redi's experiments, Lazzaro Spallanzani (1729–1799) repeated Needham's experiment, this time boiling the broth for forty-five minutes. No organisms grew in the closed container. Spallanzani's experiment should have put the kibosh on spontaneous generation, but it did not. Ninety-two years passed before the controversy was revisited. In 1860, about the same time as Darwin published "The Origin of Species by means of natural selection [69]," Pasteur showed that dust particles in air carried living micro-organisms. If boiled meat is exposed to purified air (without dust particles), bacterial growth does not occur. Nearly two centuries were required to validate Redi's early experiment, disproving spontaneous generation.

[7]**Natural selection.** The tendency for favorable heritable traits to become more common over successive generations. It is a mistake to believe that natural selection is a force of nature, like magnetic fields or gravity, or heat. Natural selection is simply a convenient term describing a somewhat abstract truism. Basically, if conditions favor the persistence of one outcome over another, then you will tend to see more instances of the favorable outcome.

Natural selection operates equally well on living matter and non-living matter. For example, let us imagine that we are observing two mountains. One mountain is composed entirely of granite, and the other is composed entirely of sand. Over time, the granite mountain is likely to persist, while the mountain of sand is likely to flatten and vanish. Natural selection favors granite over sand. Too obvious? Let us extend the metaphor to include chemical reactions. If one mountain is built from a chemical reaction that yields an insoluble product, and another mountain is built from a chemical reaction that yields a product that is slowly dissolved in water, then the insoluble mountain will persist. Moreover, if the chemical reaction that produces an insoluble product is sustained by readily available oceanic substrates, then we might expect to see new, insoluble mountains rising from the sea. If there is a geographic location where the reaction occurs more efficiently than in other areas, then we might expect to see a greater number of mountains rising in this area. If there are risen mountains wherein a novel reaction generates the catalysts and substrates for the mountain-building reaction, then we might expect to see new mountains sprouting out from such mountains; replication of sorts. The point is that natural selection favors whatever survives nature. The process of natural selection drives most, if not all, of the processes known as evolution.

[8]**Genome sizes of organisms.** The size of the genome, measured as the number of base pairs (Mega-base pairs or Giga-base pairs) of DNA.

- Escherichia coli about 5 Mbp;
- Saccharomyces cerevisiae (yeast) about 12 Mbp;
- Caenorhabditis elegans (nematode) about 100 Mbp;
- Drosophila melanogaster (fruit fly) about 120 Mbp;
- Arabidopsis thaliana (arabidopsis) about 120 Mbp;
- Mus musculus (mouse) about 2.5 Gbp;
- Homo sapiens (human) about 2.9 Gbp;
- Triticum aestivum (wheat) about 16 Gbp;

Compare the genome size of wheat (16 Gbp) with that of humans (2.9 Gbp).

Class Mimiviridae was thought to contain the largest viruses. Class Mimiviridae viruses can exceed 0.8 microns in length, hence exceeding the size of some bacteria (e. g., Mycoplasma species may be as small as 0.3 microns). The genome of species in Class Mimiviridae can exceed a million base pairs, encoding upwards of 1000 proteins. Class Megaviridae is a newly reported class of viruses, related to Class Mimiviridae, but even bigger [11].

[9]**DNA methylation.** DNA methylation is a chemical modification of DNA that does not alter the sequence of nucleotide bases. It is currently believed that DNA methylation plays a major role in cellular differentiation, controlling which genes are turned on and which genes are turned off in a cell, hence determining a cell's "type" (e. g., hepatocyte, thyroid follicular cell, neuron). Because cells of a particular cell lineage divide to produce more cells of the same lineage, DNA methylation patterns must be preserved with each somatic cell generation. About 1 % of DNA is methylated in human somatic DNA, and DNA methylation occurs preferentially on Cytosine, most often at sites where Cytosine is followed by Guanine (designated as "CpG").

DNA methylations, along with a variety of structural modifications of the DNA polymer (e. g., histone and non-histone protein attachments) are collectively known as the epigenome.

[10]**Eukaryote.** A member of Class Eukaryota, the large class of organisms whose cells contain a nucleus. There are three classes of living organisms: Class Viridae (the viruses); Class Prokaryota (organisms with no nucleus, comprising Class Bacteria and Class Archaea), and Class Eukaryota (organisms whose cells have at least one nucleus). In addition to having a nucleus, members of Class Eukaryota all seem to have evolved from organisms that contained mitochondria. Humans, and all animals, are descendants of Class Eukaryota. As descendants of Class Eukaryota, we humans are also full-fledged members of Class Eukaryota. Skeptics are urged to consider their geneology back through evolution. The gametes in humans are the result of a process can be followed iteratively up through the chain of ancestors within a species, and up through the gametes of the parent species, and on and on, until reaching the single-cell eukaryotes; organisms that are, for all practical purposes, "all gamete." Hence, we humans are dues-paying members of Class Eukaryota. This also explains why all animals have hundreds of the same core genes, give or take some acquired mutations, that are found in single-celled eukaryotic organisms.

[11]**Archaea.** For decades, archaean organisms were considered just another class of bacteria. This changed in 1977. Woese and Fox had been studying ribosomal RNA. Because ribosomal RNA is a fundamental constituent of all cellular organisms, sequence comparisons in the genes coding for ribosomal RNA are considered a reliable way estimate the degree of relatedness among organisms. In 1977, Woese and Fox surprised biologists when they demonstrated profound differences in the sequence of ribosomal RNA that distinguished archaean species from all other bacteria [269]. More shocking was their finding that the sequence of archaean ribosomal RNA was more closely related to eukaryotic cells than to other bacterial cells. Other sources have since shown

that the archaeans share with the eukaryotes a variety of features that are lacking in the bacteria. These include the presence of histones in some archaeans, the manner in which DNA is replicated and organized, and the finding that archaeans and eukaryotes share several transcription factors (proteins that bind to DNA and control the transcription of DNA to RNA) [141]. Woese and Fox proposed that the archaeans (then called archaebacteria) comprised a kingdom, separate but equal to the bacteria. That paper, and the many contributions of Woese that followed, challenged taxonomic orthodoxy and improved our understanding of the origin of terrestrial life and [173, 268].

[12]**DNA repair.** With the exception of damaged DNA, damaged cellular molecules need not be repaired. They are simply replaced with newly synthesized molecules. Because DNA is the template for its own replication, damaged DNA molecules cannot be replaced with newly synthesized DNA molecules; the damage must be repaired in situ if the organisms is to continue replicating faithfully. Repairing DNA is of such great importance to cellular life that mammals have at least seven different DNA repair pathways, each specializing in a particular type of damage repair (e. g., direct reversal repair, mismatch repair, nucleotide excision repair, homologous recombination, base excision repair, single strand break repair, non-homologous end joining, and Fanconi Anemia DNA crosslink repair). Some types of DNA lesions are substrates for more than one pathway, and the ways in which these pathways interact are complex [146].

[13]**Species.** The modern definition of species can be expressed in three words: "evolving gene pool" [76]. This elegant definition is easy to comprehend and serves to explain how new species come into existence [75, 172]. Each member of a species (i. e., each organism) has a genome constructed from the species-specific gene pool. Hence, membership within a species is immutable (e. g., a fish cannot become a cat and a cat cannot become a goat) inasmuch as the genome of any individual organism came from their species-specific gene pools. Species have a set of biological properties associated with any living entity: uniqueness, life, death, the issuance of progeny (i. e., new species), and the benefit of evolution through natural selection. Hence, we should think of a species as a biological entity; not as an artifactual abstraction created for the convenience of taxonomists.

It is estimated that 5 to 50 billion species have lived on earth, with more than 99 % of them now extinct, leaving about 10 to 100 million living species [208]. Hence, we can conclude that there is an ongoing, global process of speciation (producing new species) and extinction (disappearance of species). Because there are tens of millions of animal species living today, we can infer that the process of speciation yields diversification.

[14]**Nucleus.** In the context of biology, a nucleus is the membrane-bound organelle that contains the genome and the apparatus necessary for transcribing DNA into RNA, for preparing RNA for its role in protein translation, and for replicating the DNA in preparation for cell division.

Eukaryotes consist of all the organisms whose cells contain nuclei. In fact, the term "Eukaryote" derives from the ancient Greek, meaning "true kernel", referring to the nucleus. The prokaryotes, (eubacteria and archaeans), are organisms that do not contain a nucleus. Every known organism is either a eukaryote, or a prokaryote, or a virus. The nucleus, though necessary for cell division, is not necessary for the cell's moment-to-moment cell survival. Mature red blood cells have no nucleus, but they manage to live for about 120 days.

[15]**Retrovirus.** An RNA virus that replicates through a DNA intermediate. The DNA intermediate may become integrated into the host DNA, from which viral RNA is transcribed. When integration of the virus occurs in germ cells, the viral DNA can be inherited by the offspring. Through this mechanism, the human genome carries a legacy of retroviral DNA.

[16]**Homolog.** In the field of bioinformatics, "homolog" always refers to homologous genes. Genes from different organisms are considered homologous to one another if both descended from a gene in a common ancestral organism. If the homology was attained through speciation (i. e., the original gene persisted in the descendant species, with alterations), then the homologous genes are also orthologous. If the homology was attained through gene duplication (i. e., genetic alterations occurred in a duplicated inherited gene), then the homologous genes are also paralogous. Homologous genes of different species tend to have similar gene sequences, with sequence differences among the descendant species, and between the descendant species and the first parental species, accumulating over time.

[17]**History of the discovery of viruses.** The first inklings of the existence of viruses came in the waning years of the nineteenth century. In 1886, Mayer showed that he could reproduce the tobacco mosaic disease by injecting the juice of one tobacco plant into other plants. In 1892, Ivanowski showed the cross-infectivity occurred with juice that had passed through a fine filter, leading him to suspect that the transferrable agent was a tiny bacterium or a toxin. In 1898, Beijerinck concluded that the agent must have been something other than a bacterium or toxin, for which he coined the term "virus". In 1935, Stanley found that he could precipitate a crystalline protein from the infectious tobacco mosaic virus preparation. He inferred that the virus was some sort of protein capable of multiplying within cells. In 1936, Bawden showed that the crystalline substance was a mixture of protein and nucleic acid. A few years later, in 1939, Kausche and his coworkers viewed the virus particles using an electron microscope [4].

[18]**Types of mutation.** Some of the mutations that account for genetic diseases are: deletions (e. g., Duchenne muscular dystrophy), Frame-shift mutations (e. g., factor VIII and IX deficiencies), fusions (e. g., chronic myelogenous leukemia, hemoglobin variants), initiation and termination codon mutations (e. g. a type of thalassemia), inversions (a type of thalassemia), nonsense mutations (familial hypercholesterolemia), point mutations (e. g., sickle cell disease, glucose-6-phosphate dehydrogenase deficiency), promoter mutations (e. g., a type of thalassemia), RNA processing mutations, including splice mutations (e. g. Phenylketonuria) [258].

[19]**Evolvability.** Evolution by natural selection is not a physiological process, like respiration or replication. The theory of evolution is little more than a restatement of a

probabilistic truism, in biological terms. Namely, organisms that are most likely to survive will be the organisms most likely to reproduce.

Biologists speak in terms of evolvability, to indicate certain factors that may facilitate evolution and speciation. For the most part, these factors entail the acquisition of new or modified genetic material that may serve as the source of new genes. These would included:

– Having a large, diverse species gene pool;[i]
– Having mechanisms for horizontal gene transfer;
– Tendency toward endoduplication producing a set of non-conserved genes from which gene variants may emerge;
– Presence of pseudogenes and junk DNA, serving as "surplus parts" from which de novo genes may evolve.

[20]**Viral origin of adaptive immunity in animals.** Adaptive immunity is a somewhat specialized system wherein somatic T cells and B cells are produced, each with a unique and characteristic immunoglobulin (in the case of B cells) or T-cell receptor (in the case of T cells). Through a complex presentation and selection system, a foreign antigen elicits the replication of a B cell that produces an antibody whose unique immunoglobulin attachment site matches the antigen. Antigen-antibody complexes may deactivate and clear circulating antibodies, or may lead to the destruction of the organism that carries the antigen (e. g., a virus or bacteria).

The process of producing unique proteins requires that recombination and hypermutation take place within a specific gene region. Recombinations yield on the order of about a billion unique somatic genes, starting with one germinal genome. This process requires the participation of recombination-activating genes (RAGs). The acquisition of an immunologically active recombination activating gene, from a retrovirus, is presumed to be the key evolutionary event that led to the development of the adaptive immune system. An acquired RAG gene occurred in Class Gnathostoma (the jawed vertebrates), establishing the adaptive immune system in all extant jawed vertebrates. As one might expect, inherited mutations in RAG genes are known to cause immune deficiency syndromes in humans [276, 73].

[21]**Transposable element.** Also called transposon, and informally known as jumping gene. The name "transposable element" would seem to imply that a fragment of the genome (i. e., the transposable element) physically moves from one point in the genome to another. This is not the case. What actually happens (in the case of Class II transposons) is that a copy of the DNA sequence of the transposon is inserted elsewhere in the genome, resulting in the sequence now occupying two different locations in the genome. In the case of Class II transposons, the DNA sequence of the transposon is translated into RNA,

i A gene pool is the metaphoric name we assign to the aggregate collection of genetic material from all of the members of a population (e. g., species).

then reverse-transcribed as DNA, and reinserted at another location; likewise resulting in two of the same sequence in two locations in the genome [120]. You can see how transposable elements might bloat the genome with repeated elements.

Transposons are the ancient remnants of retroviruses and other horizonatally transferred genes that insinuated their way into the eukaryotic genome. Because much of the transposon DNA in animal genomes is not necessary for cell survival, the sequences of transposons are generally not conserved, and mutations occurring over time yield degenerate sequences that no longer function as retroviruses. As luck would have it, not all mutations to transposable elements are without benefit to the host cell. A transposon is credited with the acquisition of adaptive immunity in animals. The RAG1 gene was acquired as a transposon. This gene enabled the DNA that encodes a segment of the immunoglobulin molecule to rearrange, thus producing a vast array of protein variants [132]. A role for transposons in the altered expression of genes in cancer cells has been suggested [153].

[22]**Post-mitotic.** Refers to fully differentiated cells that have lost the ability to divide. For example, the epidermis of the skin has a basal layer of cells, each of which is capable of dividing to produce one post-mitotic cell plus another (replacement) basal layer cell. The post-mitotic cells sit atop the basal cells, where they gradually flatten out (i. e., become squamous shaped) and lose their nucleus as the cells rise through the epidermal layers, pushed up by the next generation of cells produced by the dividing basal cells.

The top layer of the epidermis sloughs off into the air and is replaced by post-mitotic epidermal cells in the next lower layer. Much of the small flakes of house dust that we see dancing in the beams of sunlight passing through our windows is composed of post-mitotic squamous cells, sloughed from our skin. This cycle of cell renewal from the bottom and cell sloughing from the top is typical of most epithelial surfaces of the body (e. g., epidermis of skin, gastrointestinal tract, and glandular organs). Aside from epithelial surfaces, post-mitotic cells arise from populations of mitotic cells that have exhausted their regenerative potential. One theory of aging holds that certain cell-types of the body (e. g., fibroblasts) have a limited number of mitotic cell cycles. When a biologically determined number of cell cycles have elapsed, such cells cannot divide further, and become post-mitotic cells that eventually wear out and die.

[23]**Differentiation.** The human body contains several hundred types of cells endowed with a set of functions and morphologic features that distinguish each cell-type from all of the other cell-types. The process by which each type of cell becomes different from the other (several hundred) cell-types is called differentiation. Every cell-type, regardless of its particular properties, has the same genome as every other cell-type. Hence, we can infer that some factors other than genome sequence controls the process of differentiation. In mammals, differentiation is determined by the epigenome, which controls the levels of expression of genes in a cell, which in turn determines how a gene looks and behaves. Epigenomic modifications are heritable from a cell to its progeny. Hence, when a hepatocyte replicates, it produces two hepatocytes and not any other cell-types.

The process by which undifferentiated cells modify their epigenomes, to produce a lineage of cell descendants having progressively greater levels of differentiation, is referred to as stem cell maturation.

[24]**Cell type.** The different cell types of the body are defined by morphologic, functional and anatomic traits that, together, distinguish cells of one type from all other types of cells.

The number of different kinds of cell in an organism varies based on how you choose to categorize and count the different cells of the body, but most would agree that there are at least 200 different cell types in the body. Because we can classify thousands of different cancers in humans, we can safely say that one cell type may correspond to many different types of cancer. For example, both follicular carcinoma of thyroid and papillary carcinoma of thyroid seem to derive from the epithelial cells that give rise to normal thyroid glands. Follicular carcinoma of the thyroid and papillary carcinoma of thyroid behave differently from each other and each has its own distinct genetic defect.

[25]**Heterokonts.** Also known as stramenopiles, are a class of single-celled eukaryotes. There are over 25,000 species of heterokonts, including some types of algae, unicellular diatoms, and single-celled components of plankton. Also includes organisms that are pathogenic in plants (late blight of potato caused by Phytophthora infestans, and sudden oak death caused by Phytophthora ramorum). Heterokonts are also known to cause two diseases in humans: blastocystosis and pythiosis. Despite the diversity and pathogenic potential of the heterokonts, biologists know shockingly little about them.

[26]**Ciliopathies.** The ciliopathies are a genetically and phenotypically diverse set of diseases that all involve inherited disorders of the primary cilium (also called the non-motile cilium). The primary cilium is a single flagellum extending from virtually every cell in vertebrates that is distinguished from motile cilia by the absence of a central pair of microtubules and outer dynein arms that are essential for normal motility. Included in the ciliopathies are: Joubert syndrome, nephronophthisis, Senior-Loken syndrome, orofaciodigital syndrome, Jeune chondrodysplasia syndrome, autosomal dominant polycystic kidney disease, recessive polycystic kidney disease, Leber congenital amaurosis, Meckel-Gruber syndrome, Bardet-Biedl syndrome, Usher syndrome, Alstrom syndrome, McKusick-Kaufman syndrome, Ellis van Creveld syndrome, cranioectodermal dysplasia (Sensenbrenner syndrome), short rib polydactyly, some forms of retinal dystrophy, and heterotaxy (including visceral situs anomalies, asplenia or polysplenia, congenital heart defects, biliary atresia, and midline defects) [188, 256]. There are only a few organs and functional systems of the human body that escape involvement by this strange collection of pathogenetically related, but clinically diverse, rare diseases.

[27]**Histone.** The major protein in chromatin (i. e., the material composing chromosomes). A segment of DNA, wound around a histone protein core, is called a nucleosome, the basic unit of DNA packaging in eukaryotes. When histones are modified by deacetylation (via deacetylases), the nucleosome tightens, reducing normal transcription by blocking transcription factors from attaching to target sites.

[28]**Mimiviridae.** Class Mimiviridae, discovered in 1992, occupies a niche that seems to span the biological gulf separating viruses from other organisms. Members of Class Mimiviridae are complex, larger than some bacteria, with enormous genomes (by viral standards), exceeding a million base pairs and encoding upwards of 1000 proteins. The large size and complexity of Class Mimiviridae exemplifies the advantage of a double-stranded DNA genome.

Class Megaviridae is a newly reported (October, 2011) class of viruses, related to Class Mimiviridae, but even larger [11]. The discovery of classes of giant viruses inspires biologists to reconsider the "non-living" status relegated to viruses and compels taxonomists to examine the placement of viruses within the classification of life [26].

[29]**Haploidy.** From Greek haplous, "onefold, single, simple". Equivalent to the chromosome set of a gamete. In humans, this would be 23 chromosomes; one set un-paired autosomes (chromosomes 1 to 22) plus one sex chromosome (X or Y chromosome). In animals, haploid cells are gametes that mature into egg cells (oocytes) or sperm cells (spermatocytes). In plants descending from Class Embryophyta (i. e., most of the organisms that we recognize as plants), haploid cells may mature into male gametophytes or female gametophytes. The somatic cells of the body (i. e., all cells that are not gametes) are diploid (i. e., having twice the haploid set of chromosomes), and somatic cells that enter meiosis produce haploid cells.

[30]**Diploidy.** Humans are diploid organisms (with the exception of haploid germ cells). In females, each chromosome is paired with a matching chromosome. In males, every autosomal chromosome (non-sex chromosome) occurs in evenly matched pairs (i. e., 2 chromosome 1s, 2 chromosome 2s, 2 chromosme 3s, etc.) while the single male sex chromosome (Y) complements a much larger X chromosome. In males and females, the diploid number of chromosomes is 46. The haploid number of chromosomes is 23.

[31]**Clonal reproduction.** The reproduction of an organism through a mechanism that does not require the contribution of genes from another organism (i. e., reproduction without sex). In the plant kingdom, where various dividing cells of the adult plant are totipotent (e. g., meristem cells), clonal growth of organisms simply involves placing meristem cells in an environment conducive to growth. Self-fertilization (in plants bearing both male and female gametophytes), grafting and coppicing are familiar plant clonal reproductive techniques. Modern agriculturalists employ plant tissue culture to clone plants in the laboratory. Fungi, protozoans, and some lower plants (all of which are capable of sexual reproduction), clonal reproduction can be achieved with spores (single-cell reproductive cells derived from the organism without fusion with another organism). In the case of animals, various methods of clonal reproduction occur naturally in non-mammalian species. A mechanism among invertebrates and certain non-mammalian tetrapods involves development of an embryo from an unfertilized egg (parthenogenesis). Insects have several additional mechanisms whereby reproduction can occur without sexual recombination. All members of a clone have the same genetic identity.

[32]**Horizontal gene transfer.** The direct transfer of genetic material between organisms, by mechanisms other than by reproduction (i. e., other than the transfer of DNA from parents to offspring). The very first eukaryotic ancestors derived their genetic material by horizontal gene transfer from prokaryotes (bacteria and archaean organisms), viruses, and possibly from other now-extinct organisms that might have preceded the eukaryotes. The early eukaryotes almost certainly exchanged DNA between one another, and we see evidence of such exchanges in modern single-celled eukaryotes and fungi [92, 134].

To an unknown extent horizontal gene transfer occurs throughout the animal kingdom. For example, tardigrades, a microscopic animal, has a genome one-sixth of which was derived from bacteria, archaeans, plants, and fungi [42].

It should be noted that many of the most significant evolutionary advances came from inter-species gene acquisitions. The primordial mitochondrion that helped to create the first eukaryotic cell was an acquisition from a bacteria. The very first chloroplast in the most primitive precursor of the plant kingdom was a pilfered cyanobacteria.

[33]**Parthenogenesis.** Reproduction from an ovum without fertilization by a sperm. Occurs in plants, invertebrates and some vertebrates, but does not occur in mammals, a class of animals that uses genomic imprinting. Mammalian imprinting may interfere with parthenogenesis. There are about 80–100 genes in mammals which are imprinted, and many of these genes are vital to normal embryogenesis and fetal development. In parthenogenesis, all genes have maternal derivation. A parthenogenic embryo would lack paternally imprinted genes. There is experimental evidence that the paternally imprinted genes are required for the normal development of a human embryo [174, 18].

[34]**Reassortment.** Often confused or used interchangeably with “recombination”, but reassortment is generally reserved for a viral event wherein two similar segmented viruses exchange part of their genomes during the co-infection of a host cell. Reassortment seems to be the major mechanism accounting for new influenza virus strains.

[35]**Endosymbiont.** An organism that lives within some other organism, often to the mutual advantage of both. An interesting example of an endosymbiont is Wolbachia pipientis, which lives happily inside most members of the filarial Class Onchocercidae [222]. Onchocerca volvulus is the filarial nematode (roundworm) that migrates to the eyes and causes river blindness, the second most common infectious cause of blindness worldwide [210]. It is the Wolbachia endosymbiont within the nematode that elicits the inflammatory response in the eye, leading to blindness.

[36]**Cyanobacteria.** The most influential organisms on earth, cyanobacteria, were the first and only organism to master the biochemical intricacies of oxygenic photosynthesis, (more than 3 billion years ago). Oxygenic photosynthesis involves a photochemical reaction, which uses carbon dioxide and water, and releases oxygen. All photosynthesizing life forms are either cyanobacteria, or they are eukaryotic cells (e. g., algae, plants) that have acquired chloroplasts. Chloroplasts, in turn, are organelles created in the distant past by endosymbiotic capture of a cyanobacteria by an early eukaryote. Hence, all oxygenic photosynthesis that has ever occurred on the planet earth

stems from a metabolic pathway that evolved in cyanobacteria. Before the emergence of oxygen-producing cyanobacteria, the Earth's atmosphere had very little oxygen and organisms were anaerobic.

[37]**Anlagen.** Embryologic precursors for fetal organs and tissues. It is important to know that the anlagen of an organ is not a miniature version of the organ, and does not contain any of the differentiated cells that will eventually appear in the developed organ. The anlagen contains stem cells growing within an architectural framework that does not have the appearance of the organ that will eventually emerge. As an example, imaginal discs are the insect equivalent of anlagen, and consist of small sacs of committed stem cells. Insect wings grow from imaginal discs that are biologically committed to develop into a wing. An insect wing does not grow from a tiny wing.

[38]**Epigenome.** The epigenome, at its simplest, consists of cell-type specific chemical modifications to DNA that do not affect the sequence of nucleotides that comprise the genome. Such modifications may control the expression of genes. There are many kinds of non-sequence modifications that fall under the rubric of "epigenome." One of the best-studied epigenomic modifications is DNA methylation. The most common form of methylation in DNA occurs on Cytosine nucleotides, most often at locations wherein Cytosine is followed by Guanine. These methylations are called CpG sites. CpG islands are concentrations of CpG sites. There are about 29,000 to 50,000 CpG islands. The patterns of methylation are inherited among cells of the same type (e. g., goblet cell to goblet cell, liver cell to liver cell). Alterations in methylation patterns are referred to as epimutations. Epimutations may persist in those specialized cells that are descendants from an epimutated cell [150].

Epigenome modifiers, other than methylation, include: histones, and non-histone nuclear proteins; conformational changes in DNA that influence gene expression (e. g., heterochromatin); protein interactions; RNA interactions; and any other physicochemical interactions that influence gene expression.

In general, the epigenome controls differentiation and the phenotype (i. e., observed biological characteristics) of the different somatic cell-types of the body [27, 28]. A simple way to think about the respective roles of genome and epigenome is as follows: The genome establishes the identity of an organism; the epigenome establishes the identity of the individual cell-types within the organism.

[39]**Totipotent stem cell.** Most stem cell are committed to producing cells of a particular type, such as skin cells, blood cells, liver cells, etc. A totipotent stem cell can produce, after cell divisions, differentiated cells of any type. This would include cells arising from any of the three embryonic layers (ectoderm, endoderm, and mesoderm), as well as germ cells, and cells of the extra-embryonic tissue (e. g., trophoblasts). A totipotent stem cell is different from a pluripotent stem cell, the latter of which cannot produce cells of extra-embryonic lineage [242].

As pertains to development in members of Class Eutheria (placental mammals), there is a short period when all of the cells are totipotent. During the so-called pre-embryonic period, corresponding to the first two weeks following fertilization, the zygote (i. e., fertilized ovum) travels to the uterus, implants, and begins to build the extra-embryonic tissues, and the blastocyst. The cells of the pre-embryo are, to the best of anyone's knowledge, totipotent. In the third week following fertilization (fifth week since the last menstrual period), the three layers of the embryo form, and the embryonic cells differentiate to the extent that they are no longer totipotent.

In the past few decades, we have come to learn that the primary difference between an embryonic stem cell and a fully differentiated cell of the body (i. e., a somatic cell) is the erasure of the epigenome. Epigenomic erasure is a process by which virtually all of the epigenomic modifications present in the mature cell are erased, and the cell is returned to its uncommitted state, much like that of a totipotent stem cell [191, 231, 235, 232].

[40]**Ectoderm.** There are three embryonic layers that eventually develop into a fully developed animal: endoderm, mesoderm, and ectoderm. The ectoderm gives rise to the skin epidermis and the skin appendages (e. g., hairs, sebaceous glands, breast glandular tissue) and to the specialized neurectoderm from which the central nervous system arises.

[41]**Endoderm.** There are three embryonic layers that eventually develop into the fully developed animal: endoderm, mesoderm, and ectoderm. The endoderm forms a tube extending from the embryonic mouth to the embryonic anus. The mucosa of the gastrointestinal tract, and the epithelial cells of the liver, pancreas, and lungs, all derive from endoderm.

[42]**Mesoderm.** The embryonic germ layer that lies between ectoderm and endoderm, giving rise to the mesenchyme, the latter consisting of most of the connective tissue, muscles and bones of the body. The kidneys and the uterus are organs that are derived entirely from mesoderm.

[43]**Germ layers.** Metazoans (i. e., members of Class Animalia) are triploblastic, meaning that the early embryo contains three layers, from which all of the tissues of the developed organism will arise. These three layers are ectoderm, endoderm, and mesoderm.

During embryogenesis, several additional germ layers arise (e. g., neuroectoderm and neural crest), but these layers are derivative of the three primary germ layers and not traditionally included when we use the term "germ layer".

When we list the primary germ layers, we traditionally omit the trophectoderm, the layer of cells that gives rise to the extra-embryonic tissues that arise early in embryogenesis but which are not incorporated into the developed organism (e. g., extraembryonic placenta, umbilicus, and amnion).

Also, the term "germ layer" must never be confused with "a layer of germ cells" (i. e., a layer of gametes and their precursors).

[44]**Diseases of the desmosome.** As with most evolutionary innovations, the arrival of the desmosome brought with it a variety of diseases. The desmosome provides a tight continuum between cells, and the gap junctions mediate the passage of electrolytes between cells. Because rhythmic contractions of the cardiac muscle require the conduction

of impulses from cell to cell, it comes as no surprise that mutations in desmosomal can result in cardiac arrhythmias; the so-called arrhythmogenic cardiomyopathies [74]. A form of arrhythmogenic right ventricular cardiomyopathy/dysplasia is caused by a heterozygous mutations in the PKP2 gene, which encodes plakophilin-2, a protein of the cardiac desmosome. In addition, desmosomes keep the epidermis from rubbing off when friction is applied to skin. Desmosomal skin disorders are characterized by blistering, acantholysis (e. g., epidermal cells falling apart from one another), and keratoses, characterized by a thickened and hardened epidermis [23, 31].

[45]**Carcinogenesis.** The cellular events leading to cancer and equivalent to the pathogenesis of cancer. Carcinogenesis in adults is a long process that involves the accumulation of genetic and epigenetic alterations that eventually produce a growing clone of malignant cells. The conjectured sequence of events that comprise carcinogenesis begins with initiation, wherein a carcinogen damages the DNA of a cell, producing a mutant clonal founder cell that eventually yields a group of cells that have one or more subtle (i. e., morphologically invisible) differences from the surrounding cells (e. g., less likely to senesce and die, more likely divide, less genetically stable, better able to survive in an hypoxic environment). After a time, which could easily extend into years, subclones of the original clone emerge that have additional properties that confer growth or survival advantages (e. g., superior growth in hypoxic conditions). The process of subclonal selection continues, usually for a period of years, until a morphologically distinguishable group of cells appear: the precancer cells. Subclonal cells from the precancer eventually emerge, having the full malignant phenotype (i. e., the ability to invade surrounding tissues and metastasize to distant sites). The entire process may take decades.

[46]**Embryonic stem cell.** In the context of tissue culture, a line of cells that has been cloned from a cell taken from the inner cell mass of a blastocyst (i. e., not from the trophectoderm and not from the thin-walled cystic area of the blastocyst). The inner cell mass consists of pluripotent stem cells that soon differentiate into an embryo composed of three germ layers. Embryonic stem cells in tissue culture can differentiate into any of the cell-types that occur in the human body, under proper laboratory conditions (Figure 11.1).

One of the most surprising advances in cell biology has been the discovery of a methodology by which differentiated cells obtained from a fully developed human, can be induced to become embryonic stem cells; thereby bypassing the traditional dependence on blastocysts as a source of embryonic stem cells for research. It seems that 4 factors (OCT4, SOX2, NANOG, and LIN28) can reprogram human somatic cells to become embryonic stem cells, capable of differentiating into all three germ layers [273].

We need to remember that the embryonic stem cell has the same genome as any of the differentiated cells of the body. In the past few decades, we have come to learn that the primary difference between an embryonic stem cell and a fully differentiated cell of the body (i. e., a somatic cell) is the erasure of the epigenome. Epigenomic erasure is a process by which virtually all of the epigenomic modifications present in the mature

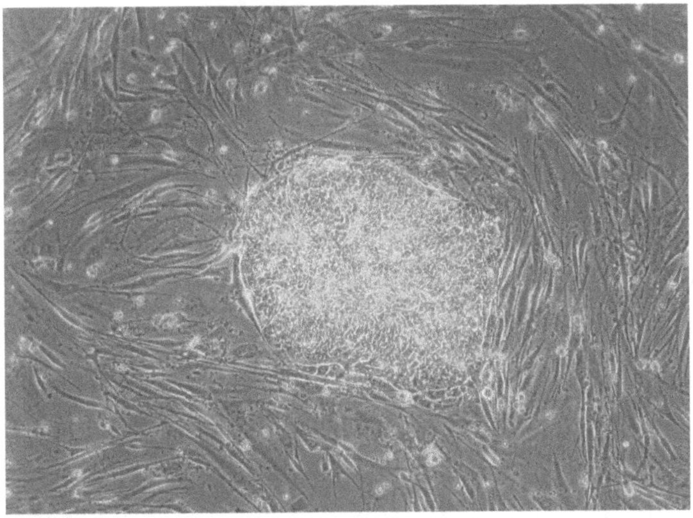

Figure 11.1: Human embryonic stem cells growing in tissue culture. The dense clump of cells in the center are stem cells. The flat cells in the periphery of the image are a layer of mouse fibroblast cells that are intended to help the embryonic stem cell colony thrive. Source, Wikipedia, from an image entered into the public domain by its author, Ryddragyn.

cell are erased, and the cell is returned to its uncommitted state, much like that of a totipotent stem cell [191, 231, 235, 232].

[47]**Progeria.** Any of several rare genetic conditions of premature aging, typically characterized by early-onset atherosclerosis and diabetes. All of the progerias, despite differences in their root genetic causes, seem to involve defects in normal cellular renewal processes (e. g., stem cell regeneration, maintenance of stem cells, DNA repair in dividing cell populations).

[48]**Telomere.** Chromosomes have a long padding sequence of repetitive DNA, at their tips, and this sequence is called the telomere. Animal cells lose a fragment of DNA from the tip of the chromosome with each cell division. This happens because one strand of DNA is replicated as sequential fragments, with each fragment requiring a template sequence beyond its end to initiate replication. Absent the telomere, the last fragment in the DNA strand has no extension template and is not replicated. By furnishing a padding at the tips of chromosomes, the telomere sequence sacrifices fragments of itself for the sake of preserving the coding sequences of the chromosome. As all good things come to an end, the telomere padding exhausts itself, after about 50 rounds of mitosis. At this time, the cell ceases further replication and eventually dies.

Cells that continually renew throughout life, such as bone marrow stem cells, skin and hair follicle basal layer cells and intestinal basal crypt cells, retain an enzyme, telomerase, that restores telomere length. If such cells contain a loss-of-function mutation in any of the genes encoding the components of the telomerase complex, then the ability of those cells to continuously divide is impaired.

A mutation in the telomerase gene is the root cause of dyskeratosis congenita, a rare inherited condition in which bone marrow failure frequently occurs [249]. Telomerase gene mutations have also been found in some cases of adults having bone marrow failure, and these late-occurring conditions are presumed to be variant forms of dyskeratosis congenita [272]. We see the clinical consequences of a telomerase deficiency in the blood forming cells of the bone marrow because these cells are the most highly replicative cells in the body [219]. A defect in the replicative capacity of dividing cells would be expected to reveal itself clinically as bone marrow failure.

Cancer cells, like bone marrow cells, continuously divide, and have high concentrations of telomerase. The ability of some cancer cells to restore their telomeres contributes to their continuous capacity to divide without limit, a phenomenon sometimes called cancer cell immortality. Telomerase insufficiency has been suggested as an possible cause of spontaneous regression in tumors, a rarely observed phenomenon [39].

[49]**Russell–Zermelo paradox.** Consider the set $\mathbb{A}$ composed of all the sets that are not members of the set $\mathbb{A}$. If a set is a member of set $\mathbb{A}$, then, by set $\mathbb{A}$'s definition, the set is not a member of set $\mathbb{A}$. If the set is not a member of set $\mathbb{A}$, then by definition it must be a member of set $\mathbb{A}$. You see the paradox.

The Russell–Zermelo paradox is one of many examples of paradox by self-reference. Set $\mathbb{A}$ references itself within itself, permitting the conditions that establish the paradox.

[50]**Pseudoscience.** A set of beliefs or theories that cannot be tested (i. e., cannot be proven wrong). Pseudoscientific beliefs are not necessarily wrong; we just have no way of knowing. Intelligent design is an example of a pseudoscientific belief. If we assert that humans were designed by an intelligent being, such as a superior but permanently absent alien life form, or an all-powerful but invisible entity, then we are making a pseudoscientific statement. There is no way to determine whether the assertion is false.

Conversely, science is built upon hypotheses that can be disproven, if false. For example, the Big Bang theory holds that our observable universe began with a singularity that occurred about 13.8 billion years ago. This theory could be disproven if we could find a single example of matter that predates the big bang (e. g., a star that has existed for 15 billion years). So far, no such instances have been found. Likewise, the theory of evolution could be disproven if we could find a single example of an animal fossil that predates its ancestor (e. g., if we found a fossilized modern bird mixed in the same rock strata that contains a fossilized extinct dinosaur). So far, the fossils in rock strata have yielded evidence consistent with our current understanding of animal evolution.

[51]**Energy.** Energy is one of those concepts that we all pretend to understand, but for which there is really no truly satisfactory definition. Often, we say that energy is the capacity to move things against a resistant force. In that case, what is the force resisting? It must be resisting our energy. Which tells us that energy is the capacity to move things against something that resists energy. Circular and not helpful.

The idea that there are different "kinds" of energy (kinetic energy, potential energy, chemical energy, nuclear energy, electromagnetic energy) indicates that there is no clear definition of our terms.[ii]

As a unifying concept, it is useful to think of energy as a quantitative property that is associated with movement. Einstein reminded us that:

> Nothing happens until something moves.
>
> – Albert Einstein

Heat energy is associated with the vibratory movement of particles. Light energy is associated with the quantized shifts in orbitals by atomic electrons. Work energy is associated with the movement of mass against resistance. The energy of the strong force is associated with the confinement of the gluons and quarks that compose neutrons and protons. In summary, wherever there is movement, there is energy.

We can derive relationships between various movements and energy. For example, the energy associated with momentum would be the integral of the momentum (mass times velocity) as velocity changes:

$$E = \int mvdv = \frac{1}{2}mv^2.$$

Conservation laws apply to energy in all of its various guises. In point of fact, we can always think of energy as whatever is being conserved when the movement (associated with its energy) occurs over time. This would apply to the energy associated with heat, light, momentum, the atomic strong force, and the atomic weak force. And, because all these forms of energy are fundamentally equivalent, we should insist that all forms of energy can be expressed in the same units or (failing that) can be converted to the same units.

[52]**Derivation of the Schrödinger wave equation.** The Schrödinger's wave equation describes the universe as a general wave function that incorporates energy and momentum. Doing so will lead us to some understanding of how reality emerged from nothing other than waves (arising from energy fluctuations).

A full discussion of this topic requires some mathematics, so we will lay out some equations in the chapter text, saving most of the details to our Glossary items. Readers with a strong aversion to mathematics are encouraged to skip forward to Section 3.5, "Watching waves."[iii]

---

ii The popular notion of "potential energy" as energy that is not causing movement (i. e., not kinetic), but which exerts itself under appropriate circumstances, is ludicrous when you come to think of it.

iii Do not panic! Yes, quantum physics relies heavily on mathematics, but most of the math is actually covered in high school level pre-calculus and linear algebra. If we pile on a little vector calculus, typically taught in the freshman year of college, we've got a lock on 90 % of modern physics.

A standing wave (also called stationary wave) is a wave that oscillates but does not move in the sense there is no succession of waves that pass a point. Examples of stationary waves are children playing with a jump rope, or a musician plucking a guitar string. Both ends of the rope and both ends of the guitar string are fixed, and the amplitude of the wave at either end is zero (because the ends are fixed and cannot raise their amplitudes). The wave oscillates up and down around an integer number of nodes having zero amplitude, with the two ends of the string being the first and the last of the zero amplitude nodes. Standing waves are always quantal insofar as there is an integer number of nodes holding an integer number of waves, with each wave oscillation being identical to every other wave oscillation (Figure 11.2).

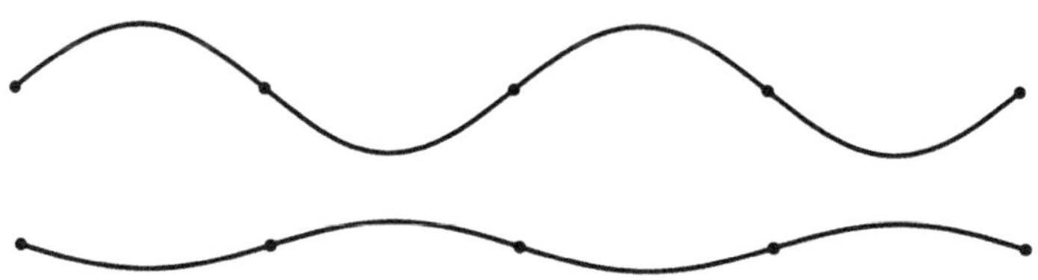

Figure 11.2: A vibrating string stretched between two fixed points will produce a standing wave. The locations of the nodes of zero amplitude stay fixed as the wave oscillates. Above, we see the wave at its peak. Below, we see the wave a moment later, when the amplitudes have fallen. Notice that the nodes of the wave do not move and that the fixed nodes at either end always have zero amplitude. There are five zero-amplitude nodes along the length of the oscillating string. The nodal configuration of a standing wave is a quantal system, with the number of waves described by an integer. Source, Wikipedia, and modified from an image released into the public domain by its author, Lucas Vieira.

Standing wave oscillation can be described with the following formula is e^{ikx} where, k is the wave number, the number of wavelengths in one revolution (i. e., $k = 2\pi \div \lambda$, where λ is wavelength); i is the unit imaginary number, and x is a distance along the wave. It is not entirely obvious that e^{ikx} has much to do with waves, until we apply Euler's formula.

$$e^{ikx} = \cos(kx) + i\sin(kx).$$

We see that the function can be represented as the sum of cosines and sines, both of which oscillate like waves, giving us a wave function. While we are at it, we can apply DeMoivre's identity[53] to find:

$$\cos(kx) + i\sin(kx) = (\cos(x) + i\sin(x))^k.$$

The wave is provided with a real component, $\cos(kx)$, and an imaginary component, $i\sin(x)$. At present, we focus on the real component. Substituting $2\pi \div \lambda$ for k, we have

$$\cos(kx) = \cos\left(\frac{2\pi}{\lambda}x\right).$$

Whenever $x = \lambda$ the wave function equals $\cos(2\pi)$, whose value is 1, the maximum value of the cosine. Therefore, the standing wave has maximum amplitude every 2π, or every full revolution around a unit circle.

This is how we can write a wave function for a standing wave. In reality, most waves do not stand still, and we need a general wave equation to describe moving waves.

Let us modify our wave function a bit, so that our waves can travel. In addition, we will give our new wave function a dignified name, ψ.

$$\psi = e^{i}(kx - \omega t).$$

Here, t is a time interval, and ω is the angular velocity of the wave, and is equal to $2\pi\nu$ where ν is the wave frequency expressed as oscillations per second.

How does this new wave equation help us? Let us consider the case when $kx-\omega t = 0$.

In this case, the wave is at a peak amplitude as $\psi = \cos(kx - \omega t) = \cos(0) = 1$, or $kx = \omega t$ or $\frac{x}{t} = \frac{\omega}{k}$.

But $\frac{x}{t}$ expresses the distance over time, or the velocity of the wave. And so, our new formula endows our wave with a velocity.

The new equation also provides a way of understanding wave velocity in terms of wave frequency and wavelength.

$$\text{velocity} = \frac{x}{t} = \frac{\omega}{k} = (2\pi\nu) \div \frac{2\pi}{\lambda} = \nu\lambda.$$

It is easy to see now that the wave velocity is the number of waves in a second, (ν), times the wavelength, (λ).

Now that we have developed general equations for a traveling wave, we can attach what we know about the physical properties of electromagnetic waves and derive a wave equation that describes quantum objects that have wave properties.

Let us look at our traveling wave function:

$$\psi = e^{ikx-\omega t}.$$

Let us take the derivative to see how the wave function changes over distance.

$$\frac{d\psi}{dx} = ike^{kx-\omega t} = ik\psi.$$

We can take the derivative of the derivative (i. e., the second derivative):

$$\frac{d^2\psi}{dx^2} = (ik)^2\psi = -k^2\psi.$$

Now we need to review Einstein's famous formulas for the energy of light in terms of either mass or frequency, where c is the speed of light, h is the Planck constant, and ν is the frequency of light.

$$E = mc^2 = h\nu.$$

Let us recall that momentum = mass times velocity. In the case of light, momentum = mass times $c = mc$. Also, we have shown that the velocity of a wave is equal to its wavelength times its frequency. Therefore, its frequency is the wave velocity divided by the wavelength. In the case of light waves, $\nu = c \div \lambda$. Substituting back into the light energy equation:

$$E = mc \times c = hc \div \lambda$$

or

$$E = pc = hc \div \lambda.$$

Therefore,

$$\lambda = h \div p.$$

We have previously shown that $k = 2\pi \div \lambda$. So, $k = 2\pi p \div h$.

By convention, we call $\frac{h}{2\pi}$ by the symbol $\hbar$. So, $k = p \div \hbar$.

Returning to our equation for the second derivative of the wave function:

$$\frac{d^2\psi}{dx^2} = -k^2\psi.$$

Substituting $p \div \hbar$ for k,

$$-\hbar^2 \times \frac{d^2\psi}{dx^2} = p^2\psi.$$

Recalling our high school physics, the total energy in a system is equal to the kinetic energy plus the potential energy, U, or $E = K.E. + U$

The kinetic energy is equal to one-half the mass times velocity squared, written as: $K.E. = \frac{1}{2}mv^2$.

The momentum of a moving object, p is mass times velocity. We'll represent momentum as p and write our equation: $p = mv$.[iv v]

---

iv Sharp readers will recognize here that momentum is just the time derivative of energy.

v You will notice that the same relationship between momentum and mass holds for solid matter and for waves, a notion developed in 1924 by Louis de Broglie [72]. A simple argument supporting de Broglie's theory is that if electrons and other subatomic particles exhibit wave/particle duality, and if all matter is composed of subatomic particles, then we would expect all matter to exhibit wave properties and to conform to the fundamental wave formula expressed in the Schrödinger wave equation.

Let us rewrite the energy equation, remembering that $p = mv$ and $v = \frac{p}{m}$, and substituting back into the original equation:

$$E = \frac{1}{2}mv^2 + U$$

or

$$E = \frac{1}{2}\frac{p^2}{m} + U.$$

Multiplying both sides of the equation by our wave function:

$$E\psi = \frac{p^2\psi}{2m} + U\psi$$

and remembering that:

$$\psi = -\frac{\hbar^2}{p^2}\frac{d^2\psi}{dx^2}.$$

We get:

$$E\psi = -\frac{\hbar^2}{2m}\frac{p^2\psi}{2m} + U\psi.$$

This is the time-independent Schrödinger's equation, and it tells us how waves evolve over distance. At the start of this topic, we took the derivative of the wave equation with respect to distance, and we derived our time-independent equation. We could have just as easily taken the derivative of the wave equation in respect to time, and that would have yielded the distance-independent Schrödinger equation.[vi] The generalized wave equation tells us how the state of any quantum object, such as an electron, will evolve over time. The purpose of this exercise is to show the reader that one of the most profound formulas in quantum physics, that describes the evolution of all waves (and matter is just another form of wave) over time and space, can be derived using simple high school mathematics.

[53]**The DeMoivre identity.** The DeMoivre identity, which is employed in the derivation of the time-independent Schrödinger's equation, can be thought of as an extension of Euler's formula.

---

vi Physicists employ a general form of the Schrödinger's equation, employing the Hamiltonian operator (upon which we shall not dwell), and it looks like this:

$$i\hbar\frac{\partial}{\partial t}|\psi(t)\rangle = \hat{H}|\psi(t)\rangle.$$

Here's DeMoivre's identity.

$$(\cos\theta + i\sin\theta)^n = \cos(n\theta) + i\sin(n\theta).$$

The proof follows easily from Euler's formula, which states,

$$e^{i\theta} = \cos\theta + i\sin\theta.$$

Let us raise both sides of Euler's formula to the nth power.

$$(e^{i\theta})^n = (\cos\theta + i\sin\theta)^n.$$

The left-hand side can be represented as,

$$(e^{i\theta})^n = e^{i(n\theta)}.$$

So, we can just substitute $n\theta$ as our angle in Euler's formula, to yield the DeMoivre identity.

$$e^{i(n\theta)} = \cos(n\theta) + i\sin(n\theta) = (\cos\theta + i\sin\theta)^n.$$

DeMoivre's identity can give us the nth roots of unity, also known as DeMoivre's numbers. Let us set $x = \frac{2\pi}{n}$, then,

$$\left(\cos\frac{2\pi}{n} + i\sin\frac{2\pi}{n}\right)^n = \cos 2\pi + i\sin 2\pi = 1.$$

The general rule (not developed here) for generating DeMoivre numbers is:

$$\xi_n^k = \cos\left(2\pi\frac{k}{n}\right) + i\sin\left(2\pi\frac{k}{n}\right),$$

while $0 \le k \le n$, where the greatest common divisor of k and n is 1.

The nth roots of unity form an irreducible representation of any cyclic group of order n. DeMoivre's simple and unassuming identity has significance in quantum physics, where spatial symmetries of particles have irreducible representations.

[54]**Wave parameters.** There are several simple equations that express the common measurements that apply to waves, and light waves in particular.

Wave speed = $\nu \times \lambda$, the wave frequency times the wavelength.

In the case of light, the wave speed is a constant, c, so $c = \nu \times \lambda$.

Alternately, the wavelength of light is $\lambda = \frac{c}{\nu}$.

The angular frequency of a wave, ω is $2\pi\nu$ or the number of oscillations that would occur in a unit circle per second.

The wave number, k is the number of waves that pass in a unit distance. By convention, k is expressed as the number of waves that would pass a point in a circle of unit radius. Accordingly, $k = \frac{2\pi}{\lambda}$ or $k = \frac{2\pi}{c} \times \nu$ or $k = \frac{2\pi}{c} \times \frac{\omega}{2\pi}$ or $k = \frac{\omega}{c}$.

The period of a wave is the time required for one wavelength to pass a point, or the reciprocal of frequency. Thus, the period of a light wave is $\frac{1}{\nu} = \frac{\lambda}{c}$.

In the case of light waves, the energy is directly related to frequency, wavelength, or period. $E = h\nu$, where h is Planck's constant. Energy in terms of wavelength is $E = h\frac{c}{\lambda}$.[vii]

[55]**Proving the convolution theorem.** The convolution theorem states that the Fourier transform of the convolution of two functions equals the product of the Fourier transforms of the functions.

In mathematical terms:

$$\mathcal{F}(f \star g) = \mathcal{F}(f)\mathcal{F}g = \hat{f} \cdot \hat{g}.$$

Here, $\mathcal{F}$ is the Fourier transform and $\hat{f}$ or $\hat{g}$ is a shorthand way of expressing the Fourier transform of f or g. As noted previously, the Fourier transform takes a function in the time domain and yields a function in the frequency domain. The inverse Fourier transform does the reverse. Using the equation above, we can take the inverse Fourier transform of $\hat{f}\hat{g}$, to yield the convolution of f and g.

Consequently,

$$\mathcal{F}^{-1}(\hat{f} \cdot \hat{g}) = f \star g.$$

The convolution theorem is often summarized by saying that convolution in one domain (time or frequency) is equivalent to multiplication in the other domain.

Before we begin our proof of the convolution theorem, let us review the general equations for a Fourier transform and its inverse, and the convolution of two functions.

The Fourier transform of a function:

$$\mathcal{F}(f(x)) = \int_{-\infty}^{\infty} f(x)e^{-i\omega x}dx = \hat{f}(\omega).$$

The inverse Fourier transform of the same function

$$\mathcal{F}^{-1}(\hat{f}(\omega)) = \frac{1}{2\pi}\int_{-\infty}^{\infty} \hat{f}(\omega)e^{i\omega x}d\omega = f(x).$$

vii Sharp-eyed readers may have noticed that the equation for the energy of light is dependent only upon the speed of light, and the frequency of light waves, but not on the amplitude of the light waves at a given frequency. The amplitude of light is determined by the number of quanta (i. e., photons) moving. For simplicity sake, the wave equations are derived for a single quantum.

The definition of a convolution,[viii]

$$f \star g = \int_{-\infty}^{\infty} f(x-\xi)g(\xi)d\xi.$$

The three listed equations are all that we need for our proof.

$$\begin{aligned}
\mathcal{F}^{-1}(\hat{f}\hat{g}) &= \frac{1}{2\pi}\int_{-\infty}^{\infty} \hat{f}(\omega)\hat{g}(\omega)e^{i\omega x}d\omega \\
&= \frac{1}{2\pi}\int_{-\infty}^{\infty} \hat{f}(\omega)\left(\int_{-\infty}^{\infty} g(y)e^{-i\omega y}dy\right)e^{i\omega x}d\omega \quad \text{Evaluate the Fourier transform of } \hat{g}(\omega) \\
&= \int_{-\infty}^{\infty} g(y) \times \frac{1}{2\pi}\int_{-\infty}^{\infty} \hat{f}(\omega)e^{i\omega(x-y)}d\omega dy. \qquad (11.1)
\end{aligned}$$

But $\frac{1}{2\pi}\int_{-\infty}^{\infty}\hat{f}(\omega)e^{i\omega(x-y)}d\omega$ is the inverse of the Fourier transform of $f(x-y)$, yielding back itself. So, we can substitute $f(x-y)$ back into our last equation above:

$$\begin{aligned}
&= \int_{-\infty}^{\infty} g(y)f(x-y)dy \\
&= (g \star f) \\
&= (f \star g) \quad \text{This completes our proof.}
\end{aligned}$$

The very last step in the proof assumes that convolution is commutative, which is easily provable.

[56]**The impulse response convolution theorem.** Waves have all the properties of linear systems, and one such property tells us that there are functions that, when applied to a linear system, yield a result that is also a linear system. Let us suppose that $\mathcal{H}$ is such a function. Hence, all of the properties of linear systems apply to the products of the function.

Linear system homogeneity accounts for the scaling rule:

$$\mathcal{H}[ax(t)] = a\mathcal{H}[x(t)].$$

Linear system additivity accounts for the additive rule:

$$\mathcal{H}[x_1(t) + x_2(t)] = \mathcal{H}[x_1(t)] + \mathcal{H}[x_2(t)].$$

---

viii We will not discuss convolution more than to define it mathematically.

Together, the scaling rule and the additive rule justify the superpositioning rule, which we will use in our proof:[ix]

$$\mathcal{H}[ax_1(t) + bx_2(t)] = a\mathcal{H}[x_1(t)] + b\mathcal{H}[x_2(t)].$$

We add that time-shift invariance is a special form of shift invariance and accounts for the following:

$$\mathcal{H}[x(t - t_0)] = y(t - t_0) \quad \text{Here, } y() \text{ is the result of } \mathcal{H}[x()]$$

There is one more equality to remember. From the sifting property of the impulse function (not derived here), we know that:

$$x(t) = \int_{-\infty}^{\infty} x(\tau)\delta(t - \tau)d\tau.$$

Here, δ is the Dirac impulse function.

Now that we have dispensed with the preliminaries, we can easily prove impulse response convolution theorem.

$$\begin{aligned}
y(t) &= \mathcal{H}[x(t)] \quad \text{Here, } y(t) \text{ is the } \mathcal{H} \text{ operation on } x(t)\\
&= \mathcal{H}\left[\int_{-\infty}^{\infty} x(\tau)\delta(t - \tau)d\tau\right] \quad \text{Substitute the sifting identity for } x(t)\\
&= \int_{-\infty}^{\infty} \mathcal{H}\left[x(\tau)\delta(t - \tau)d\tau\right] \quad \text{Linear additivity property}\\
&= \int_{-\infty}^{\infty} x(\tau)\mathcal{H}[\delta(t - \tau)]d\tau \quad \text{Scaling property}\\
&= \int_{-\infty}^{\infty} x(\tau)h(t - \tau)d\tau \quad \text{Here, } h(t) \text{ is } \mathcal{H}[\delta(t)]\\
&= (x \star h)(t) \quad \text{The convolution of } x \text{ with } \mathcal{H}[\delta].
\end{aligned}$$

Using just the fundamental properties of linear systems, and the sifting property, we have shown that to find a transformation of any component of the linear system (or any combination of components of the linear system), all we need to do is to convolve

ix Of the six properties of linear systems, the most important is "wave superpositioning"), insofar as any system for which superpositioning applies will always fulfill the minimal requirements for a linear system (i. e., additivity and homogeneity), and will nearly always fulfill all of the remaining properties of a linear system.

the component with the transformation of the Dirac delta function. This property, along with Fourier analysis (to which it is closely related) forms the core of all modern signal analysis. It is appears frequently whenever linear systems are analyzed, and this would include all waves and harmonic oscillators of the Standard Model.

[57] **The canonical commutation relation.** In the process of deriving the Heisenberg's uncertainty principle, we also derived another fundamental property of an operator for particular types of physical systems. The general equation for the canonical commutation operator is shown here.

$$\{\hat{x}, \hat{p_x}\} = \hat{x}(\hat{p_x}(\psi(x))) - \hat{p_x}(\hat{x}(\psi(x))).$$

For complex conjugates:

$$\langle f \mid g \rangle = \int_{-\infty}^{\infty} f^*(x) \cdot g(x)\, dx,$$

and

$$\langle g \mid f \rangle = \int_{-\infty}^{\infty} g^*(x) \cdot f(x)\, dx.$$

The asterisk in the equation denotes the complex conjugate. At this point, we can insert the wave equations and evaluate.

$$\langle f \mid g \rangle - \langle g \mid f \rangle = \int_{-\infty}^{\infty} \psi^*(x)\, x \cdot \left(-i\hbar \frac{d}{dx}\right) \psi(x)\, dx \\ - \int_{-\infty}^{\infty} \psi^*(x) \left(-i\hbar \frac{d}{dx}\right) \cdot x\, \psi(x)\, dx.$$

This reduces to:

$$\begin{aligned} \langle f \mid g \rangle - \langle g \mid f \rangle &= i\hbar \cdot \int_{-\infty}^{\infty} \psi^*(x)\psi(x)\, dx \\ &= i\hbar \cdot \int_{-\infty}^{\infty} |\psi(x)|^2\, dx \\ &= i\hbar. \end{aligned}$$

The value *ih* is a fixed value that describes a particular relationships between any two physical properties that happen to be the Fourier transforms of one another (i. e., Fourier conjugates). This relationship, as shown here for locations (x) and momentum (p) is known as the canonical commutator and shown:

$$\{\hat{x}, \hat{p}_x\} = i\hbar.$$

Incidentally, the (absolute) average of the possible outcomes of position and location, gives us $\hbar/2$, their Heisenberg uncertainty.

[58]**Derivation of Heisenberg's uncertainty principle.** The Heisenberg's uncertainty principle can be constructed using two clever mathematical theorems: the Cauchy–Schwarz inequality, and the canonical commutation relation. Heisenberg's uncertainty principle applies to any two conjugate variables, (in this case, variables that are the Fourier conjugates of each other), so the Uncertainty principle applies to either dual measurements for location and momentum or for energy and time. Let us consider position (x) and momentum (p).

The Cauchy–Schwarz inequality for complex conjugate variables yields the following formula for the standard deviations (σ) of x and y.

$$\sigma_x^2\sigma_p^2 = \langle f \mid f\rangle \cdot \langle g \mid g\rangle \geq |\langle f \mid g\rangle|^2.$$

When we let

$$z = \langle f|g\rangle$$

and

$$z^* = \langle g \mid f\rangle.$$

We can show that

$$|\langle f \mid g\rangle|^2 \geq \left(\frac{\langle f \mid g\rangle - \langle g \mid f\rangle}{2i}\right)^2.$$

For complex conjugates:

$$\langle f \mid g\rangle = \int_{-\infty}^{\infty} f^*(x) \cdot g(x)\, dx,$$

and

$$\langle g \mid f\rangle = \int_{-\infty}^{\infty} g^*(x) \cdot f(x)\, dx.$$

The asterisk in the equation denotes the complex conjugate.

At this point, we can insert the wave equations and evaluate.

$$\langle f \mid g\rangle - \langle g \mid f\rangle = \int_{-\infty}^{\infty} \psi^*(x)\, x \cdot \left(-i\hbar\frac{d}{dx}\right) \psi(x)\, dx$$

$$- \int_{-\infty}^{\infty} \psi^*(x) \left(-i\hbar \frac{d}{dx}\right) \cdot x\,\psi(x)\,dx.$$

This reduces to:

$$\begin{aligned} \langle f \mid g\rangle - \langle g \mid f\rangle &= i\hbar \cdot \int_{-\infty}^{\infty} \psi^*(x)\psi(x)\,dx \\ &= i\hbar \cdot \int_{-\infty}^{\infty} |\psi(x)|^2\,dx \\ &= i\hbar. \end{aligned}$$

Plugging this into the above inequalities, we get:

$$\sigma_x^2\sigma_p^2 \geq |\langle f \mid g\rangle|^2 \geq \left(\frac{\langle f \mid g\rangle - \langle g \mid f\rangle}{2i}\right)^2 = \left(\frac{i\hbar}{2i}\right)^2 = \frac{\hbar^2}{4}$$

or taking the square root

$$\sigma_x\sigma_p \geq \frac{\hbar}{2}.$$

This is the general Heisenberg's uncertainty principle and can be applied to any pair of wave conjugates.

[59]**Inverse square law.** The inverse square law, dictating that the force emanating from a point source diminishes as the inverse square of the distance from the source, is a consequence of two conditions:
- That force moving through empty space is conserved. If you have a force (e. g., light, or gravity) that emanates from a point (the so-called point source) then the force generated will be conserved throughout space.
- The spread of force from a point source is spacetime invariant, so that the force is distributed (diluted) equally through all points extending from the source.

Let us imagine a point source. The force emanating through empty space from the point source can be envisioned as an enlarging sphere, the sphere representing the total symmetry of the expanding force arriving with equal strength at all points. Due to conservation of force, the total force moving through a spherical shell at radius "r" will the equal to the total force moving through a spherical shell at radius "$2r$" and at radius "$4r$" and so on. If the force arrives at different times (as would the force associated with a spreading point light source), it makes no difference, because the emanation is spacetime invariant.

The area of a 3-dimensional sphere is $4\pi r^2$. This means that the total force moving through a sphere of radius r will be equally distributed over the area $4\pi r^2$. As the force expands outward, the same total force moving through a sphere of radius 2r will be distributed over an area of $4\pi(2r)^2$, or 4 times the total area of the smaller radius. This means that the passing force measured per unit surface area is reduced by 1/4, when we double the distance from the original point source (i. e., as an inverse square relationship).

Generalizing, the expanding force reduces as the inverse square of the distance from the emanating point source, in 3-dimensional space. We could extend the same line of reasoning to any n-dimensional space, as shown:

$$I \propto \frac{1}{r^{n-1}}.$$

[60]**Inertial frame of reference.** Any reference frame that is not accelerating. This would include any reference frame moving at a constant velocity from zero up to (but not including) the speed of light in a vacuum. Any inertial reference frame can be matched with another reference frame moving at the same speed, so that each reference frame appears stationary to the other.

The principles underlying special relativity are:

- The laws of physics are the same for all inertial frames of reference
- The velocity of light is the same for all observers in inertial frames of reference

In this book, whenever we mention "reference frames", we will be referring to initial frames of reference.

[61]**Proving that all massless particles move at exactly the speed of light.** We can show, in separate proofs that massless particles cannot move more slowly than light speed; and that massless particles cannot move faster than light speed. Therefore, massless particles must move constantly at the speed of light. Let us begin by proving that massless particles cannot move more slowly than light.

The currently recognized massless particles are photons and gluons.[x] If we wanted to be cute, we could argue that photons must always move at the speed of light (and not slower than the speed of light) because photons are light, and light always moves at the speed of light. We can construct a more general proof to cover all massless particles (known or unknown), as follows:

1. Imagine that we had a massless particle that moves slower than the speed of light.
2. If the particle is moving slower than the speed of light, then there is a frame of reference in which the particle is motionless
   This is equivalent to saying that an observer could match its speed, at which point the observed velocity of the particle would be zero.

---

x In addition, one type of neutrino may be massless, and the as yet unconfirmed graviton particle may also be massless.

3. Furthermore, all frames of reference are equally valid.
   That is to say that measurements of the massless particle are valid within its zero-velocity reference frame.
4. In the zero-velocity reference frame, the massless particle would have no energy.
   Energy is what makes particles move. If the velocity of the particle is zero, and there is no movement, and there is no mass, then there is no energy.
5. A particle with no size, no mass, no movement, and no energy is a non-existent particle.[xi]
   Therefore,
6. Our initial premise is impossible, and there can be no frame of reference that a moving observer can match.
7. The speed that cannot be matched by a moving observer is the speed of light.
   Therefore,
8. All massless particles move at the speed of light (at the very least).

Now, we will prove that massless particles cannot move faster than the speed of light.

- Argument 1. As a particle moves faster and faster, its clock slows down (as per special relativity). When it reaches the speed of light, time stops completely and the particle exists in the so-called "eternal now".[xii] It is meaningless to imagine a particle moving at a speed greater than the speed of light because once time has stopped, there is no way in which time can be further stopped, and greater-than-light speed becomes meaningless.
- Argument 2. If a massless particle were to exceed the speed of light, then it would become unobservable (by outpacing the light by which they could be observed). An unobservable object, with no mass and no size (fundamental particles have no size) cannot arguably exist (there is nothing there, really). Therefore, if a particle were to move faster than the speed of light, it would exit existence.[xiii]
- Argument 3. Objects moving faster than the speed of light would violate the dictum that observations are valid in all reference frames. Observations would not be repeatable in different reference frames because the light that captures an event could never reach the observer traveling at greater than light speed.

xi This calls to mind one of my favorite limericks: "As I was sitting in my chair, I saw the bottome was not there; nor legs nor back; but I just sat; ignoring little things like that."

xii The term "eternal now" seems to have made its first appearance in Karl Pearson's *The Grammar of Science*, published in 1900 by Adam and Black. Pearson's book was studied by Albert Einstein, and it is likely that the notion of the "eternal now" played some role in Einstein's development of the theory of special relativity.

xiii This suggests yet another reason why massive objects cannot exceed the speed of light. A massive object is composed, in part, by massless fundamental particles (quarks, gluons, and photons). An aggregate object cannot move faster than the fastest moving particle of which it is composed.

If massless particles cannot move slower or faster than light, then they must move at the speed of light.

[62]Derivation of the time dilation formula of special relativity.

> You can observe a lot by watching.
>
> – Yogi Berra

All of special relativity stems from two fundamental conditions of spacetime:
- The laws of physics are the same in all inertial frames of reference
- The velocity of light is the same for all inertial observers

The second condition was shown to be true experimentally by Michelson and Morley, in 1886 (for discussion, see Section 7.10, "Fundamental constants save the ether."). But anyone could have inferred the second condition from the first, with a little thought.
1. If the laws of physics are the same in all inertial frames of reference.
2. The observables must be the same in all inertial frames of reference.
3. What we observe is dependent upon the speed of light.
4. Therefore, the speed of light must be the same for all inertial observers.

So, knowing only that the laws of physics are the same in all inertial frames of reference, scientists had all they needed to develop the principles of special relativity.[xiv] This being the case, it is something of a mystery as to why the world needed to wait until 1905 for Albert Einstein to enlighten us.

In point of fact, there are several ways in which the special relativity could have been derived, without resorting to modern advances in physics and mathematics. One such way depends on nothing more than an understanding of the Doppler effect, discovered in 1842 by Christian Doppler. Here, we shall derive the time dilation formula of special relativity, using Christian Doppler's mid-nineteenth century observation.[xv]

We will begin with a light source is moving at a constant velocity toward one observer and away from a second observer.

---

xiv Note on terminology. What we call special relativity is a misnomer, as it suggests a brand of relativity that applies only in special circumstances; while classical relativity (the relativity of Galileo Galilei and of Isaac Newton) would apply under normal circumstances. Of course, special relativity applies in all circumstances, and classical relativity applies only when dealing with objects in slow motion; and then only approximately. When we think about it, everyday motions of atoms, electrons, planets, stars, and galaxies all occur at so-called relativistic speeds. Clearly, the relativity of Newton and Galileo is special, and the relativity of Einstein is general. Unfortunately, the "general" term, as applied to relativity, is reserved for gravity, so we are stuck with a less than accurate terminology.

xv Marco Moriconi derived the special relativity formulas for time dilation, length contraction, addition of velocities, and the mass-energy relation using nothing more than his understanding of the Doppler effect. In this section, we generally follow Moriconi's work, "Special Theory of Relativity through the Doppler Effect" in: arXiv:physics/06052040v1, May 24, 2006.

Let us look at our variables:

- Observer Toward = the observer to which the light is approaching.
- Observer From = the observer from which the light is receding.
- v is the velocity of the light source.
- ω is the frequency of light.
- ω_0 is the frequency of the emitted light when stationary.
- ω_v is the frequency of the emitted light moving at velocity, v.
- λ is the wavelength of light.

We will soon be needing to evaluate the ratio of ω_v/ω_0 and although we do not know what this equals, we do know that the ratio will be a dimensionless number. Therefore, we can infer that the ratio can be represented by a function of v/c, a dimensionless number computed from the velocity of the emitter and the velocity of light. That is to say: $\omega_v = \omega_0 \times g(\beta)$, where $\beta = v/c$ and $g(\beta)$ is some function of β.

Δt is the time between two pulses from the light source. Note that since the observers are receiving two pulses of light, the time between pulse 1 and pulse 1 represents the time for light to travel one the distance of between the two pulses, which the observers would perceive as one wavelength.

As perceived by Observer Toward, the Doppler effect compresses the wavelength, which in this case is the distance measured between the first pulse and the second pulse. The wavelength measured by Observer Toward is:

$$c\Delta t - v\Delta t = (c - v)\Delta t = \lambda_{\text{Observer Toward}}.$$

The wavelength perceived by Observer From will be broadened as the light source recedes, as:

$$c\Delta t + v\Delta t = (c + v)\Delta t = \lambda_{\text{Observer From}}.$$

f_v is the frequency of light observed by Observer Toward. f_{-v} is the frequency of light observed by Observer From

The general formula for the speed of any wave is:

wavelength times the frequency of the wave = wave speed.

Now that we have assembled all of our variables, plus the general formula for the speed of a wave, plus our knowledge that both observers must find that the measured speed of light must equal the constant, c, we can compare the findings of the two observers (Observer Toward and Observer From).

Observer Toward finds that the speed of light is:

$$\lambda_{\text{Observer Toward}} \times \omega_v = ((c - v)\Delta t)\omega_v = c.$$

In the case of Observer From, the frequency of the light is the frequency occurring at $-v$ because the wave is moving away from the observer. Observer From finds that the speed of light is:

$$\lambda_{\text{Observer Toward}} \times \omega_{-v} = ((c + v)\Delta t)\omega_{-v} = c.$$

Dividing the first equation by the second and making our substitutions, (i. e., $g(\beta) = \omega_v/\omega_0$ and $\beta = v/c$, we find:

$$\frac{g(\beta)}{g(-\beta)} = \frac{1+\beta}{1-\beta}.$$

The next part of the derivation is a bit tricky, requiring us to imagine an observer moving toward the emitter at speed v (this being equivalent to the emitter moving toward the observer at speed v).

The equation for the frequency of light observed by the observer at rest is:

$$\omega_v = \omega_0 \times g(\beta).$$

The emitter "looking" at the observer at rest will receive light at a frequency of $\omega_v \times g(-\beta)$ or $\omega_0 \times g(\beta) \times g(-\beta)$.

But both the observer and the emitter are in the same rest frame and will obviously both observe light at the same frequency, v_0.

So,

$$\omega_0 = \omega_0 \times g(\beta) \times g(-\beta).$$

Therefore,

$$1 = g(\beta) \times g(-\beta).$$

Returning to our prior equation:

$$\frac{g(\beta)}{g(-\beta)} \times (g(\beta) \times g(-\beta)) = \frac{1+\beta}{1-\beta}$$

or,

$$g(\beta)^2 = \frac{1+\beta}{1-\beta}$$

and,

$$g(\beta) = \left(\frac{1+\beta}{1-\beta}\right)^{1/2}.$$

Plugging our derived value for $g(\beta)$ back into our relationship between doppler frequency and emitter frequency, we have the formula for the relativistic Doppler effect.

$$\omega_v = \omega_0 \times g(\beta) = \omega_0 \times \left(\frac{1+\beta}{1-\beta}\right)^{1/2}.$$

Now, we can proceed to derive the relativistic time dilation formula. Let us go back to our original gedanken experiment, in which a pulse of light is emitted by a light source moving toward an observer (Observer Toward) at speed v. Remember, between these two pulses, one wave is emitted and received by the observer. The formula for the speed of light, as perceived by Observer Toward, must equal the constant speed of light (our second postulate).

$$\lambda_{\text{Observer Toward}} \times \omega_v = ((c-v)\Delta t)\omega_0\left(\frac{1+\beta}{1-\beta}\right)^{1/2} = c.$$

Using the same formulaic approach, we can compute the perceived speed of light from the light-source frame of reference, in which $v = 0$, and $\beta = v/c = 0$ and $\Delta\tau$ is the perceived time interval between pulses.

$$c\Delta\tau\omega_0 = c.$$

So,

$$(c-v)\Delta t\omega_0\left(\frac{1+\beta}{1-\beta}\right)^{1/2} = c\Delta\tau\omega_0.$$

Or,

$$\Delta t = \frac{\Delta\tau}{(1-\beta^2)^{1/2}}.$$

Substituting back $\beta = v/c$, we obtain the relativistic time dilation formula.

$$\Delta t = \frac{\Delta\tau}{\sqrt{1-v^2/c^2}}.$$

[63]**Fine structure constant.** Also known as Sommerfeld's constant. Early in the 20th century, the fine structure of the hydrogen atom spectrum was observed to produce two narrowly separated lines, where only one line was predicted by the Bohr theory of the atom. The fine structural separation of spectral lines was quantified by Sommerfeld as a dimensionless value, 1/137, and was named the fine structure constant (1916). Since then, numerous interpretations for the physical meaning of the fine structure constant have been established, and values for the constant have been more precisely established from

experimental measurements and from predictions based on quantum electrodynamics theory.[xvi]

The current value for the fine structure constant is 7.297352566417 × 10^{-3} or 1/137.035999206. It can be expressed in the following relationship:

$$\alpha = \frac{1}{4\pi\varepsilon_0}\frac{e^2}{\hbar c} \approx \frac{1}{137},$$

where, α is the fine structure constant; ε_0 is the electric constant, e is the elementary charge, $\hbar$ is the reduced Planck constant, and c is the speed of light in a vacuum.

[64]**Transformations.** A transform is a mathematical operation that takes a function or a time series (e. g., values obtained at intervals of time) and transforms it into something else. An inverse transform takes the transform function and produces the original function. Transforms are useful when there are operations that can be more easily performed on the transformed function than on the original function.

Some examples might be helpful. In 1614, John Napier (1550–1617), a Scottish mathematician, published tables of logarithms and antilogarithms. The primary purpose of these tables was to facilitate multiplication, division and exponentiation. Napier had transformed multiplication with numbers into addition with the logarithms of those numbers, as shown:

If

$$a = b \times c,$$

then

$$\log(a) = \log(b \times c) = \log(b) + \log(c)$$

and

$$\text{antilog}(\log(a)) = \text{antilog}(\log(b) + \log(c))$$

or

$$a = \text{antilog}((\log(b)) + (\log(c)).$$

---

xvi The cosmic significance of the fine structure constant has been a topic of great speculation, at times exceeding the bounds of orthodox analysis. The following story has been told by Max Dresden. "The great Arthur Eddington gave a lecture about his alleged derivation of the fine structure constant from fundamental theory. Goudsmit and Kramers were both in the audience. Goudsmit understood little, but recognised it as far-fetched nonsense. After the discussion, Goudsmit went to his friend and mentor Kramers and asked him, 'Do all physicists go off on crazy tangents when they grow old? I am afraid.' Dramers answered, 'No Sam, you don't have to be scared. A genius like Eddington may perhaps go nuts, but a fellow like you just gets dumber and dumber."'

Simply put, we can multiply together any collection of numbers by transforming them to their logs, and replacing multiplication with summation; then determining the antilog (the inverse transform) of the summed logs. Hence, logarithms transform multiplication into summation. Division works much the same way, with the logarithms of the divisors subtracted, not added.

The most useful transform in mathematical analysis is the Fourier transform, which represents signals from the time domain as a sum of frequency components (i. e., converts from the time domain to the frequency domain). The Fourier transform is easily computed with great speed on modern computers, using a modified form known as the fast Fourier Transform. Periodic functions (i. e., functions with repeating trends in the data, including waveforms and periodic time series data) can be represented as the sum of oscillating functions (i. e., functions involving sines, cosines, or complex exponentials). The summation function is the Fourier series. Often, just a the first few components of the Fourier series serve as a suitable approximation of the original signal (Figure 11.3). Operations on the transformed function can sometimes eliminate repeating artifacts or frequencies that occur below a selected threshold (e. g., noise). The transform can be used to find similarities between two signals. When the operations on the transform function are complete, the inverse of the transform can be calculated and substituted for the original set of data.

Forward and inverse Fourier transformations:

$$\hat{f}(\xi) = \int_{-\infty}^{\infty} f(x) e^{-2\pi i x \xi}\, dx,$$

$$f(x) = \int_{-\infty}^{\infty} \hat{f}(\xi)\, e^{2\pi i x \xi}\, d\xi.$$

The figure below demonstrates how a function may be approximated by just a few components of a Fourier series.

[65]**Group.** A group is a set having the properties of closure and associativity under an operation; having an identity element; and having an inverse element for every member of the set.

Specifically, for some operation (which we will call "x" in this example, "closure" tells us that for all a, b in G, the result of the operation, $a \times b$, is also in G.

Associativity tell us that for all a, b and c in G, $(a \times b) \times c = a \times (b \times c)$.

To have an identity element, there must exist an element e in G such that, for every element a in G, the equation $e \times a = a \times e = a$ holds. The identity element is unique within a group, and we speak of "the" identity element.

To have an inverse element, there must exist for each a in G, an element b in G, such that $a \times b = b \times a = e$, where e is the identity element.

A simple group is a group G that does not have any normal subgroups except for the trivial group and G itself.

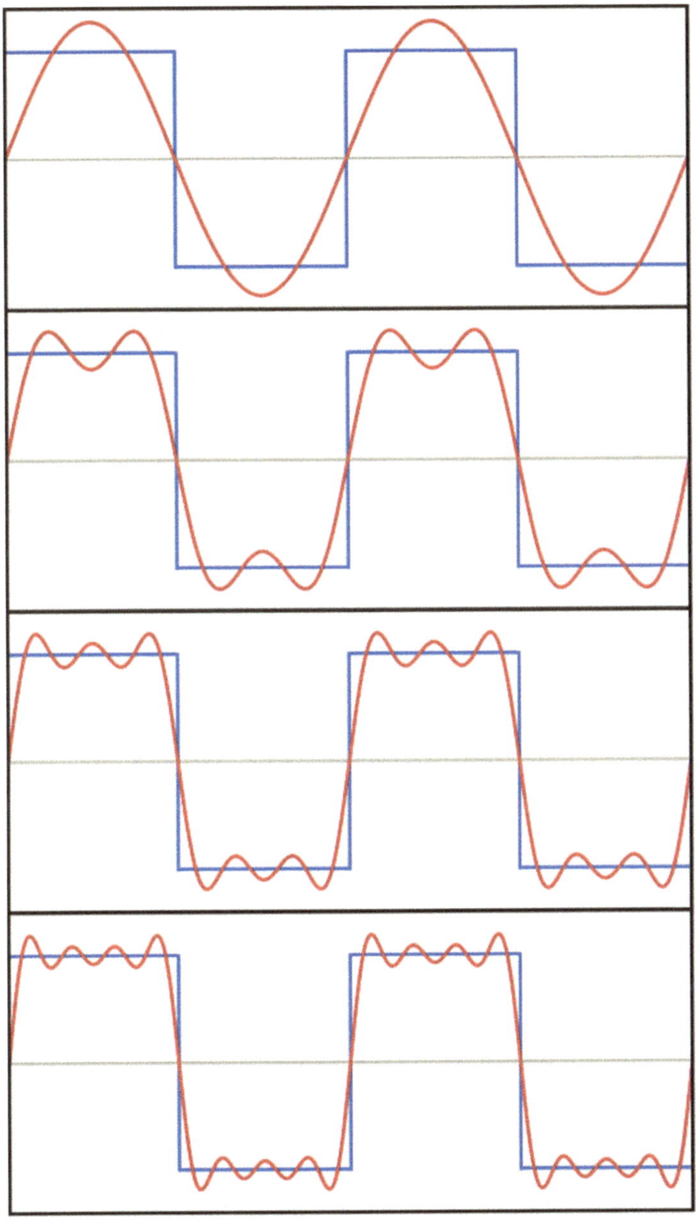

Figure 11.3: A square wave is approximated by a single sine wave, the sum of two sine waves, three sine waves, and so on. As more components are added, the representation of the original signal or periodic set of data, is more closely approximated. Source, Wikipedia, from an image intered into the public domain by its author, Jim.belk.

[66]**Point group.** The point group is formed from all of the symmetry transformations that can be performed on a geometric shape, about a fixed point.

Point groups play a role in the symmetry groups of crystals. There are 32 different point groups in crystals and 14 Bravais lattices [40]. Together they account for all 230 theoretical (and empirically observed) space groups of crystals.

[67]**Crystal classes.** All crystals can be assigned to one of 32 classes, with each class consisting of a point symmetry group. To appreciate the meaning of a point symmetry group, we need to know a few things about crystals. Every crystal has an irreducible structure, the so-called unit cell of the crystal, that repeats itself to form the visible crystal. There are 7 unit cell structures, and these are known as Bravais lattices [47]. Hence a crystal that can be held in the hand has the same structure as a crystal that is viewed under the microscope. The unit cell lattices of different crystals can all be transformed by any combination of symmetry operations (i. e. transformation that leave the structure unaffected; and these operations for unit cells include reflection through particular planes, or inversion with respect to the center, or rotations around an axes through the center). The angles of rotational symmetry transformations are limited to $\frac{2\pi}{n}$, where n must be either 2,3,4, or 6. Point groups are the various combinations of symmetry transformations that can be performed around a fixed point on a unit cell. As mentioned, there are 32 crystal point groups. Using the point symmetry groups and the translational symmetries, we find a total of 230 space groups for crystals. Furthermore, all of the crystal symmetry groups can be represented by matrix algebras, greatly facilitating their mathematical analysis. This information laid the groundwork for x-ray crystallographic analysis, a field that did not yet exist when the classification of crystals was being finalized [79].

[68]**Equivalence of fundamental particles and wave packets.**

There are no facts, only interpretations.

– Friedrich Wilhelm Nietzsche (1844–1900)

On a simplistic level, we can prove the equivalence of fundamental particles and wave packets by specifying the definitions of each, and recognizing that the definitions are equivalent.

- Let us define what we mean by a “wave packet”. A wave packet is a wave composed of a collection of superpositioned waves whose frequencies cluster around a single value. For example, the wave packet known as an electron is a wave whose composite wave frequencies are tightly clustered around a value that is the characteristic frequency of the electron. The electron wave-packet behaves in a manner that is characteristic of its role as the quantum of the electromagnetic field.
- A physical particle is a set of field properties at a point in space. As such, it has no size, and its energy is determined by the mechanics of its motion. Because fundamental particles have no size, we cannot “see” an individual particle. We can only detect the presence of a particle due to its interaction with physical systems (i. e.,

sensors). Our functional definition of a particle is that it is whatever is detected by a sensor tuned to its energy.
- A physical detector cannot distinguish a wave packet (characteristic energy at a location) from a particle (characteristic energy at a location).
- There being no physical way to distinguish a wave packet from a particle, then the two must be equivalent.

Of course, our "proof" does not tell us very much about the nature of wave packets or of particles. We really need to step back and ask ourselves what it means to be a particle composed as a wave packet.

A single wave has a pure sinusoidal pattern (Figure 11.4). What happens when we superposition multiple waves? As we add sinusoidals together, we get a combined wave envelope. Sound engineers refer to the combined wave envelope as the "beat". When many combined waves, all with nearly the same frequency, are combined, they yield a single prominent moving peak, surrounded by small ripples at all other locations. The attenuation of the combined wave outside of its sharp peak is due to the inability of many waves to stay in phase outside of the point where the peak occurs. Once out of phase, they interfere with one another, canceling there combined amplitude. Beyond the peak, the likelihood that the component waves will get back into the same phase again (forming additional peaks) drops off to something very small (Figure 11.5). The single moving peak is known as the wave packet, and the speed of the moving wave packet may be less than the speed of its component waves.

Wave packets can be detected by instruments designed to register point-like pulses of energy, so-called particle detectors. We cannot see fundamental particles (they have

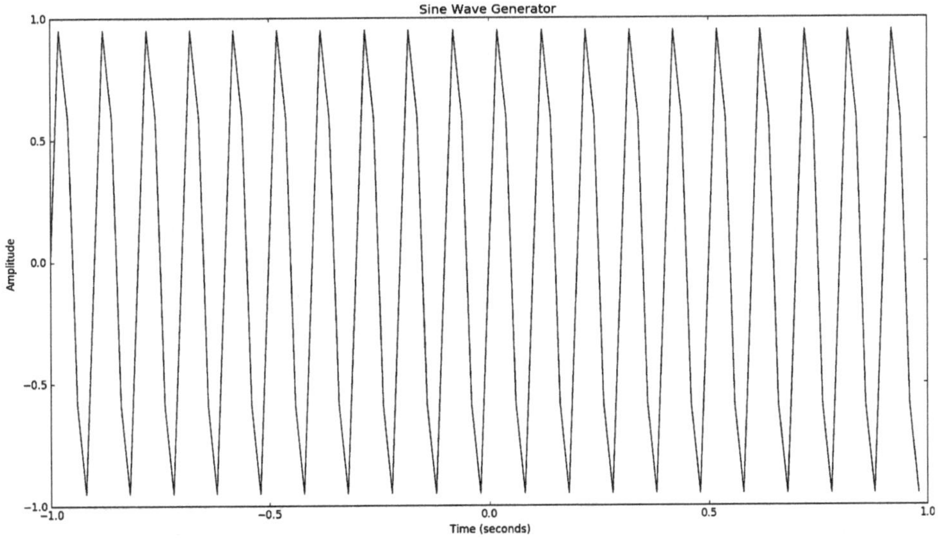

Figure 11.4: A sine wave, oscillating at 10 cycles per second with a peak amplitude of 1.0. Source, author.

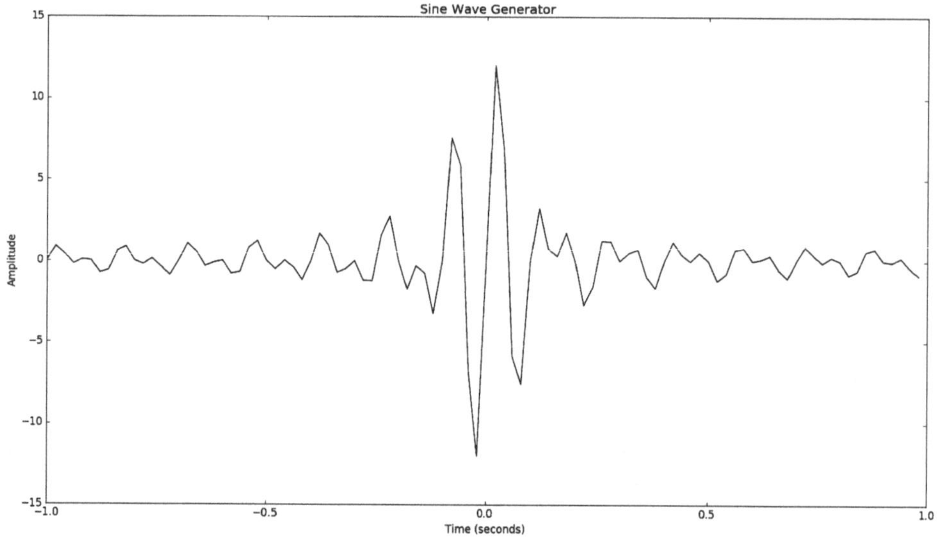

Figure 11.5: Something akin to a wave packet formed from adding together 14 sinusoidal waves to the wave shown in the figure above, all with nearly the same frequency and amplitude. There is one central peak formed (i. e., the wave packet), with only small waves everywhere else. Source, author.

no size), so we depend upon our particle detectors to tell us when and where a particle has hit the sensor grid. The wave packet appears to us as a detected particle (i. e., a quantal object with the energy to register on the sensor, at a determined location). Hence, as detected by the sensor, there is no observable distinction between a wave packet and a particle.

[69]**Emergent properties.** The emergent properties arise as the physical consequences of initial conditions interacting. textAs such, they are dependent upon pre-existing properties and are not strictly fundamental. The familiar emergent properties occur when multiple objects resulting from initial conditions interact with one another to produce secondary phenomena. For example, a single particle does not have a temperature, When we have trillions of particles all vibrating, then temperature emerges as an observable property. Likewise, the familiar forms of matter arise from emergent properties. We know that H_2O can exist as a liquid (water), solid (ice) or gas (steam). If there were only one molecule of water in the universe, then none of these forms would apply. They are all emergent features of a system that permits physical interactions among numerous molecules of H_2O.

Concepts such as "wave" and "solid particles" are emergent properties drawn from everyday perceptions. The duality of light as "wave" and "particle" may be an artifact of our crude way of seeing only emergent properties, when a single and fundamental property sets the conditions upon which the dual nature of light emerges. Likewise, we are at risk of dividing quantum objects into different types when they are one and the same thing on an individual level and only become different things when observed as emergent properties. When we generalize further, and we allow ourselves to imagine that all the properties that we observe to be "fundamental" (e. g., forces and fields) emerged from higher, more fundamental properties, we can begin to see why grand unification theories are worth pursuing. Indeed, theorists have suggested models of reality from which various emergent properties arise that were otherwise conceived as fundamental. For example, physicists view gravity as one of the four fundamental forces of the universe (along with the weak force, the strong force, and electromagnetic force). Tejinder P. Singh has proposed that gravity is an emergent property, "precipitated by the spontaneous localisation of highly entangled fermions" [221].

If we pushed the point, we might imagine that "time" is an emergent property of a universe that contains matter (i. e., fermions moving at less than the speed of light). We can define time as a measurable duration (e. g., in ticks of a clock) between fermionic interactions, establishing an ordered sequence. If there were no fermions in the universe (or only a single fermion), then there would be no interactions, no observable time intervals, and thus no "time". Why did we limit our definition of time to interactions among fermions? The non-fermions (i. e., photons and other bosons) travel at the speed of light, and are effectively timeless; existing in the so-called "eternal now".

[70]**Tempo and mode.** In the context of this book, "tempo and mode" refer to two features of evolution. Tempo is the rate of evolution at any give epoch. Paleontologists have noticed periods of acceleration or deceleration in the rate of evolution of particular classes of species at particular periods in the history of evolution. Mode is the method by which evolutionary changes in classes or species of organisms are acquired [220]. Recently, the same two terms, "tempo and mode" have been applied to molecular genetics, to address the rate of change in genes over time (i. e., the tempo of the molecular clock) and the mechanisms that drive those changes [91].

[71]**HOX gene diseases.** At least ten different HOX genes have been found to cause rare diseases in humans, and each affected gene may be associated with several different syndromes. In general, the reported cases of HOX gene diseases are syndromes of structural anomalies. For example, the Bosley-Hand-foot-genital syndrome, associated with a mutation in HOXA13, is characterized by both limb and urogenital malformations. Human HOX diseases have very similar phenotypes to the mouse diseases resulting from mutations in homologous genes [206]. This would suggest that HOX genes have very similar functions throughout the animal kingdom. This view is strengthened by the startling finding that Pax6 (a HOX-like gene that controls eye development) in insects can be swapped for its inserted human homolog, and still produce functioning eyes [84]. Furthermore, because the insect eye is structurally different from the human eye, this observation indicates that the genes that command that eye development must begin are different from the genes that implement the command.

[72]**Homeobox.** Genes that code for transcription factors involved in anatomic development in animals, fungi, and plants. Specialized homeobox (abbreviated as HOX) genes found in metazoans determine the axial relationship of organs. Mutations of homeobox genes are associated with remarkably specific, often isolated, anatomic alterations.

Examples are:
- MSX2 homeobox gene mutation, which produces enlarged parietal foramina.
- PITX1 homeobox gene mutation, which produces Rieger syndrome (hypodontia and malformation of the anterior chamber of the eye including microcornea and muscular dystrophy).
- PITX3 homeobox gene mutation, which produces anterior segment dysgenesis of the eye, moderate cataracts, and anterior segment mesenchymal dysgenesis.
- NKX2.5 homeobox gene, which produces atrial septal defect and atrioventricular conduction defects.
- SHOX homeobox (short stature homeobox) gene mutation which is the root cause of Leri–Weill dyschondrosteosis (deformity of distal radius, ulna, and proximal carpal bones as well as mesomelic dwarfism).

The reason why homeobox mutations tend to produce malformations in isolated anatomic locations or involve some specific function, probably results from the restricted regulatory activity of the individual homeobox genes. For example, one gene might regulate the synthesis of a group of proteins exclusively involved in growth of particular skull bones; another homeobox gene might regulate proteins involved in insulin production.

[73]**Syndrome.** A syndrome is a constellation of tissue findings, producing more than one type of pathological lesion, that typically occur at about the same time.

The common cold is a disease, with a specific cause, that produces a pattern of headache, sniffles, cough, and malaise. In the case of the common cold, all these symptoms arise at about the same time after a viral infection, and all the symptoms clear at about the same time. The common cold is not generally considered a syndrome because a single pathogenesis characterizes the disease, in all of the affected tissues.

Diverse expressions of a physical disorder, occurring in multiple body systems are more likely to be syndromes than diseases, when the different conditions are inconsistently present or separated by intervals of time, or have different pathogenetic mechanisms in the different affected tissues. or affect different cell types in the same affected tissue.

There are no hard or fast rules, but the distinction between a disease and a syndrome has therapeutic and diagnostic consequences. A true disease has one pathogenesis, despite its multi-organ manifestations and can be potentially prevented, diagnosed, or treated by targeting events and pathways involved in the development of the disease. A syndrome is a confluence of clinical findings in multiple tissues, that typically have a root cause in common, but for which the pathogenesis diverges in different tissues.

[74]**Valence.** The number of electrons that an atom uses when bonding with another atom. In general, the valence can be determined by examining the configuration of the outer shell of electrons. If the outer shell is full, then the atom will not bond with other atoms. Predictions of bonding are based upon the ability of two atoms to complete their

respective outer shells by sharing electrons. In doing so, one electron may seem to occupy the outer shell of two atoms, at once. For example, an atom with one electron in its outer shell might form a bond with an atom that needs one electron to complete its outer shell. The shared electron completes the outer shell of the first atom by seemingly moving away from the atom, thus emptying its shell , and moving towards the second atom, where it seemingly completes the second atom's outer shell. Because the two atoms stay close to one another (i. e., in a chemical bond), sharing an electron that resides in a zone overlapped by both atoms is credible.

[75]**Gedanken.** Gedanken is the German word for "thought." A gedanken experiment is one in which the scientist imagines a situation and its outcome, without resorting to any physical construction of a scientific trial. Albert Einstein, a consummate theoretician, was fond of inventing imaginary scenarios, and his use of the term "gedanken trials" has done much to popularize the concept. The scientific literature contains many descriptions of gedanken trials that have led to fundamental breakthroughs in our understanding of the natural world and of the universe [27].

[76]**Non-Mendelian inherited genetic disease.** Refers to genetic diseases that do not exhibit the patterns of inheritance described by Gregor Mendel. Specifically, diseases with an autosomal dominant inheritance pattern would be expected in half the offspring of a couple in which one parent was affected. A recessive inheritance pattern would be expected in one quarter of offspring when each parent carries one disease-causing recessive gene. Deviations from either of these patterns would suggest that the inheritance pattern is non-Mendelian.

Polygenic diseases (i. e., in which altered variants must occur in multiple genes for the expression of disease) are never Mendelian. In addition, monogenic diseases that might be expected to display a Mendelian inheritance pattern may appear to be non-Mendelian due to a variety of confounding factors (e. g., allelic heterogeneity, epistasis, expressivity, genetic heterogeneity, incomplete dominance, lethal genotypes, penetrance, phenocopies,pleiotropy). As a general rule, Mendelian inheritance is monogenic; Non-Mendelian inheritance is polygenic; de novo diseases are not inherited (i. e., have neither Mendelian inheritance nor non-Mendelian inheritance). Genetic diseases that have Mendelian patterns of inheritance tend to be less common than diseases with polygenic inheritance.[xvii] [xviii]

[77]**Multifactorial diseases.** These are diseases that arise as the result of multiple factors, which may include both environmental conditions as well as genetic variations. Consider a patient with no known underlying medical condition who is stung by a bee and immediately succumbs to anaphylactic shock. It is tempting to say that the root

xvii Polygenic diseases such as diabetes, obesity, and hypertension, have a high occurrence rate because the sets of genes that cause these diseases are common variants within the general population.

xviii De novo genetic diseases arise from an alteration in a gene of the affected person's zygote or the earliest embryonic cells.

cause is the bee sting, but we know that most individuals who are stung by a bee do not develop an anaphylactic response. Clearly, some underlying condition must have predisposed the patient to develop shock. You want to blame a gene passed to the child from a parent, but if there were no parental or familial history of anaphylaxis, then it would be hard to put the blame on one inherited gene. It is possible that a de novo mutation (a mutation originating in the patient's zygote), but such mutations are rare, while anaphylactic responses are not particularly uncommon. In such instances, we look toward a polygenic explanation, wherein multiple gene variants together produce a physiological condition that predisposes the individual to anaphylactic shock. Of course, we cannot be certain that we are correct until we identify all of the modified genes and demonstrate the biological mechanism by which they exert their effect. That's a very tall order. In the meantime, we work under the tentative assumption that we are correct. That is science.

[78]**Non-quantitative trait.** Non-quantitative traits can be described as being present or absent, without quantitative measures, that occur in individuals of a species, but which are not defining properties of the species. An example would be the trait of having blue eyes.

Unlike the quantitative traits, that always seem to be polygenic, with a non-Mendelian inheritance pattern, the non-quantitative traits can have a monogenic origin. An example is the horse's tölt. Some Icelandic horses have a natural ability to use a specialized gait known as the tölt. This gait is common in a certain breed of Icelandic horse and rare amongst all other horse breeds. Most horses, no matter how hard they try, cannot tölt. Recently, the tölt was found to come from a single gene mutation [107].

It should be noted that the term "trait" has a second meaning when it is used by geneticists to indicate that an individual carries a recessive disease mutation on a single allele. For example, sickle cell disease occurs when an individual carries two recessive alleles of the characteristic hemoglobin mutation. Individuals with only one allele affected, who may develop some tendency toward red blood cell sickling, but who do not develop the complete sickle cell disease phenotype, are said to carry the sickle cell trait.

[79]**Gravitational radiation.** Because gravity is a curvature in space, it is not entirely clear that gravity can radiate energy. For some number of years, Einstein himself was convinced that gravitational energy did not exist. The idea that a gravity field did not radiate was difficult for physicists to accept, inasmuch as all other types of physical fields were known to radiate energy and it seemed difficult to imagine any field in which radiation did not fit into the scheme of things. To the general relief of physicists, a solution was offered by Richard Feynman (in 1957) in the form of his "sticky bead" detector. Imagine a rod piercing two beads that could slide freely over the rod, producing some friction along the way. If the rod were placed in the path of a gravitational wave, the length of the rod would not change (we're supposing that the rod is rigid) but the distance between the two beads will oscillate (as predicted by general relativity), causing the beads to rub against the rod, producing heat (i. e., the agitation of molecules that

requires energy). Therefore, gravity radiates energy, and the "sticky bead" experiment detects gravitational waves.[xix]

[80]**Proving the net energy in the universe is zero.** Let us use the principle of energy conservation to prove that the net amount of energy in the universe is zero and that negative energy must exist.

1. Energy is conserved over time.
2. All of the observable energy in the universe today was present at the moment of the Big Bang.
   In fact, most of the electromagnetic energy observed in the universe today is found in the cosmic background radiation released sometime after the Big Bang, and most of the atoms in the universe today are the same atoms that were produced when the cosmic background radiation was released.
3. Under the assumption that nothing existed before the Big Bang, we can infer that the net energy prior to the Big Bang was zero.
4. Energy conservation requires that there was the same amount of energy in the universe before the Big Bang as there was during the Big Bang and after the Big Bang.
   Therefore,
5. The net amount of energy in the universe today is zero.
   Therefore,
6. There must be negative energy in the universe to balance the observable positive energy.

[81] **The impossibility of negative mass.** Sometimes it seems as though everything in our universe has an opposite counterpart (e. g., positive and negative charges, positive and negative energy, positive and negative numbers, left and right-handed molecules, yin and yang, and so on). Is it possible that there is both positive and negative mass? Actually, no. Here is the proof.

1. Positive mass distorts space so that objects fall towards it (i. e., attraction). Conversely, negative mass distorts space so that objects move away (i. e., repulsion).
2. If a negative mass were placed near a positive mass, the positive mass would attract the negative mass and the negative mass would repel the positive mass, causing the positive mass to run away from the negative mass, with the negative mass chasing the positive mass, forever.
3. This would make possible perpetual motion machines (an impossibility).
   Therefore,
4. The universe must be composed of all positive mass or all negative mass.
5. But we have positive mass.
   Therefore,

xix It is now accepted that a strong gravitational force (such as we find in black holes) is a source of radiating energy. It is the dissipation of gravitational radiation that causes black holes to shrink [116].

6. There can be no negative masses.

[82]**Kalam argument.** Intuitively, it pleases us to think that everything that exists must have a beginning, springing from a time preceding its existence. Because the universe exists, then we can assume that the universe must have a beginning. But can we really make any assumptions about something as inscrutable as our universe? Let's stop a moment and think a bit about a theory of the universe that is credited to an 11th century Persian scholar named Al-Ghazali. This theory is known today as the Kalam argument, named for a medieval Muslim school of philosophy. The theory roughly follows the following line of thinking, with my comments attached:

1. Everything that begins must have a cause.
   Causality, as we discussed at length earlier, is a tenuous concept. Modern scientists fare much better with the concept of sets of initial conditions from which a sequence of events may unfold. What we see as a "cause" is just an emergent event that occurs proximate to some effect.
2. The universe began.
   We do not know that the universe began. The universe may be eternal, and eternity does not have an ending and need not have a beginning. It is easy to invent a model of the universe that is timeless, with our human concept of time providing us with a convenient way of putting sequences in order.
3. Therefore, the universe has a cause.
   The preceding two steps are far from settled, rendering the inference (i. e, step 3) unproven.
4. If the universe has a cause, then a powerful God must have created the universe.

Of course, this leaves open the obvious objections, such as "If God exists, then who or what caused God and when did God begin?" We might also ask "If God operates under the physical laws of his creation, wouldn't he be limited by the speed of light?" To rebuff these questions, and many others, philosophers have stipulated that God, the Creator of the Universe must be causeless, without beginning, changeless, immaterial, timeless, and without location. Basically, the Kalam argument ascribes to God the same properties that some modern physicists assign to our Universe (i. e., something eternal, without a beginning and with unchanging properties).

[83]**Ciliate life cycle.** The tactic of harboring two organisms having the same genetic material, but living very different lives, is not an exclusive characteristic of multicellular organisms (e. g., plants and animals). A division between germ and soma (body) was developed by an ancient class of single-celled eukaryotes known as the ciliates (i. e., members of Class Ciliophora). These organisms are morphologically distinguished by a profusion of cilia visible on their surfaces (Figure 11.6). Each ciliate organism is endowed with two nuclei, with each nucleus having its own peculiar genomic architecture and cellular purpose. One nucleus is large and polyploid (the macronucleus), and serves the general physiologic needs of the cell (e. g., producing RNA and maintaining

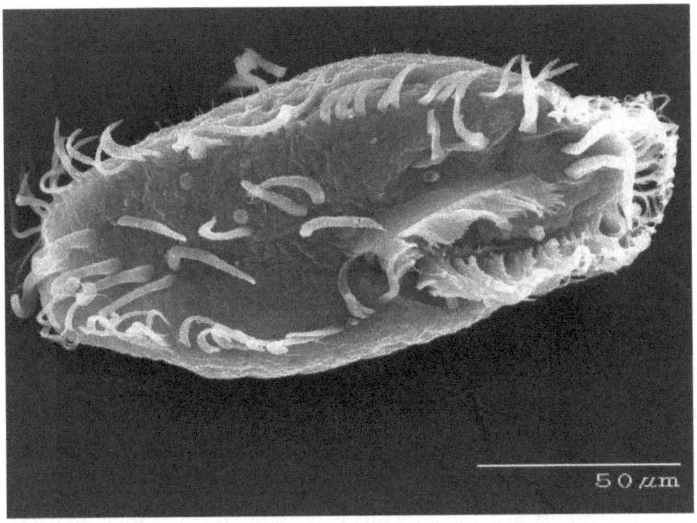

Figure 11.6: Oxytricha trifallax, a typical ciliate, with ovoid shape and plentiful cilia. Source, Wikipedia, from a US government public domain work.

the cell's phenotype). The other nucleus (the micronucleus) is tiny and does not express its genes. It serves as the genetic repository preserving the germline of the organism. The macronucleus divides by a process known as amitosis, with each amitotic division reducing its viability. After about 200 such divisions, macronucleus is no longer capable of amitosis, and a new macronucleus is generated by the micronucleus.

Within a single cell, the ciliates maintain a somatic entity (determined by the macronucleus) and a separate germinal entity (the micronucleus). Aside from regenerating the somatic macronucleus, the micronucleus has an active sex life. On occasion, two ciliates may couple, and their micronuclei exchange DNA via a conjugal mechanism that involves the meiotic production and exchange of haploid chromosomes. The ciliates many to orchestrate the basic reproductive activities, such as we see in animals and plants, within a single-cell organism. Remarkable.

[84]**Germ cell line.** The germ cell line is one very specialized lineage of cells that produces fully differentiated gametes (ova in females and sperm in males). The primitive cells of the germ cell line have the ability to erase the epigenome, thus producing cells that may become totipotent (i. e., able to differentiate into any cell type).

The term "germ cell line" needs to be distinguished from "germline", the latter being the first cells to arise from the zygote, from which all of the cells of the embryo, including the cells of the germ cell line, will develop.

[85]**Germline.** The germline consists of the cells that derive from the fertilized egg of an organism. All of the somatic cells (i. e., the cells composing the body), as well as the germ cells of the body (i. e., oocytes and spermatocytes) arise from the same germline. The extra-embryonic cells (e. g., placental cells) have the same germline as the somatic

cells. An inherited condition can be described as being in the germline; meaning that the mutation is in every cell that derives from the fertilized egg.

The word "germline" has confused many students, who use the term "germline cell" interchangeably with "germ cell" or with "germ cell line". In fact, if you search for a definition of germline on the web, you're apt to find the somewhat recursive definition that the germline constitutes all of the cells in the offspring that are produced by the germ cell lines of the parents. True, but apt to mislead.

It is best to think of a germline mutation by its functional definition, a mutation found in every cell in an organism.

[86]**Germ cell.** A cell in the lineage that produces gametes (i. e., oocytes in females and sperm cells in males). Equivalent to a cell belonging to the germ cell line."

[87]**Dichotomous fungal nomenclature.** Fungal organisms have two options for reproduction: sexual and asexual. Both these forms of reproduction have their own morphologic appearances, in the same species of organism. Factors that determine the mode of reproduction for a cultured fungus are a mystery. It is possible for one mycologist to observe a fungus that reproduces exclusively asexually, while another mycologist, observing the same species in a culture dish or growing in the wild, may observe sexual reproduction (e. g., fruiting bodies). Depending on the phase of reproduction observed, and ignorant of the existence of an alternate morphologic form, taxonomists have assigned different names (sexual and asexual) to the same organism. Rather than harmonizing a dichotomous nomenclature under one preferred name, the ICBN (International Code of Botanical Nomenclature) had ruled, in the past, that it is acceptable to assign two different binomials to an organism: a sexual (also called teleomorphic, perfect, or meiotic form), and an asexual (also called anamorphic, imperfect, or mitotic form). For instance, two binomials legitimately apply to the same organism: Filobasidiella neoformans (the teleomorphic form) and Cryptococcus neoformans (the anamorphic form). Clinical mycologists typically use the asexual name, because it is the asexual form that grows in human tissues.

The unbridled craziness of one species having several different names could not continue forever. The 18th International Botanical Congress assembled in 2011 and abolished longstanding provisions permitting the separate names for the different morphologic forms of a pleomorphic fungal species (i. e., names for anamorphs and teleomorphs) [45].

[88]**Amyloid world.** We tend to think of proteins as individual molecules; not as large aggregates. It seems intuitively obvious that any protein that serves as a catalyst (i. e., any enzyme) would be in the best position to bind to its intended substrate molecules if it were free-floating and untethered. All the introductory biochemistry books have graphic illustrations of enzymes that seem to float across the page, to participate in well-defined metabolic pathways.

As it happens, polypeptides and proteins (large polypeptides) can form large fibrillar aggregates, known as amyloid. Fibrils of amyloid, on very close inspection, have the form of pleated protein sheet (the so-called beta-pleated sheet). Occasionally, amyloid is produced in large quantities, layering extracellular spaces and interrupting the normal function of organs. The term "amyloid" applies to any protein that has malformed into a beta-pleated sheet. The clinical presentation of amyloidosis will depend on the cellular source of the amyloid (e. g., immunocytes, neurons) and the molecular identity of the aggregate (immunoglobulin light chains, transthyretin, alpha-synuclein), the quantity of amyloid produced, and the its location in tissues (e. g., brain, heart, connective tissue, vessels).

It is easy to dismiss amyloid as a wholly pathogenic material serving no benefit to living organisms, but in the epoch preceding the life as we know it, it is quite possible that amyloid protein conformations may have played a vital role in the evolution of life [109]. Amyloid aggregates are structurally stable and resistant to most proteolytic metabolic pathways [65]. Amyloid proteins can serve as enzymatic catalysts, and they can act as templates in their own replication. If things had gone differently, ancient organisms may have evolved into an amyloid world. Among currently extant organisms, amyloid assists in the production fimbriae in some types of bacteria, as an epigenomic constituent in fungi, and in hormone release in humans [241].

89 **Amyloidosis.** A term that includes at least 30 different subtypes of disease each characterized by the accumulation of misfolded proteins that cannot be digested by proteolytic enzymes [65]. The term "amyloidosis" is reserved for conditions in which amyloid proteins accumulate in the extracellular spaces [114]. Apart from amyloidosis, amyloid can also accumulate in intracellular inclusion bodies, as is observed in several neurodegenerative diseases (Huntington disease, Alzheimer disease, Parkinson disease, and amyotrophic lateral sclerosis). There are inherited and acquired forms of amyloidosis, producing dysfunction in the tissues where amyloid deposits.

90 **Codon.** A codon is a sequence of three nucleotides that can be translated to an amino acid or as a stop instruction. Synonym: nucleotide triplet.

Messenger RNA is translated into protein by successively matching sequential triplets of mRNA against the amino acids for which they code. Each RNA nucleotide coding triplet is known as a codon.

As the mRNA sequence is being processed, successive triplets are attached to a triplet-specific tRNA (transfer RNA) molecule in the ribosome that binds to the amino acid matching the triplet, and the amino acid is then added to the protein sequence that is being built. The process of reading the codons is accomplished at the same time as the amino acid sequence of its encoded protein is assembled.

There are 64 possible nucleotide triplets, and only about 20 naturally occurring proteogenic amino acids. Hence, there is considerable redundancy in the code. Among the 64 triplets, 3 are stop codons.

91 **Ribozymes.** Ribozymes, a term conflated from "ribonucleic acid" and "enzymes" function as part of the large subunit ribosomal RNA, linking amino acids during protein synthesis. In addition, ribozymes act as catalysts in a variety reactions within the ribosome (e. g., including RNA splicing, transfer RNA synthesis, and viral replication) [147]. The finding that RNA molecules may act as genetic material and as enzymes has

raised the possibility that ancient life forms relied upon molecules much like ribozymes, that evolved to yield the various types of specialized RNA molecules that are essential to modern cells.

[92]**Neutrophils.** Circulating blood cells are either red (red blood cells) or white (white blood cells). More accurately, blood cells are either red or they are unpigmented. The red blood cells greatly outnumber the white blood cells, but there is a large variety of circulating white cells, dominated by lymphocytes, platelets, monocytes, and granulocytes (neutrophils, eosinophils, basophils). The granulocytes can be distinguished from one another by the dyes that can be absorbed by the different types of their contained granuloes. Eosinophils retain a pink dye. Basophils retain a blue dye. Neutrophils do not retain dyes of either color. The most numerous type of white cell is the neutrophil, and, among the granulocytes, the neutrophil is the most important defender against infections.

[93]**Congenital disorder.** Includes all the congenital anomalies (i. e., structural or anatomic deformities) plus the genetic diseases, including metabolic disorders, and acquired diseases of the embryo or fetus, that are present at birth.

[94]**Classification system versus identification system.** It is important to distinguish classification systems from an identification systems. An identification system matches an individual objects (e. g., organism, disease) with its assigned name. In the case of organisms, identification is based on finding several features that, taken together, can help determine the species name of an organism. For example, if you have a list of characteristic features: large, hairy, strong, African, jungle-dwelling, knuckle-walking; you might correctly identify the organisms as a gorilla. These identifiers are different from the phylogenetic features that were used to classify gorillas within the hierarchy of organisms (Animalia: Chordata: Mammalia: Primates: Hominidae: Homininae: Gorillini: Gorilla). Specifically, you can identify an animal as a gorilla without knowing that a gorilla is a type of mammal. Conversely, you can classify a gorilla as a member of Class Gorillini without knowing that a gorilla happens to be large.[xx]

One of the most common mistakes in the biological sciences is to confuse an identification system with a classification system. The former simply provides a handy way to associate an object with a name; the latter is a system of relationships among objects.

[95]**Parent class.** The immediate ancestor, or the next-higher class (i. e., the direct superclass) of a class. For example, in the classification of living organisms, Class Vertebrata is the parent class of Class Gnathostomata. Class Gnathostomata is the parent class of Class Teleostomi.

---

xx A class is a collection of members that share a property that is absent from members not in the class or its descendant classes. The word "class", lowercase, is used as a general term. The word "Class", uppercase, followed by an uppercase noun (e. g. Class Animalia), represents a specific class within a formal classification.

In a classification, which imposes single class inheritance, each child class has exactly one parent class; whereas one parent class may have several different child classes. Furthermore, some classes, in particular the bottom class in the lineage, have no child classes (i. e., a class need not always be a superclass of other classes). A class can be defined by its properties, its membership (i. e., the instances that belong to the class), and by the name of its parent class. When we list all of the classes in a classification, in any order, we can always reconstruct the complete class lineage, in their correct lineage and branchings, if we know the parent class of each class.

[96]**Subclass.** A subclass is a descendant class of its direct line of ancestral classes. The members of a subclass inherit the properties and methods of the ancestral classes in its direct lineage. For example, all subclasses of mammals have mammary glands because mammary glands are a defining property of the mammal class. In addition, all subclasses of mammals have vertebrae because the class of mammals is a subclass of the class of vertebrates. The subclass that is the immediate descendant of a class is called the child class. In common parlance, when we speak of the subclass of a class, we are referring specifically to its child class.

[97]**Unclassifiable objects.** Classifications create hierarchical collections of classes, and taxonomies assign each and every named object to its correct class. This means that a classification is not permitted to contain unclassified objects; a condition that puts fussy taxonomists in an untenable position. Suppose you have an object, and you simply do not know enough about the object to confidently assign it to a class. Or, suppose you have an object that seems to fit more than one class, and you can't decide which class is the correct class. What do you do? Historically, scientists have resorted to creating a "miscellaneous" or "problematica" or "incertae sedis" (uncertain placement) class into which otherwise unclassifiable objects are given a temporary home, until more suitable accommodations can be provided [127].

The promiscuous creation of "miscellaneous" classes have proven to be a huge impediment to the advancement of the biological sciences. In the case of the classification of living organisms, the class of protozoans stands as a case in point. Ernst Haeckel, a leading biological taxonomist in his time, created the Kingdom Protista (i. e., protozoans), in 1866, to accommodate a wide variety of simple organisms with superficial commonalities. Haeckel himself understood that the protists were a blended class that included unrelated organisms, but he believed that further study would resolve the confusion. In a sense, he was right, but the process took much longer than he had anticipated; occupying generations of taxonomists over the ensuing 150 years. Today, the protozoans have been reassigned to positions among the animals, plants and fungi and separately named single-celled eukaryotes. In the meantime, therapeutic opportunities for eradicating so-called protozoal infections, using class-targeted agents, have no doubt been missed [26]. For practical reasons, textbooks still use the term "protozoan", but strictly speaking, Kingdom Protista no longer exists [214].

You may think that the creation of a class of living organisms, with no established scientific relation to the real world, was a rare and ancient event in the annals of biology, having little or no chance of being repeated. Not so. A special pseudoclass of fungi, deuteromycetes (spelled with a lowercase "d", signifying its questionable validity as a true biologic class) has been created to hold fungi of indeterminate speciation (i. e., fungi whose status as a distinct species, or whose species name, has not been established). At present, there are several thousand such fungi, sitting in a taxonomic limbo, waiting to be placed into a definitive taxonomic class [111, 26].

[98]**Fundamental particles of the Standard Model.** The fundamental particles predicted from the Standard Model are:

- Six flavors of quarks (up, down, strange, charm, bottom, and top).
- Six types of leptons (electron, electron neutrino, muon, muon neutrino, tau, tau neutrino).
- Twelve gauge bosons (force carriers): the photon of electromagnetism, the three W and Z bosons of the weak force, and the eight gluons of the strong force.
- The Higgs boson.

[99]**Least action, Lagrangians, and the Euler–Lagrange equation.** Here is the Euler–Lagrange equation:

$$\frac{\mathrm{d}}{\mathrm{d}t}\left(\frac{\partial L}{\partial \dot{x}_0}\right) = \frac{\partial L}{\partial x_0}.$$

On first inspection, this equation does not seem to bear any profound signifi cance, but physicists rely upon this equation whenever they model the laws and the symmetries that represent physical reality. For us to appreciate the relevance and power of the Euler–Lagrange equation, we first need to introduce the Lagrangian, and its so-called action integral. The Lagrangian is a mathematical function that encapsulates the forces operating within a space. It is suited for solving sets of equations where the forces are conservative (i. e., energy differences at the beginning and the ending of a forced motion are path-independent).

In mathematical terms, the Lagrangian is the difference between the kinetic energy of the system (i. e., the sum of the kinetic energy of all the parts of a system) and the potential energy of the system. In effect, the Lagrangian describes the physical system.[xxi]

$$L = T - V \quad \text{The Lagrangian, } T = \text{kinetic energy, } V = \text{potential energy.}$$

xxi We are taught that potential energy is the "stored" energy of a system. This definition is highly misleading. In technical terms, potential energy is the difference between the energy of an object in its location and its energy at a reference location. In practical terms, potential energy is the energy that would be spent to return the object to its original state after some kinetic event. In the case of a rock dropped from the top of a cliff, the potential energy is the energy required to lift the rock from the cliff bottom to the cliff top. It might be of value to replace the term "potential energy" with "restorative energy", the energy needed to restore the system to its original state. In a fully reversible process, the restorative energy equals the kinetic energy.

When we speak of symmetries under Noether's theorem, we are referring to symmetries of the Lagrangian.[119] That is to say that we are dealing with those symmetries that leave the Lagrangian unchanged. The "action" of the Lagrangian is the Lagrangian's integral over time.

$$S = \int_{t_1}^{t_2} L(x, \dot{x}, t)\, \mathrm{d}t \quad S = \text{Action.}^{\text{xxii}}$$

The action has an extreme value (i. e., a maximum, a minimum, or a saddlepoint) at locations where an infinitesimal variation in the action is zero.[xxiii] As shown here,

$$\delta S = \delta \int_{t_1}^{t_2} L(x, \dot{x}, t) dt = 0 \quad \text{Least action.}$$

Why is this the case? When a function (or a functional, in the case of the Action) is at a minimum or maximum, or saddlepoint, it is at a stable position, where there is no accelerated movement, and this calmness extends to the values in its immediate vicinity. So an infinitesimal variation in the Action at its extremum will not change the Action value (i. e., its derivative is zero).

To describe the laws of motion in a system having conservative forces, we must find the function $x_0(t)$ that produces least action (i. e., a function whose action will be an extremum, and whose derivative is zero). To review, at the action's extremum, the action will not change over small variations in time, so that its derivative, $\delta S = \delta \int_{t_1}^{t_2} L(x, \dot{x}, t) dt$ will equal zero. Where the action does not change over time its least action becomes a conserved quantity.[xxiv] We can also point out that when we discuss physical systems, actions involve motions (changes of position, the "x" variable), over time (the "t" variable). So we can intuit here that discovering $x_0(t)$, the function that produces the least action, will lead us to the equations of motion in a conservative system.

Physicists find $x_0(t)$ by using the Euler–Lagrange equation.

$$\frac{\mathrm{d}}{\mathrm{d}t}\left(\frac{\partial L}{\partial \dot{x}_0}\right) = \frac{\partial L}{\partial x_0} \quad \text{Euler–Lagrange equation.}$$

---

xxii Note that S is a functional, not a function. Its value is determined by the values of all functions in the integrand, over all points of time in the evaluated interval.

xxiii To appreciate the concept of a saddlepoint, imagine that we have a saddle on a horse, and we are placing a small pebble on the saddle. There may be points of stability on the surface of the saddle from which small nudges in position will not cause a pebble to fall down or rise up. These would be the saddlepoints.

xxiv Remember, a conserved quantity is anything that does not change over time.

Before we get ahead of ourselves, we must prove that the Euler–Lagrange equation is true.[xxv]

Let us begin by assuming that $x_0(t)$ is a function that produces a least action for S (i. e., wherein $\delta S = 0$). Now consider some other function $x_a(t)$ such that:

$$x_a(t) = x_0(t) + a\beta(t).$$

Here, a is a number and $\beta(t)$ is a function whose endpoints at t_1 and t_2 yield $\beta(t_1) = \beta(t_2) = 0$.

Now, let us return to our definition of the action:

$$S = \int_{t_1}^{t_2} L(x, \dot{x}, t)\,\mathrm{d}t \quad S = \text{Action}.$$

We recall that the action is a functional (not a function) and depends upon all of the possible functions of $x_{\text{anything}}(t)$ that satisfy the Lagrangian equality. So we can take the partial derivative of the action, for the changing functions, using a variable, a, in $x_a(t)$ like so:

$$\frac{\partial}{\partial a} S[x_a t] = \frac{\partial}{\partial a} \int_{t_1}^{t_2} L(x, \dot{x}, t) dt = \int_{t_1}^{t_2} \frac{\partial L(x, \dot{x}, t)}{\partial a} dt.$$

But L is a function with three variables $L(x, \dot{x}, t)$, so in our equation, we have:

$$\partial L = \partial L(x_a) + \partial L(\dot{x_a}) + \partial L(t).$$

But we are looking for some $x_a t$ that conserves the Lagrangian over time, so we can drop the last term of the right side of the equation to give us,

$$\partial L = \partial L(x_a) + \partial L(\dot{x_a}).$$

Plugging this into our equation,

$$\frac{\partial L(x, \dot{x}, t)}{\partial a} = \frac{\partial L(x_a)}{\partial a} + \frac{\partial L(\dot{x_a})}{\partial a}$$

The chain rule[100] tells us that if $y = f(x)$ and $x = g(t)$, then the derivative of y with respect to t is,

xxv Here, we loosely follow the proof described by David Morin in "Introduction to Classical Mechanics With Problems and Solutions", Cambridge University Press, Cambridge, UK, 2008.

$$\frac{dy}{dt} = \frac{dy}{dx} \times \frac{dx}{dt}.$$

Applying the chain rule to the right side of the equation,

$$\frac{\partial}{\partial a} S[x_a t] = \int_{t_1}^{t_2} \frac{\partial L(x, \dot{x}, t)}{\partial a} dt,$$

which equals,

$$\int_{t_1}^{t_2} \left(\frac{\partial L}{\partial x_a} \frac{\partial x_a}{\partial a} + \frac{\partial L}{\partial \dot{x_a}} \frac{\partial \dot{x_a}}{\partial a} \right) dt.$$

Let us go back to see how we originally conceived of $x_a(t)$:

$$x_a(t) = x_0(t) + a\beta(t),$$

where a is a number and $\beta(t)$ is a function whose endpoints at t_1 and t_2 yield $\beta(t_1) = \beta(t_2) = 0$.

Let us take the partial differential with respect to a

$$\frac{\partial x_a}{\partial a} = \beta.$$

Differentiating again:

$$\frac{\partial \dot{x_a}}{\partial a} = \dot{\beta}.$$

Substituting back into our prior equation:

$$\frac{\partial}{\partial a} S[x_a t] = \int_{t_1}^{t_2} \left(\frac{\partial L}{\partial x_a} \beta + \frac{\partial L}{\partial \dot{x_a}} \dot{\beta} \right) dt.$$

Now it is time to integrate by parts (see note in Chapter 11 on integration by parts[101]). Integration by parts states that:

$$\int u dv = uv - \int v du.$$

Applying integration by parts to the far-right term in the integral, where $u = \frac{\partial L}{\partial \dot{x_a}}$ and $dv = \dot{\beta} dt$ we have:

$$\int \frac{\partial L}{\partial \dot{x_a}} \dot{\beta} dt = \frac{\partial L}{\partial \dot{x_a}} \beta - \int \left(\frac{d}{dt} \frac{\partial L}{\partial \dot{x_a}} \right) \beta dt.$$

We will substitute the right-hand side of the equation back into the third term of the prior expression for $\frac{\partial}{\partial a}S[x_a t]$. This gives us:

$$\frac{\partial}{\partial a}S[x_a t] = \int_{t_1}^{t_2} \left(\frac{\partial L}{\partial x_a} - \frac{d}{dt}\frac{\partial L}{\partial \dot{x}_a} \right)\beta dt + \frac{\partial L}{\partial \dot{x}_a}\beta \Big|_{t_1}^{t_2}.$$

Now, all that remains is to evaluate the terms in the equation based upon our original goal (i. e., to find an $x_0(t)$ that yielded an action derivative equal to zero) and the constraints that we set (i. e., $\beta_{t_1} = \beta_{t_2} = 0$).

By definition, for $x_0(t)$, the derivative of the action is zero, so:

$$0 = \int_{t_1}^{t_2} \left(\frac{\partial L}{\partial x_a} - \frac{d}{dt}\frac{\partial L}{\partial \dot{x}_a} \right)\beta dt + \frac{\partial L}{\partial \dot{x}_a}\beta \Big|_{t_1}^{t_2}.$$

If $\beta_{t_1} = \beta_{t_2} = 0$, then $\frac{\partial L}{\partial \dot{x}_a}\beta|_{t_1}^{t_2} = 0$ and:

$$0 = \int_{t_1}^{t_2} \left(\frac{\partial L}{\partial x_a} - \frac{d}{dt}\frac{\partial L}{\partial \dot{x}_a} \right)\beta dt.$$

But if the integral evaluates to zero, then the quantity within the parentheses must be zero or β must be zero. We previously stipulated that β could be any function satisfying $x_a(t) = x_0(t) + a\beta(t)$, so we know that *beta* is not zero. Therefore the quantity within the parentheses is zero, and we have:

$$\frac{d}{dt}\frac{\partial L}{\partial \dot{x}_0} = \frac{\partial L}{\partial x_0}.$$

This is the Euler–Lagrange equation. Here is an oversimplified outline of how physicists employ the equation:

1. Compose a symmetry in terms of the Lagrangian equation (i. e., compose a Lagrangian that preserves an invariance under transformations of the Poincaré group).[xxvi]
2. Compose the Euler–Lagrange equation for the Lagrangian.
3. Solve the Euler–Lagrange equation to produce a function that yields an Action derivative of zero
4. Formulate the laws of motion and the symmetry of the system.

---

xxvi Named for Henri Poincaré (1854–1912) who was one of the first mathematical physicists to find the relevance of symmetries in physical laws under various transformations.

The principle of least action holds that physical systems do not change the value of the action. The symmetry transformation groups that obey the principle of least action (i. e., whose transformations leave the Lagrangian stationary) are the symmetry groups that represent the physical laws of the universe.

[100]**Proving the chain rule.** The chain rule tells us that if $y = f(x)$ and $x = g(t)$, then the derivative of y with respect to t is:

$$\frac{dy}{dt} = \frac{dy}{dx} \times \frac{dx}{dt}.$$

There are several available proofs of the chain rule, but we will discuss the shortest.

1. If $x = g(t)$, then there is infinitesimal Δt such that

$$\Delta x = g(t + \Delta t) - g(t).$$

2. In the same way, for $y = f(x)$, there is infinitesimal Δx such that

$$\Delta y = f(x + \Delta x) - f(x).$$

3. But $\frac{\Delta y}{\Delta t} = \frac{\Delta y}{\Delta t} \cdot \frac{\Delta x}{\Delta x}$
4. Rearranging, $\frac{\Delta y}{\Delta t} = \frac{\Delta y}{\Delta x} \cdot \frac{\Delta x}{\Delta t}$
5. Applying the infinitesimal for each Δ value

$$\frac{dy}{dt} = \frac{dy}{dx} \cdot \frac{dx}{dt} \quad \text{The chain rule.}$$

[101]**Proving integration by parts.** The method of integration by parts requires that,

$$\int u dv = uv - \int v du.$$

The proof of integration by parts comes directly from the product rule [102]. Applying the product rule to two functions of x,

$$\frac{d(uv)}{dx} = v\frac{du}{dx} + u\frac{dv}{dx}.$$

Integrating both sides with respect to x we have:

$$\int d(uv)dx = \int v\frac{du}{dx}dx + \int u\frac{dv}{dx}dx.$$

But the integral on the left side is just the anti-derivative of the derivative of uv, so:

$$uv = \int v du + \int u dv$$

or

$$\int udv = uv - \int vdu \quad \text{Integration by parts.}$$

[102]**The product rule.** Here is the product rule:

$$\frac{d(f(x)g(x))}{dx} = \left(\frac{df(x)}{dx}\right)g(x) + f(x)\left(\frac{dg(x)}{dx}\right).$$

Expressed alternately,

$$(fg)' = f'g + g'f.$$

There are several simple proofs for the product rule, but one of the simplest involves using the previously proven equation for the derivative of the natural logarithm, namely:

$$\frac{1}{y} = \frac{d(ln(y))}{dy}$$

or,

$$\left(\frac{1}{y}\right)dy = d(ln(y)).$$

Or, in an even simpler form,

$$\frac{y'}{y} = (ln(y))'.$$

Let us let y equal the product of $f(x)$ and $g(x)$

$$y = f(x)g(x).$$

Or, in simple form,

$$y = fg.$$

Substituting for y,

$$\frac{(fg)'}{fg} = (ln(fg))' = ln(f)' + ln(g)'.$$

Multiplying both sides by fg, and remembering that $\frac{y'}{y} = (ln(y))'$, we substitute again,

$$(fg)' = (fg)\left(\frac{f'}{f} + \frac{g'}{g}\right)$$

and,

$$(fg)' = f'g + g'f \quad \text{The product rule.}$$

[103]**Gluon paradox.** Gluons are elementary particles that bind quarks together to form hadrons (i. e., protons and neutrons). Eight gluons are found in the protons and neutrons. Gluons happen to be bosons, and, like all bosons, gluons travel at the speed of light. This being the case, it would seem impossible for gluons, traveling at the speed of light, to be confined within the nucleus of an atom traveling more slowly than the speed of light.

The paradox is readily explained when we consider the life-expectancy of a gluon. As it happens, the time interval between the emission and the absorption of a gluon within the nucleus is about 10^{-24} seconds. This tells us that a gluon can only exist for 10^{-24} seconds, which corresponds to the time required for an object moving at the speed of light to traverse a proton (which has a diameter of about 10^{-15} meters). In fact, the reason that a proton is the size that it is, is determined by the distance traversed by a gluon before absorption. If the gluon could move further, then the proton would be larger.

[104]**Proving fundamental particles do not decay.** Radioactive isotopes decay, and neutrons decay. Should not the fundamental particles also decay. Actually, we have good reason to believe that fundamental particles such as electrons, neutrinos and quarks, are endowed with eternal stability and do not decay.[xxvii]

We posit the following:

- A particle may transition to an available lower energy state.
- A particle that has no lower energy state to which it can transition is stable and cannot decay.
- Such particles, in the absence of intervening events, are eternal.

It may surprise you, but most of the electrons, protons, and photons that were created soon after the Big Bang are still in existence, and are likely to remain in existence for a very long time. For example, Protons are estimated to have a lifespan of at least 10^{34} years. Electrons are thought to have a mean lifetime exceeding 10^{24} years. At this point, readers may interject that every particle has a well-described path to destruction: electrons can be annihilated by positrons; protons can be annihilated by anti-protons; photons can be absorbed by opaque matter; protons and electrons can collapse together to form neutrons (as we see in neutron star formation). It is all true, but none of these aforementioned processes are examples of decay; they are allowed activities between

xxvii The word "decay", as used in this section, only applies to processes observed in particles that are not the result of forces external to the particle. For the purposes of our discussion, results of particle interactions in which one particle ceases to exist and is replaced by other particles, are not "decay" reactions, though they are sometimes referred to as such in the literature.

consenting particles. In addition to not being decay processes, they are also rare. Nearly all of the matter in our current universe is found in the persistent remains of interstellar Hydrogen and helium formed a few hundred thousand years after the Big Bang. The cosmic background radiation accounts for most of the light traveling freely through the universe and was released from a primordial hot plasma at the same moment when atomic Hydrogen and helium atoms were created.

Some of the fundamental particles have very simple explanations for their stability.

- Electrons cannot decay because they carry charge, which is conserved.
  An electron cannot simply "decay" without causing a drop in the total charge present in the universe, and this would be a violation of the conservation of energy.[xxviii]
- Protons, like electrons, carry charge, and cannot decay without violating a conservation law.
  Neutrons, unlike protons, do not carry charge and can decay. A free neutron decays quickly, with an expected lifetime of under 15 minutes.[xxix] Because a neutron carries no charge, neutron decay does not violate charge conservation.[xxx]
- Photons, being timeless, cannot decay inasmuch as decay is always a **time-dependent** phenomenon.
  Photons the quanta of electromagnetic waves. Photons live in the "eternal now" attained by any particle traveling at the speed of light. For a photon, one moment in time is not observably different from any other moment in time. Accordingly, at light-speed, internal events that play out over time (such as decay) do not occur (they just can not find the time).

[105]**Entropy.** Entropy is one of those terms that has been endowed with much more significance that it deserves. Entropy is often discussed under any of the following constructs:

1. As the measure of a system's disorder (in turn, a measure of the combinations of ways that the contents of a system can be rearranged). A crystal has relatively lit-

xxviii An exception would be the balanced annihilation of one electron and one oppositely charged positron. As mentioned, an electron-positron annihilation is not an example of the decay in that it is not spontaneous and involves more than one particle.

xxix For a discussion of neutron lifespan, see Pieter Mumm's article, "Resolving the neutron lifetime puzzle", in Science 360:605–606, 2018.

xxx When a neutron decays, it yields a proton (positive charge) plus an electron (negative charge) plus an anti-neutrino. Some forms of neutron decays may yield high energy gamma radiation (i. e., high frequency electromagnetic waves) and highly energetic electrons. Because a great deal of energy may be released when neutrons decay, the reverse reaction (neutron formation) is energetically unfavorable. Other than in neutron star formation, we would not expect to find protons and electrons coming together to form neutrons.

tle entropy, compared to a gas, and a compressed gas has less entropy than an expanded gas.

2. As the amount of work that can be extracted from a system (by moving heat energy around).
3. As the numeric embodiment of the second law of thermodynamics (thermal reactions that are irreversible always increase the system's entropy).
4. As the information needed to describe a system (in a the quantum states of every atom in a system). The information, as formulated below, takes all of the states of a system, x_i, determines the probability of each state, $P(x_i)$ and expresses the summation for all n possible states:

$$\mathrm{H}(X) = -\sum_{i=1}^{n} \mathrm{P}(x_i) \log_b \mathrm{P}(x_i) \quad \text{Information Entropy formula.}$$

5. As a measure of the possible outcomes of an event. An event that has only one or several outcomes is a low entropy event. An event that has many possible outcomes is a high entropy event.
6. As the reason for the forward direction of time (insofar as both entropy and time may seem to be continuously increasing).
7. As the measure of a thermodynamic state, determined by the following formula, wherein S is entropy, Q is heat, and T is temperature

$$\mathrm{d}S = \frac{\delta Q}{T}.$$

A simple and practical way of thinking of a system's entropy is as a number that we can calculate that tells us something about how much we would need to know if we wanted to describe the components of the system. Increasing the number of components of a system would increase its entropy, as would increasing the degrees of freedom by which those components may move, as would increasing the space in which we might find those components. Compressing the universe to a single point in space, containing no distinguishable objects, would decrease the entropy of the universe. Processes that are reversible are not associated with a change in entropy insofar as the state of the system must be equivalent before and after the process is completed (that is what it means to be reversible). In a static universe, wherein every process is reversible and the size of the universe is fixed, the number associated with the entropy of the universe would be constant.

Entropy has no direct relationship with the arrow of time [21]. Regarding symmetries, we can say that entropy is conserved by reversible processes. However, in the case of irreversible processes, entropy always increases (as per the second law of thermodynamics).

We can get a practical taste of entropy by playing with the calculated entropies of image files. An near-empty image file has a calculated entropy very close to zero (Figure 11.7).

Figure 11.7: A simple image (black background with a smear of white) requires very little information to describe the image. Consequently, it has a calculated entropy near zero (actually, 0.0229745). Source, author.

An image composed of pixels each having a randomly selected gray level between 0 and 255 will require a great deal of information to fully describe the full image (Figure 11.8). It is calculated entropy will be very close to the maximal, 1 (actually, 0.999893).

When we combine two distinguishable images by overlaying one atop the other, we always see an increase in the calculated entropy (i. e., the entropy of the combined image is always greater than that of either of the individual images) (Figure 11.9). Overlaying one image atop another is an irreversible process (unless we add information to the system).

However, when we combine an image with itself the entropy of the combined image is exactly equal to the entropy of each identical image alone. Overlaying an image on itself is a reversible process. Combining an image with itself is analogous to the mixing experiment wherein two volumes of identical atoms are combined (described in Section 6.9, "Making a point of identicality.") There is no change in the entropy of the system.

[106]**Spherical harmonics.** Just as the wave function can be transformed via the two-dimensional Fourier transform, a three-dimensional wave function can be transformed by the three dimensional Laplacian (a function very similar to the Fourier transform) and evaluated with radial coordinates (i. e., a sphere, with radius r and two angular

Figure 11.8: An image of randomly selected grey-value pixels. It is calculated entropy is high, being nearly 1.0 (actually, 0.999893). Source, author.

coordinates). Solution of the 3-D Laplacian provides a volume of equal probability distribution (i. e., wherein ψ^2 is equal everywhere within the surface-limited volume).

$$\psi_{n\ell m}(r,\theta,\varphi) = \sqrt{\left(\frac{2}{na_0}\right)^3 \frac{(n-\ell-1)!}{2n[(n+\ell)!]}}, \tag{11.2}$$

$$e^{-r/na_0}\left(\frac{2r}{na_0}\right)^{\ell} L_{n-\ell-1}^{2\ell+1}\left(\frac{2r}{na_0}\right) \cdot Y_{\ell}^{m}(\theta,\varphi), \tag{11.3}$$

where

$$a_0 = \frac{4\pi\varepsilon_0\hbar^2}{m_q q^2}$$

is the Bohr radius, and

$$L_{n-\ell-1}^{2\ell+1}(\cdots)$$

are the generalized Laguerre polynomials of degree $n-\ell-1$, and n, ℓ, m are the principal, azimuthal, and magnetic quantum numbers respectively, which take the values $n = 1, 2, 3, \ldots, \ell = 0, 1, 2, \ldots, n-1, \ell = 0, 1, 2, \ldots, n-1$.

Figure 11.9: Two images superimposed. The entropy of the combined image is greater than that of either of the component images alone. Source, author.

The surface-limited volumes for the different values of the quantum states of a hydrogen electron can be computed and represented as shapes expanded around a central nucleus (Figure 11.10).

We need not get bogged down in the mathematics. It is enough to know that the various shapes correspond to the locations wherein an electron would exist with equal probability, for its various combinations of increasing quantum numbers (corresponding to the increasing energy levels of shells and orbitals around the nucleus). Observe that each increasing harmonic volume actually subsumes (occupies the same space) portions of the preceding volumes. This means that two electrons (if we were dealing with more than one electron) could occupy the same location at the same moment, consistent with their spatial probability distribution. Furthermore, an electron may occupy the same location as its nucleus! How is this possible? We must remember that quan-

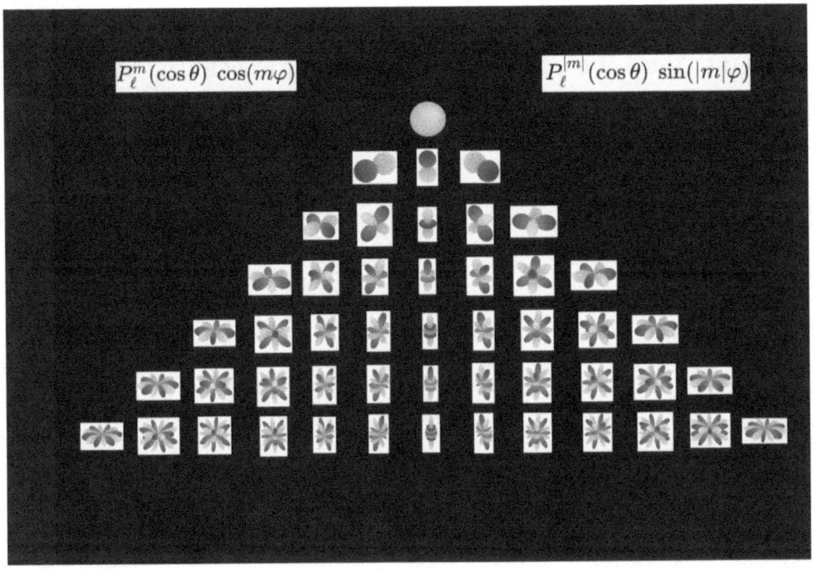

Figure 11.10: The spherical harmonic solutions for the Schrödinger wave probability for Hydrogen. The distributions at ascending sets of quantum numbers provide the location probabilities for orbiting electrons. Source, Wikipedia, as provided to the public domain by 3*eϕp*.

tum objects are waves, and waves have the property of superposition (i. e., two quantum objects can occupy the same location at the same time). The only restriction on superposition occurs when two identical orbiting fermions, such as two electrons, have the same quantum state (including spin). Otherwise, it is acceptable for two electrons to superpose on one another. It is always acceptable for an orbiting electron to superpose on its nucleus.

[107]**Proving Euler's formula and Euler's identity.** Imagine a function, $f(\theta)$, with the following property:

$$f(\theta) = e^{-i\theta}(\cos\theta + i\sin\theta).$$

We will use the product rule[102] to differentiate $f(\theta)$.

$$f'(\theta) = e^{-i\theta}(i\cos\theta - \sin\theta) - ie^{-i\theta}(\cos\theta + i\sin\theta).$$

Evaluating and combining terms,

$$f'(\theta) = e^{-i\theta}(i\cos\theta - \sin\theta - i\cos\theta + \sin\theta).$$

In the equation above, we see that everything on the left cancels out, indicating that $f'(\theta) = 0$, which tells us that $f(\theta)$ must be a constant.

We can determine the constant value of $f(\theta)$ by substituting any value of θ into the original equation (that's one of the perks of being a constant). Let us substitute zero as our value of θ.

$$f(0) = e^{-i0}(\cos 0 + i \sin 0) = 1.$$

Evaluating the left-hand side of the equation, we find that $f(\theta)$ always evaluates to 1. Thus,

$$1 = e^{-i\theta}(\cos\theta + i\sin\theta).$$

Dividing both sides by $e^{-i\theta}$,

$$e^{i\theta} = \cos\theta + i\sin\theta \quad \text{Euler formula.}$$

When we substitute π as θ in Euler's formula, we have,

$$e^{i\pi} = \cos\pi + i\sin\pi.$$

Evaluating, we get,

$$e^{i\pi} = -1 + i0.$$

Rearranging,

$$e^{i\pi} + 1 = 0 \quad \text{Euler's identity.}$$

Euler's identity, widely credited as the most beautiful equation in mathematics, manages to include virtually every important mathematical symbol in one compact relationship ($e, i, \pi, 1, 0$, and $=$).

Now that we have Euler's formula, we can derive a few useful variations. Substituting kx for x,

$$e^{ikx} = \cos(kx) + i\sin(kx).$$

The substitution of kx for x is equivalent to raising e^{ix} to the kth power.

$$e^{ikx} = (e^{ix})^k = (\cos(x) + i\sin(x))^k = \cos(kx) + i\sin(kx).$$

This latter equivalence is known as the DeMoivre identity.[53]

[108] **Compton wavelength.** Imagine for a moment that you are a particle such as an electron, or a proton, or a neutron. Now, ask yourself, "If I were a light wave, rather than the particle that I am, then what would be my wavelength? The answer to this wildly hypothetical proposition is "The Compton wavelength"!

Every particle has a specific resting mass, and this resting mass is associated with a specific energy. Likewise, every electromagnetic wave, though it has no mass, is associated with a specific mass-equivalent energy. The Compton wavelength is the wavelength of light whose energy that is the rest energy of the particle. Expressing these relationships in mathematical terms, we see that a particle of mass m has a rest energy of $E = mc^2$, Einstein's famous equation. Furthermore, the energy of an electromagnetic wave is $h\nu$ (i. e., its frequency, ν, times the Planck constant, h or $6.62607015 \times 10^{-34} \frac{\mathrm{m^2 kg}}{\mathrm{s}}$).

When we look for an electromagnetic frequency that has an energy equal to the energy of a particle with mass m, we have:

$$E = h\nu = mc^2.$$

Therefore, when the energy of an electromagnetic wave and a particle are equal, then the corresponding frequency of the wave must equal $\frac{mc^2}{h}$. Furthermore, since the frequency of an electromagnetic wave is related to its wavelength by the equation $\nu = \frac{c}{\lambda}$, then the corresponding wavelength (i. e., the Compton wavelength) is $\frac{h}{mc}$.[xxxi]

So what? As it happens, we can learn a great deal from having the Compton wavelength of different particles. With it, we can determine the particle's radius, and we can learn how photons were scattered in the early years following the Big Bang, when the universe was confined to a dense hot plasma. Just for fun, let's show that we can never determine the position of a particle to within half of the particle's Compton wavelength.

To refresh, Heisenberg's uncertainty principle asserts that we can never measure a change of momentum $\triangle p$ and a change of position $\triangle x$ to within $\frac{\hbar}{2}$, where $\hbar$ is the reduced Planck's constant.

Expressed as an equation, we have:

$$\triangle x \triangle p \geq \frac{\hbar}{2},$$

or,

$$\triangle x \geq \frac{\hbar}{2 \triangle p}.$$

Let us imagine that the momentum of the particle is as great as it can possibly be, and maybe a tad more. This would mean that the particle moved at the speed of light, giving it an upper limit momentum of mc. Let us put this into the equation, as the upper limit for momentum, to see what this does to our limit on finding the position of the particle, with certainty.

xxxi When I was barely surviving undergraduate physics, a popular joke at the time involved greeting a classmate with the question, "What's new?", and the classmate would reliably answer $\frac{E}{h}$. Not so funny, in retrospect, but it serves as a sturdy mnenomic for the relation between frequency, ν and the energy of electromagnetic radiation.

$$\Delta x \geq \frac{\hbar}{2mc} = \frac{1}{2} \times \frac{\hbar}{mc}.$$

But we have previously shown that $\frac{\hbar}{mc}$ is just the Compton wavelength.

So, the position of a particle can only be measured to a precision within one-half of the Compton wavelength of the particle. This is just one example that demonstrates how we can use fundamental relationships and constants established when the symmetries of our universe revealed themselves, to answer fundamental questions about the behavior of matter, light, and energy. This conclusion helps us to understand some of the properties of dark matter, hypothesized to account for the clustering of galaxies and the increase in the rotational speed of galaxies over time.[xxxii] The predicted mass of a hypothetical dark matter particle is $10^{-22}eV$. Not much. With this mass, a dark matter particle will have a Compton wavelength of about 1.3 lightyears, meaning that we could never know the position of a dark matter particle to anything under 0.65 light years. For all practical purposes, dark matter particles have no discernible position, no structure, and a limited ability to interact with just about anything. It is not surprising that we have so far not detected dark matter particles.[xxxiii]

[109]**Accuracy versus precision.** Accuracy measures how close your data comes to being correct. Precision provides a measurement of reproducibility (i. e., whether repeated measurements of the same quantity produce the same result). Data can be accurate but imprecise. If you have a 10 pound object, and you report its weight as 7.2376 pounds, on every occasion when the object is weighed, then your precision is remarkable, but your accuracy is dismal [29].

[110]**Proving that light travels at finite speed.** Nineteenth century scientists were aware that the speed of light, though fast, was not infinitely fast. A few casual observations served as proof. If the speed of light were infinite, then all manner of physical impossibilities might unfold. For example, imagine a planet in another galaxy that emits a death ray aimed at earth. Earth scientists, anticipating the attack, have launched a space armada intent on intercepting the death ray before it reaches our planet. Unfortunately, the death ray is traveling at infinite speed, and reaches the earth at the very same time as it reaches the armada and at the very same time it left the enemy planet. The death ray cannot be intercepted because it has already arrived at its target! With a little imagination, we can see that causality, time, and space are three fundamental concepts that depend upon the finite speed of light.

For any unimaginative skeptics, there is another simple proof that light must move at finite speed. We know that light is a wave, and the speed of any wave is determined

---

xxxii Not all physicists are confident of the existence of dark matter. A theory known as Modified Newtonian Dynamics, first developed by Mordehai Milgram, predicts galaxy condensation without hypothesizing the existence of dark matter [183].

xxxiii Incidentally, the term "dark matter" is a misnomer. Dark matter cannot block out light waves and is not dark. A better name would be "transparent matter".

by its frequency (i. e., waves passing a point in unit time) multiplied by its wavelength. If the speed of light were infinite, then either the frequency of light must be infinite, or the wavelength of light must be infinite, or both must be infinite. If the frequency of light were infinite, then light would carry infinite energy (because the energy of light is a function of frequency). If the wavelength of light were infinite, then all light would have the same wavelength and there would be no colors in the rainbow. Therefore, the frequency and the wavelength of light must be finite, leading us to conclude that the speed of light must be finite.

111 **Entanglement.** When a quantum object with spin (e. g., an electron) is created, another quantum object must be created having the opposite spin; otherwise angular momentum would not be conserved. When one of the two quantum objects commits to a particular spin value (e. g., when its spin is measured), then we can be certain that its twin must have the opposite spin. This would be true regardless of the location of the two particles. This feature is known as quantum entanglement.

112 **Auto-correlation.** Statisticians have devised clever ways to determine the degree to which one moment's measurements depended upon some previous moment's value. The auto-correlation function is a method for producing a measurement indicating the dependence of data elements on prior data elements. Long-range dependence occurs when a value is dependent on many prior values. Long-range dependence is determined when the serial correlation (i. e., the auto-correlation over multiple data elements) is high when the number of sequential elements is large [78].

113 **Protein sizes.** Typical proteins have a molecular weight about 50,000 Daltons, roughly equivalent to a protein containing 454 amino acids. Many proteins are constructed to lie within a cellular membrane (e. g., receptor proteins), and proteins in the 50,000 Dalton range are a good fit. As an example, rhodopsin, a protein found in the membrane of photoreceptor cells, has a molecular weight of 38,893 Daltons.

The largest protein in human cells is titin, with a length of 27,000 to 33,000 amino acids (or 2 to 3 million Daltons). Very large proteins have a tough time being exported into the plasma. Von Willebrand factor (500 to 20,000 kiloDaltons, depending on the number of subunits) is an example of a protein that is much too large to pass through the cell membrane. Yet Von Willebrand factor is an important blood constituent. How does it travel from the inside of the cell, where proteins are synthesized, to the blood plasma? Von Willebrand factor is synthesized in two types of cells: vascular endothelial cells and megakaryocytes. Each of these two cell types has its own way of extruding Von Willebrand factor into circulation. In endothelial cells, Von Willebrand factor is packaged ito Weibel–Palade bodies, which are extruded from the endothelial cell, when needed. In megakaryocytes, the factor is packaged into alpha granules that enter the circulation when platelets are pinched from the megakaryocyte's cytoplasm.

114 **Acquistion of new genes.** New genes often occur in duplicated genes, in which one of the two genes is conserved, and the other gene undergoes slow alterations over time, eventually yielding a conserved gene variant having a new or improved function. Gene duplication occurs frequently in nature, and is particularly common in plants wherein

whole sets of chromosomes duplicate to yield plants with stable polyploid genomes (i. e., genomes consisting of some multiple of the diploid size). Mutation in a duplicated gene seldom causes the death of the organism (because the twin gene compensates). Consequently, mutations occurring in germ line cells can be "kept" in the species gene pool. When mutated duplicate genes happen to provide a benefit to the organism, they are conserved and a "new" gene becomes available to the gene pool. It was once thought that all "new" genes in animals arose through the mutation of duplicated genes.

The assumption that new genes derive from old genes, exclusively following gene duplication events, has been discredited by the observation of recently-evolved genes having no homologs to ancient genes. Whole genome sequencing has uncovered genes that are species-specific; hence, not inherited from ancestral organisms and not belonging to gene families. These species-specific genes were, at first, called "orphan genes" (i. e., having no parental origin), and are now called "de novo genes" (i. e., new to a species). It is hypothesized that de novo genes arise from existing non-coding gene sequences that mutate to include a start codon sequence that can initiate RNA transcription and protein synthesis (thus creating an open reading frame). We assume that most such events are either deleterious or of no advantage to the cell. However, if the new protein fragment serves some useful purpose for the cell, then the new coding sequence may be conserved [216]. The emergence of de novo genes is not particularly rare [185]. Scientists have found 142 genes that are present in some species of flies, but not in others, a sign that these genes arose de novo.[xxxiv] The emergence of de novo genes is yet another example of the value of junk DNA.[xxxv]

[115]**Open reading frame.** Abbreviation: ORF. A sequence within the genome that can be translated (via an mRNA intermediate) into protein. As such, an ORF will have a start codon (usually AUG) and a stop codon (usually UAA, UAG, UGA).

[116]**Law of sequence conservation.** If a sequence is conserved through evolution (i. e., if we can find a closely similar sequence that is present in various animal species having a shared ancestral class), then that sequence must perform a useful function for the organism. Furthermore, sequences that are highly conserved (i. e., with very little difference among class members, over time), are likely to have a very important function. This law is so useful and so fundamental to genomics and to gene-related computational algorithms that it may as well be known as the first law of bioinformatics.

---

xxxiv See Carl Zimmer's article, "The continuing evolution of genes" in the April 28, 2014 edition of The New York Times.

xxxv Junk DNA is DNA that does neither codes nor regulates genes, and has no discernible function. The majority of our genome consists of non-conserved "junk" DNA. Why do we have junk DNA? If there were no junk DNA, then all of our DNA would be conserved (i. e., preserved in the gene pool without mutation). If all of our genes were conserved, then new variants of genes would not arise in the population, and evolution of the species would cease. But we observe that all species evolve continuously. Therefore, all species must contain junk DNA.

The corollary to the law is that genomic sequences that degenerate over time, and for which there are large variations in closely related species, must not have a very important function in the organism.

[117]**Time symmetry.** In our spacetime universe, time is a dimension through which we move, much the same way as we move through the familiar *x*-, *y*-, and *z*-dimensions of Euclidean space. Euclidean space permits us to move reversibly through any dimension. We can move forward or backward in the *x*-, *y*-, and *z*-dimensions. A forward motion is equivalent to a backward motion, played in reverse.

When we think about forward and backward motions in time, we encounter some practical limits to time reversibility. For example, as we watch a movie depicting a high-octane car chase culminating in a fiery crash, we know that the movie is progressing forwards in time. If the movie were played in reverse, with a car suddenly being assembled from the shattered remnants of an explosion, then instantly speeding backwards down the street, we would all agree that what we were seeing was physically impossible. We would conclude that time, unlike the Euclidean *x*-, *y*-, and *z*-coordinates, cannot be reversed. Hence, time is asymmetric, meaning that it proceeds in one direction but not the other.

Let us imagine another movie, in which there are only a few particles moving in space, occasionally bumping into one another. In this case, if we watched a movie of those particles, and played the movie backwards, we would have no way of distinguishing whether the movie was playing forwards or backwards (i. e., in reverse). When the number of particles is small, assigning forward time and backward time is arbitrary and, hence, reversible and symmetric.

There seem to be small-scale (quantum) truths that are different from large-scale truths. The fundamental laws of particle interactions are compatible with the notion that quantum events have time symmetry.

[118]**Proving the speed of light is constant.** Imagine a universe wherein light of different wavelengths traveled at different speeds. For example, blue light might travel at a speed 10 % greater than the speed of red light. If this were the case, white light, which is composed of all the different colors (i. e., wavelengths) of visible light, would quickly lose its coherence and its “whiteness”. Emitted white light would reach us as blue light first, and the red light would come in some time later. We never observe this kind of behavior in our light.

Let us imagine another scenario entirely, in which the speed of light depended upon the location at which the light was emitted. We might have a situation in which a pulse of light leaves a star at a distance of 1 billion miles from Earth. Along the way, half of the light is absorbed and re-emitted by a mirror located in a region of space where emitted light travels at 10 times the speed of light. In this case, we would see the pulse from the star arriving at one time (from the fast-moving re-emitted source) and the slower light from the same pulse would arrive sometime later. Again, such occurrences are at odds with reality.[xxxvi]

Actually, there is a very simple way to prove that the speed of light must be constant. Everyone knows Einstein's relationship between the rest mass of an object and the speed of light.

$$E = mc^2.$$

Suppose the speed of light were not constant, changing from moment to moment. For example, suppose the speed of light suddenly doubled. Then we would have, pretending for the sake of argument that light has mass,

$$E = m(2c)^2 = 4mc^2.$$

When the speed of light doubles, the associated energy increases 4-fold, without adding any energy to the system. This is a clear violation of the law of conservation of energy. Therefore, the speed of light must be constant. The constancy of the speed of light was demonstrated experimentally in the Michelson–Morley experiment, discussed in some detail in Section 7.10, "Fundamental constants save the ether".

Let us look at another proof of the constancy of the speed of light, based on simple gedanken experiment, credited to William de Sitter. Imagine that we have our telescope trained on a luminous body rotating around a point in distant space. We observe it going around and around in its orbit, as a single point of light tracing a smooth ellipse. Now imagine that the speed of the light emanating from the body is variable. One moment it travels at the traditional speed of light; the next moment it travels a bit faster, and at other moments it travels a bit slower. In that case, the light received at our telescope will consist, at any given moment, of a combination of light beams originating at different moments, at different positions in the elliptic, but "catching up" to one another when they reach the telescope. What we would see might be multiple luminous bodies, spaced at different locations in the elliptic, or we might have moments when our image goes totally blank, because the light from the body has arrived previously or has yet to arrive. As we observe, the continuous reception of single point of light tracing a continuous elliptic curve informs us that the light emitted from the orbiting body must have traveled at a constant velocity.

[119]**Proving Noether's theorem.** Put in simplest terms, Noether's theorem states that **wherever we have an invariance, we have a symmetry, and wherever we have**

---

xxxvi Readers may object here that gravity lensing, a phenomenon that occurs when light from a distant star passes through an high gravity area of space will arrive on Earth at different moments of time, and will seem to come from different points in space. This phenomenon is due to gravity's distortion of space (sometimes referred to as the bending of light by gravity), causing portions of the emitted light to follow different paths. In this instance, the speed of light is not changed, but the angle of the path, and its distance to Earth, are subtly altered.

a symmetry, there is a conservation law. This straightforward theorem is key to our understanding of the fundamentals of the physical world.[xxxvii] Physicists interpret Noether's theorem as an expression of the symmetries of the Lagrangian. It permits us to conceptualize physical laws as symmetries, and to model our symmetry relations as Lagrangians that satisfy the Euler–Lagrange equation. Through these steps, physicists develop mathematical models of the operating system of the universe. Accordingly, every scientist should understand the meaning of Noether's theorem, and should have some notion of its proof. Unfortunately, a rigorous proof of Noether's theorem demands a bit more mathematics than we can feasibly provide herein. We present here a stripped-down mathematical proof of Noether's theorem that preserves the most important steps.

We need to prove that for each symmetry of the Lagrangian, there is a conserved quantity. This is equivalent to saying that when we change the coordinates of the system, then there is a quantity described by the Lagrangian that does not change. We shall need a solid understanding of the meaning of the Lagrangian, plus familiarity with the chain rule of differentiation.[100 xxxviii]

We can think of the Lagrangian as an expression of the energy state of a system as it moves incrementally through small changes in its coordinate system.

So, if q_i is one of n coordinates, a small change in the coordinate may be expressed as:

$$q_i \to q_i + \epsilon K_i(q).$$

Here, ϵ is a small number and $K_i(q)$ is a function that applies to the ith coordinate that may depend on the values of some or all of the total number of coordinates.

So, a change in coordinates for the Lagrangian will alter the Lagrangian (assuming the Lagrangian is variant under a change of coordinates) as a function of all the changed coordinates and their first-order derivatives:

$$L = L(q_i, \dot{q}_i).$$

Because we stipulate that the Lagrangian does not change with a small increment in coordinates, we know that:

$$0 = \frac{\partial L}{\partial \epsilon} = \sum \frac{\partial L(q_i)}{\partial \epsilon} + \frac{\partial L(\dot{q}_i)}{\partial \epsilon}.$$

xxxvii The theorem was published by Emmy Noether, in 1918, as "Invariante Variationsprobleme," in: Nachrichten von der Gesellschaft der Wissenschaften zu Gottingen, Mathematisch-Physikalische Klasse 1918:235–257, 1918.

xxxviii As in the prior proof of the Euler–Lagrange Equation, we loosely follow the steps described by David Morin in "Introduction to Classical Mechanics With Problems and Solutions", Cambridge University Press, Cambridge, UK, 2008.

The chain rule tells us that if $y = f(x)$ and $x = g(t)$, then the derivative of y with respect to t is:[100]

$$\frac{dy}{dt} = \frac{dy}{dx} \times \frac{dx}{dt}.$$

Applying the chain rule, we have:

$$0 = \frac{\partial L}{\partial \epsilon} = \sum \frac{\partial L}{\partial q_i}\frac{\partial q_i}{\partial \epsilon} + \frac{\partial L}{\partial \dot{q}_i}\frac{\partial \dot{q}_i}{\partial \epsilon}.$$

But we know that:

$$q_i \rightarrow q_i + \epsilon K_i(q).$$

So,

$$\frac{\partial q_i}{\partial \epsilon} = K_i(q).$$

Substituting back into our equation,

$$0 = \frac{\partial L}{\partial \epsilon} = \sum \left(\frac{\partial L}{\partial q_i} K_i(q) + \frac{\partial L}{\partial \dot{q}_i} \dot{K}_i(q) \right).$$

Now we come to the point where we can apply the Euler–Lagrange equation. To review, when there is in invariance of the Lagrangian, the Euler–Lagrange equation specifies:

$$\frac{d}{dt}\frac{\partial L}{\partial \dot{x_0}} = \frac{\partial L}{\partial x_0}.$$

In this case,

$$\frac{\partial L}{\partial q_i} = \frac{d}{dt}\frac{\partial L}{\partial \dot{q}_i}.$$

Substituting back into our equation:

$$0 = \frac{\partial L}{\partial \epsilon} = \sum \left(\frac{d}{dt}\frac{\partial L}{\partial \dot{q}_i} K_i(q) + \frac{\partial L}{\partial \dot{q}_i} \dot{K}_i(q) \right).$$

But, $\dot{K}_i(q) = \frac{d}{dt} K_i$. Substituting back into our equation:

$$0 = \frac{\partial L}{\partial \epsilon} = \frac{d}{dt} \sum \left(\frac{\partial L}{\partial \dot{q}_i} K_i(q) \right).$$

Therefore, when we have a Lagrangian that does not change with incremental alterations in the coordinates, we ALWAYS have a quantity, $\frac{\partial L}{\partial \dot{q}_i} K_i(q)$ that does not change over time. A quantity that does not change over time is, by definition, a conserved quantity. Therefore, we have proven that when there is a symmetry in the Lagrangian, there must be a conserved quantity. This is Noether's theorem.[xxxix]

[120]**Lie groups.** Named for Sophus Lie (rhymes with tophus knee), who developed this area of mathematics. A Lie group, G, is a group that satisfies the following:

1. The is a function which can be called multiplication, such that:
   $G \times G \to G$: $(a, b) \to ab$ (i. e., when any two members of the group are multiplied, they yield another member of the group)
2. There is an inverse operation
   The inverse of $G \to G$: $a \to$ The inverse of a (every element in G has an inverse in G)
3. G is smooth (i.e, G is differentiable everywhere and is continuous)

The condition of smoothness confers upon the Lie Group the qualities of infinitesimals (such as those used in the familiar operations of calculus). As it happens Lie Groups can be represented by corresponding Lie Algebras, which are linear objects amenable to the familiar methods of linear algebra. Consequently, Lie groups abstractly combine the two worlds of the continuous and the discrete. Because the laws of our universe encompass continuous and discrete operations (e. g., continuous motion through space and time, and quantized (discrete) states of particles), Lie groups are often chosen as the symbolic representations of reality. Examples are the Lorentz group of special relativity, the spin groups (relativistic and non-relativistic), and the gauge groups for quantum electrodynamics describing the weak and strong nuclear forces [266].

[121]**Classification of finite simple groups.** Mathematicians have recently provided proof that we have achieved a complete classification of all possible finite simple groups, allowing us to find mathematical and physical relations among different finite simple groups [136, 108]. One of the most complex of the symmetry groups listed in the classification theorem of finite simple groups is the vividly named monster group. Freeman Dyson, a physicist and a science visionary, predicted that even the monster group would eventually be found to represent the behavior of some physical laws in the real world. His prediction seems to have been validated. Today, study of the monster group had led to a deeper understanding of string theory and black holes [44]. The completeness of the classification theorem for the finite simple groups opens the possibility that all of the operational principles of the universe can be represented within the classification of related finite simple groups [34].

[122]**Potential energy.** The term "potential energy" is usually defined as the energy available to a system that is not expended during some action on an object (e. g., falling

---

xxxix In the section titled "Proving the canonical commutation relation," we discussed the mathematical relationship between properties that happen to be the Fourier transforms of one another (i. e., paired Fourier conjugates). Noether's theorem extends to the Fourier conjugates in the following way. When there is an invariance that applies to one of the paired conjugates, then the other conjugate will be conserved.

off a cliff and landing on a beach). This definition is essentially meaningless. What we think of as the potential energy is the energy that would be spent to return the object to its original state (the state prior to the expenditure of kinetic energy). In the case of a rock dropped from the top of a cliff, the potential energy is the energy required to lift the rock from the cliff bottom to the cliff top. It might be of value to replace the term "potential energy" with "restorative energy", the energy needed to restore the system to its original state. In a reversible process, the restorative energy equals the kinetic energy.

[123]**Quantum particle.** The characteristic particle of a field has been defined in various ways, all of which are functionally equivalent:

- The minimal observable excitation of a field
  Each field has a minimal energy below which field interactions cannot be observed,[xl] and this exact minimal energy is the quantum of the field.
- The quantum of a field
  The quantum of a field is detected as a discrete particle that has no size and no material contents, and which is absolutely identical to every other particle of the same kind (e. g., all electrons are identical to one another; all photons are identical to one another).
- A wave packet.
- A probability distribution for field properties that can be conveniently modeled by a wave packet.
- A collapsed wave function (the result of wave model at the moment when it fails).
- Whatever is detected by a particle detector (i. e., the detectable effect of the wave packet).
- A wave that has lost its coherence with other waves (thus losing its full complement of wave properties), causing it to revert to a particle.
- A conserved property (such as a charge, in the case of the electron) near a point, having a non-stationary distribution.
- A wave having rotational motion.
- An observable capable of entanglement (e. g., wherein conserved angular momentum dictates that the spin of one particle determines the spin of its pair).
- A manifestation of the vibrational state of a quantum string.

xl If an event is unobservable, then we can infer it has no effect in our universe. We know this because if an unobservable event had an effect, then the event would have been observable via the effect it produced.

Bibliography

[1] Adamo SA. "Octopus: multiple minds or just a slow thinker?" In: *Animal Sentience* 2019 (2019), p. 278.

[2] Adams FC and Laughlin G. "A dying universe: the long-term fate and evolution of astrophysical objects." In: *Rev Mod Phys* 69 (1997), pp. 337–372.

[3] Agrelo R et al. "Inactivation of the lamin A/C gene by CpG island promoter hypermethylation in hematologic malignancies, and its association with poor survival in nodal diffuse large B-cell lymphoma." In: *J Clin Oncol* 23 (2005), pp. 3940–3947.

[4] Agrios G. *Plant Pathology*. Academic Press, Orlando, FL, 1978.

[5] Aharonov Y, Albert DZ, and Vaidman L. "How the result of a measurement of a component of the spin of a spin-1/2 particle can turn out to be 100." In: *Phys Rev Lett* 60 (1988), pp. 1351–1354.

[6] Ahuja R et al. "Relativity and the lead-acid battery." In: *Phys Rev Lett* 106 (2011), p. 018301.

[7] Akopyants NS et al. "Demonstration of genetic exchange during cyclical development of Leishmania in the sand fly vector." In: *Science* 324 (2009), pp. 265–268.

[8] Anderson PW. "More is different." In: *Science* 177 (1972), pp. 393–396.

[9] Angly FE et al. "The marine viromes of four oceanic regions." In: *PLoS Biol* 4 (2006), p. 4.

[10] Archibald JD. Eutheria (Placental Mammals), In: *Encyclopedia of Life Sciences*. MacMillan, London, pp. 1–4, 2001.

[11] Arslan D et al. "Distant Mimivirus relative with a larger genome highlights the fundamental features of Megaviridae." In: *PNAS* 108 (2011), pp. 17486–17491.

[12] Bai L, Baker DR, and Rivers M. "Experimental study of bubble growth in Stromboli basalt melts at 1 atm." In: *Earth Planet Sci Lett* 267 (2008), pp. 533–554.

[13] Baldauf SL. "An overview of the phylogeny and diversity of eukaryotes." In: *J Syst Evol* 46 (2008), pp. 263–273.

[14] Ball WWR. *Mathematical Recreations and Essays*. Fourth edition. MacMillan, Company, London, and New York, 1905.

[15] Bamford DH. "Do viruses form lineages across different domains of life?" In: *Res Microbiol* 154 (2003), pp. 231–236.

[16] Bandin I and Dopazo CP. "Host range, host specificity and hypothesized host shift events among viruses of lower vertebrates." In: *Vet Res* 42 (2011), p. 67.

[17] Bartholdi D et al. "Epigenetic mutations of the imprinted IGF2-H19 domain in Silver-Russell syndrome (SRS): results from a large cohort of patients with SRS and SRS-like phenotypes." In: *J Med Genet* 46 (2009), pp. 192–197.

[18] Barton SC, Surani MAH, and Norris ML. "Role of paternal and maternal genomes in mouse development." In: *Nature* 311 (1984), pp. 374–376.

[19] Basto R et al. "Flies without centrioles." In: *Cell* 125 (2006), pp. 1375–1386.

[20] Bazin P et al. "Phytoplankton diversity and community composition along the estuarine gradient of a temperate macrotidal ecosystem: combined morphological and molecular approaches." In: *PLoS ONE* 9 (2014), p. 94110.

[21] Ben-Naim A. "Entropy and time." In: *Entropy* 22 (2020), p. 430.

[22] Benton MJ and Donoghue PCJ. "Palaeontological evidence to date the tree of life." In: *Mol Biol Evol* 24 (2006), pp. 26–53.

[23] Berman J. *Precision Medicine, and the Reinvention of Human Disease*. Academic Press, Cambridge, MA, 2018.

[24] Berman JJ. "Informatics research using publicly available pathology data." In: *J Pathol Inform* 2 (2011), p. 5.

[25] Berman JJ. "Post-informatics pathology." In: *J Pathol Inform* 2 (2011), p. 18.

[26] Berman JJ. *Taxonomic Guide to Infectious Diseases: Understanding the Biologic Classes of Pathogenic Organisms*, First Edition. Academic Press, Cambridge, MA, 2012.

[27] Berman JJ. *Armchair Science: No Experiments, Just Deduction*. Amazon Digital Services, 2014.

https://doi.org/10.1515/9783111498690-012

[28] Berman JJ. *Rare Diseases and Orphan Drugs: Keys to Understanding and Treating Common Diseases*. Academic Press, Cambridge, MA, 2014.
[29] Berman JJ. *Repurposing Legacy Data: Innovative Case Studies*. Morgan Kaufmann, Waltham, MA, 2015.
[30] Berman JJ. *Data Simplification: Taming Information With Open Source Tools*. Morgan Kaufmann, Waltham, MA, 2016.
[31] Berman JJ. *Evolution's Clinical Guidebook: Translating Ancient Genes into Precision Medicine*. Academic Press, Cambridge, MA, 2019.
[32] Berman JJ. *Taxonomic Guide to Infectious Diseases: Understanding the Biologic Classes of Pathogenic Organisms*, Second Edition. Academic Press, Cambridge, MA, 2019.
[33] Berman JJ. *Logic and Critical Thinking in the Biomedical Sciences, Volume I: Deductions Based Upon Simple Observations*. Academic Press, Cambridge, MA, 2020.
[34] Berman JJ. *Classification Made Relevant: How Scientists Build and Use Classifications and Ontologies*. Academic Press, San Diego, CA, 2022.
[35] Bianciardi G et al. "Complexity analysis of the Viking labeled release experiments." In: *Int J Aeronaut Space Sci* 13 (2012), pp. 14–26.
[36] Bianconi E et al. "An estimation of the number of cells in the human body." In: *Ann Hum Biol* 40 (2013), pp. 463–471.
[37] Blanco-Melo D, Gifford RJ, and Bieniasz PD. "Reconstruction of a replication-competent ancestral murine endogenous retrovirus-L." In: *Retrovirology* 15 (2018), p. 34.
[38] Bloom BJ et al. "An optical lattice clock with accuracy and stability at the 10–18 level." In: *Nature* 506 (2014), pp. 71–75.
[39] Bodey B. "Spontaneous regression of neoplasms: new possibilities for immunotherapy." In: *Expert Opin Biol Ther* 2 (2002), pp. 459–476.
[40] Bonolis L. "From the rise of the group concept to the stormy onset of group theory in the new quantum mechanics: a saga of the invariant characterization of physical objects, events and theories." In: *Riv Nuovo Cimento* 27 (2004), pp. 1–110.
[41] Booch G. *Object-Oriented Analysis and Design With Applications*, Second edition. Addison-Wesley, Boston, MA, 1994.
[42] Boothby TC et al. "Evidence for extensive horizontal gene transfer from the draft genome of a tardigrade." In: *PNAS* 112 (2015), pp. 15976–15981.
[43] Borgwardt S et al. "Why are psychiatric imaging methods clinically unreliable?" In: *Behav Brain Funct* 8 (2012), p. 46.
[44] Boya LJ. *Introduction to Sporadic Groups*. Proceedings of the Workshop, Supersymmetric Quantum Mechanics and Spectral Design, Benasque, Spain, 2010.
[45] Braun U. "The impacts of the discontinuation of dual nomenclature of pleomorphic fungi: the trivial facts, problems, and strategies." In: *IMA Fungus* 3 (2012), pp. 81–86.
[46] Braunstein SL and Pati AK. "Quantum information cannot be completely hidden in correlations: implications for the black-hole information paradox." In: *Phys Rev Lett* 98 (2007), p. 080502.
[47] Bravais A. "Memoire sur les systemes formes par les points distribues regulierement sur un plan ou dans l'espace." In: *J Éc Polytech* 19 (1850), pp. 1–128.
[48] Brekke D. "Quiet passing of unlikely hero." In: *Wired* (Jan. 22, 2000).
[49] Britten RJ. "Almost all human genes resulted from ancient duplication." In: *PNAS* 103 (2006), pp. 19027–19032.
[50] Brocks JJ et al. "Archaean molecular fossils and the early rise of eukaryotes." In: *Science* 285 (1999), pp. 1033–1036.
[51] Brocks JJ et al. "Early sponges and toxic protists: possible sources of cryostane, an age diagnostic biomarker antedating Sturtian Snowball Earth." In: *Geobiology* 14 (2016), pp. 129–149.
[52] Bromham L. "What can DNA tell us about the Cambrian explosion?" In: *Integr Comb Biol* 43 (2003), pp. 148–156.

[53] Brunet T et al. "The evolutionary origin of bilaterian smooth and striated myocytes." In: *eLife* 5 (2016), p. 19607.

[54] Burke DL et al. "Positron production in multiphoton light-by-light scattering." In: *Phys Rev Lett* 79 (1997), p. 1626.

[55] Burri L et al. "Microsporidian mitosomes retain elements of the general mitochondrial targeting system." In: *PNAS* 103 (2006), pp. 15916–15920.

[56] Burt A. "Perspective: sex, recombination, and the efficacy of selection–was Weismann right?" In: *Evolution* 54 (2000), pp. 337–351.

[57] Cammack R, Rao KK, and Hall DO. "Metalloproteins in the evolution of photosynthesis." In: *Biosystems* 14 (1981), pp. 57–80.

[58] Cavalier-Smith T. "Coevolution of vertebrate genome, cell, and nuclear sizes." In: *Symposium on the Evolution of Terrestrial Vertebrates* (1991), pp. 51–86.

[59] Chan EF et al. "A common human skin tumour is caused by activating mutations in beta-catenin." In: *Nat Genet* 21 (1999), pp. 410–413.

[60] Chang K. "A hobbyist challenges papers on growth of dinosaurs." In: *NY Times* (Dec. 16, 2013).

[61] Chang K. "A spat over the search for killer asteroids." In: *NY Times* (May 23, 2016).

[62] Chatrchyan S and CMS Collaboration. "Search for supersymmetry at the LHC in events with jets and missing transverse energy." In: *Phys Rev Lett* 107 (2011), p. 221804.

[63] Chen J, Odenike O, and Rowley JD. "Leukemogenesis: More than mutant genes." In: *Nat Rev Cancer* 10 (2010), pp. 23–36.

[64] Chung E. "Stardust citizen scientist finds first dust from outside solar system: dust collected by NASA spacecraft a decade ago finally identified and analyzed." In: *CBC News* (Aug. 14, 2014).

[65] Ciechanover A and Kwon YT. "Degradation of misfolded proteins in neurodegenerative diseases: therapeutic targets and strategies." In: *Exp Mol Med* 47 (2015), p. 47.

[66] Cockell CS. "The laws of life." In: *Phys Today* 70 (2017), pp. 42–48.

[67] Coulson CA. "The contributions of wave mechanics to chemistry." In: *J Chem Soc* 2 (1955), pp. 69–84.

[68] Dacks J and Roger AJ. "The first sexual lineage and the relevance of facultative sex." In: *J Mol Evol* 48 (1999), pp. 779–783.

[69] Darwin C. *Origin of the Species by Means of Natural Selection, or the Preservation of Favoured Races in the Struggle for Life*. John Murray, London, 1859.

[70] Davies PCW. *Superforce: The Search for a Grand Unified Theory of Nature*. Simon and Schuster, New York, 1985.

[71] Dawkins R. *The Greatest Show on Earth: The Evidence for Evolution*. Free Press, New York, 2009.

[72] de Broglie. *The Theory of Measurement in Wave Mechanics, Usual Interpretation and Causal Interpretation*. Gauthier-Villars, Editeur-Imprimeur-Libraire, Paris, 1957.

[73] de Villartay JP et al. "A novel immunodeficiency associated with hypomorphic RAG1 mutations and CMV infection." In: *J Clin Invest* 115 (2005), pp. 3291–3299.

[74] Delmar M and McKenna WJ. "The cardiac desmosome and arrhythmogenic cardiomyopathies from gene to disease." In: *Circ Res* 107 (2010), pp. 700–714.

[75] DeQueiroz K. "Ernst Mayr and the modern concept of species." In: *PNAS* 102 (2005), pp. 6600–6607.

[76] DeQueiroz K. "Species concepts and species delimitation." In: *Syst Biol* 56 (2007), pp. 879–886.

[77] Dodd MS et al. "Evidence for early life in Earth's oldest hydrothermal vent precipitates." In: *Nature* 543 (2017), pp. 60–64.

[78] Downey AB. *Think DSP: Digital Signal Processing in Python*. Green Tea Press, Needham, MA, 2014.

[79] Dunitz JD. "Symmetry arguments in chemistry." In: *Proc Natl Acad Sci USA* 93 (1996), pp. 14260–14266.

[80] Durruthy-Durruthy J et al. "The primate-specific noncoding RNA HPAT5 regulates pluripotency during human preimplantation development and nuclear reprogramming." In: *Nat Genet* 48 (2016), pp. 44–52.

[81] Emerman M and Malik HS. "Paleovirology: modern consequences of ancient viruses." In: *PLoS Biol* 8 (2010), p. 1000301.

[82] Epstein D. "The DIY Scientist, the Olympian, and the mutated gene: How a woman whose muscles disappeared discovered she shared a disease with a muscle-bound Olympic medalist." In: *ProPublica* (Jan. 15, 2016).

[83] Erb TJ and Zarzycki J. "A short history of RubisCO: the rise and fall of nature's predominant CO2 fixing enzyme." In: *Curr Opin Biotechnol* 49 (2018), pp. 100–107.

[84] Erwin DH. "The origin of body plans." In: *Am Zool* 39 (1999), pp. 617–629.

[85] Erwin DH, Valentine JW, and Jablonski D. "The origin of animal body plans." In: *Am Sci* 85 (1997), pp. 126–137.

[86] Faustino NA and Cooper TA. "Pre-mRNA splicing and human disease." In: *Genes Dev* 17 (2003), pp. 419–437.

[87] Fikes BJ. "The patient from the future, here today." In: *The San Diego Union Tribune* (Mar. 5, 2014).

[88] Filee J. "Multiple occurrences of giant virus core genes acquired by eukaryotic genomes: the visible part of the iceberg?" In: *Virology* 466 (2014), pp. 53–59.

[89] Finsterer J. "Mitochondriopathies." In: *Eur J Neurol* 11 (2004), pp. 163–186.

[90] Fisher RA. "The correlation between relatives on the supposition of Mendelian inheritance." In: *Trans R Soc Edinb* 52 (1918), pp. 399–433.

[91] Fitch WM and Ayala FJ (eds). *Tempo and Mode in Evolution*. National Academy Press, Washington, DC, 1995.

[92] Fitzpatrick DA. "Horizontal gene transfer in fungi." In: *FEMS Microbiol Lett* 329 (2012), pp. 1–8.

[93] Fogli A et al. "Ovarian failure related to eukaryotic initiation factor 2B mutations." In: *Am J Hum Genet* 72 (2003), pp. 1544–1550.

[94] Fort A et al. "Deep transcriptome profiling of mammalian stem cells supports a regulatory role for retrotransposons in pluripotency maintenance." In: *Nat Genet* 46 (2014), pp. 558–566.

[95] Forterre P. "The two ages of the RNA world, and the transition to the DNA world: a story of viruses and cells." In: *Biochimie* 87 (2005), pp. 793–803.

[96] Forterre P. "Three RNA cells for ribosomal lineages and three DNA viruses to replicate their genomes: A hypothesis for the origin of cellular domain." In: *PNAS* 106 (2006), pp. 3669–3674.

[97] Forterre P. "Defining life: the virus viewpoint." In: *Orig Life Evol Biosph* 40 (2010), pp. 151–160.

[98] Forterre P and Gaia M. "Giant viruses and the origin of modern eukaryotes." In: *Curr Opin Microbiol* 31 (2016), pp. 44–49.

[99] Fowler RH. "On dense matter." In: *Mon Not R Astron Soc* 87 (1926), pp. 114–122.

[100] Frederic MY et al. "Identification of 526 conserved metazoan genetic innovations exposes a new role for cofactor E-like in neuronal microtubule homeostasis." In: *PLoS Genet* 9 (2013), p. 9.

[101] Garcia-Bellido A. "Symmetries throughout organic evolution." In: *Proc Natl Acad Sci USA* 93 (1996), pp. 14229–14232.

[102] Gibbs JW. *On the Equilibrium of Heterogeneous Substances, Originally published in Connecticut Acad Sci, 1875–1878, Selected Papers on Thermodynamics and Statistical Physics*. Cambridge University Press, London, 1951.

[103] Gilbert W. "Origin of life: The RNA world." In: *Nature* 319 (1986), p. 618.

[104] Glansdorff N, Xu Y, and Labedan B. "The last universal common ancestor: emergence, constitution and genetic legacy of an elusive forerunner." In: *Biol Direct* 3 (2008), p. 29.

[105] Glusman G et al. "The complete human olfactory subgenome." In: *Genome Res* 11 (2001), pp. 685–702.

[106] Godel K. "Uber formal unentscheidbare Satze der Principia Mathematica und verwandter Systeme, I, On formally undecidable propositions of Principia Mathematica and related systems." In: *Monatshefte Math Phys* 38 (1931), pp. 173–198.

[107] Goldberg R. "Horses' ability to pace is written in DNA." In: *NY Times* (Sept. 10, 2012).

[108] Gorenstein D. "Classifying the finite simple groups." In: *Bull Am Math Soc* 14 (1986), pp. 1–98.

[109] Greenwald J, Kwiatkowski W, and Riek R. "Peptide amyloids in the origin of life." In: *J Mol Biol* 430 (2018), pp. 3735–3750.

[110] Griffiths DJ. "Endogenous retroviruses in the human genome sequence." In: *Genome Biology 2, Reviews* (2001), p. 1017.
[111] Guarro J, Gene J, and Stchigel AM. "Developments in fungal taxonomy." In: *Clin Microbiol Rev* 12 (1999), pp. 454–500.
[112] Hardison RC. "Evolution of hemoglobin and its genes." In: *Cold Spring Harb Perspect Med* 2 (2012), p. 011627.
[113] Haring M et al. "Virology: independent virus development outside a host." In: *Nature* 436 (2005), pp. 1101–1102.
[114] Hassan W et al. "Amyloid heart disease: new frontiers and insights in pathophysiology, diagnosis, and management." In: *Tex Heart Inst J* 32 (2005), p. 32.
[115] Hawking SW. "Gravitationally collapsed objects of very low mass." In: *Mon Not R Astron Soc* 152 (1971), pp. 75–78.
[116] Hawking SW. "Black hole explosions?" In: *Nature* 248 (1974), pp. 30–31.
[117] Hildenbrand A et al. "A potential cephalopod from the early Cambrian of eastern Newfoundland, Canada." In: *Commun Biol* 4 (2021), p. 388.
[118] Hochner B. "An embodied view of octopus neurobiology." In: *Curr Biol* 22 (2012), pp. 887–892.
[119] Holland HD. "The oxygenation of the atmosphere and oceans." In: *Philos Trans R Soc, Biol Sci* 361 (2006), pp. 903–915.
[120] Holmes I. "Transcendent elements: whole-genome transposon screens and open evolutionary questions." In: *Genome Res* 12 (2002), pp. 1152–1155.
[121] Horie M et al. "Endogenous non-retroviral RNA virus elements in mammalian genomes." In: *Nature* 463 (2010), pp. 84–87.
[122] Horike S et al. "Loss of silent chromatin looping and impaired imprinting of DLX5 in Rett syndrome." In: *Nat Genet* 32 (2005), pp. 31–40.
[123] Horowitz NH. "On the evolution of biochemical syntheses." In: *PNAS* 31 (1945), pp. 153–157.
[124] Huang W and Ferris JP. "One-step, regioselective synthesis of up to 50-mers of RNA oligomers by montmorillonite catalysis." In: *J Am Chem Soc* 128 (2006), pp. 8914–8919.
[125] Huggins W. "On the photographic spectra of stars." In: *Philos Trans R Soc Lond* 171 (1880), pp. 669–690.
[126] International Consortium for Blood Pressure Genome-Wide Association Studies. "Genetic variants in novel pathways influence blood pressure and cardiovascular disease risk." In: *Nature* 478 (2011), pp. 103–109.
[127] Jenner RA and Littlewood TJ. "Problematica old and new." In: *Philos Trans R Soc B* 363 (2008), pp. 1503–1512.
[128] Jern P, Sperber GO, and Blomberg J. "Use of Endogenous Retroviral Sequences (ERVs) and structural markers for retroviral phylogenetic inference and taxonomy." In: *Retrovirology* 2 (2005), p. 50.
[129] Johnston WK et al. "RNA-catalyzed RNA polymerization: accurate and general RNA-templated primer extension." In: *Science* 292 (2001), pp. 1319–1325.
[130] Jung CG. *Memories, Dreams, Reflections*. Vintage Books, New York, 1961.
[131] Kalghatgi S et al. "Bactericidal antibiotics induce mitochondrial dysfunction and oxidative damage in mammalian cells." In: *Science* 5 (2013), p. 192.
[132] Kapitonov VV and Jurka J. "RAG1 core and V(D)J recombination signal sequences were derived from Transib transposons." In: *PLoS Biol* 3 (2005), p. 181.
[133] Kasliwal MM et al. "Illuminating gravitational waves: a concordant picture of photons from a neutron star merger." In: *Science* 358 (2017), pp. 1559–1565.
[134] Keeling PJ and Palmer JD. "Horizontal gene transfer in eukaryotic evolution." In: *Nat Rev, Genet* 9 (2008), pp. 605–618.
[135] Kingsbury DW. "Species classification problems in virus taxonomy." In: *Intervirology* 24 (1985), pp. 62–70.
[136] Kishore B. "Review of symmetry and the monster, by Mark Ronan." In: *Math Intell* (2009), pp. 76–78.

[137] Koester D and Chanmugam G. "Physics of white dwarf stars." In: *Rep Prog Phys* 53 (1990), pp. 837–915.

[138] Koga Y et al. "Archaeal and bacterial cells arise independently from noncellular precursors? A hypothesis stating that the advent of membrane phospholipid with enantiomeric glycerophosphate backbones caused the separation of the two lines of descent." In: *J Mol Evol* 46 (1998), pp. 54–63.

[139] Kokotas H, Petersen MB, and Willems PJ. "Mitochondrial deafness." In: *Clin Genet* 71 (2007), pp. 379–391.

[140] Koonin EV. "How many genes can make a cell: The minimal-gene-set concept." In: *Annu Rev Genom Hum Genet* 1 (2000), pp. 99–116.

[141] Koonin EV and Galperin MY. *Sequence, Evolution, Function: Computational Approaches in Comparative Genomics*. Kluwer Academic, Boston, 2003.

[142] Koonin EV and Starokadomskyy P. "Are viruses alive? The replicator paradigm sheds decisive light on an old but misguided question." In: *Stud Hist Philos Biol Biomed Sci* 59 (2016), pp. 125–134.

[143] Koonin EV and Starokadomskyy P. "Studies in history and philosophy of biological and biomedical sciences." In: *Stud Hist Phil Sci Part C* 59 (2016), pp. 125–134.

[144] Koonin EV and Wolf YI. "Evolution of microbes and viruses: a paradigm shift in evolutionary biology?" In: *Front Cell Infect Microbiol* 2 (2012), p. 119.

[145] Koster MJ, Snel B, and Timmers HT. "Genesis of chromatin and transcription dynamics in the origin of species." In: *Cell* 161 (2015), pp. 724–736.

[146] Kothandapani A et al. "Epistatic role of base excision repair and mismatch repair pathways in mediating cisplatin cytotoxicity." In: *Nucleic Acids Res* 41 (2013), pp. 7332–7343.

[147] Kruger K et al. "Self-splicing RNA: Autoexcision and autocyclization of the ribosomal RNA intervening sequence of tetrahymena." In: *Cell* 31 (1982), pp. 147–157.

[148] Lahr DJ et al. "The chastity of amoebae: re-evaluating evidence for sex in amoeboid organisms." In: *Proceedings: Biol Sci* 278 (2011), pp. 2081–2090.

[149] Lamb WE and Retherford RC. "Fine structure of the hydrogen atom by a microwave method." In: *Phys Rev* 72 (1947), pp. 241–243.

[150] Lancaster AK and Masel J. "The evolution of reversible switches in the presence of irreversible mimics." In: *Evolution* 63 (2009), pp. 2350–2362.

[151] Lane N. *Life Ascending: The Ten Great Inventions of Evolution*. Profile Books, London, 2009.

[152] Leger A and Puget JL. "Identification of the "unidentified" IR emission features of interstellar dust?" In: *Astron Astrophys* 137 (1984), pp. L5–L8.

[153] Lerat E and Semon M. "Influence of the transposable element neighborhood on human gene expression in normal and tumor tissues." In: *Gene* 396 (2007), pp. 303–311.

[154] Levin-Zaidman S et al. "Ringlike structure of the Deinococcus radiodurans genome: a key to radioresistance?" In: *Science* 299 (2003), pp. 254–256.

[155] Lifton RP, Gharavi AG, and Geller DS. "Molecular mechanisms of human hypertension." In: *Cell* 104 (2001), pp. 545–556.

[156] Liu Y et al. "Direct evidence that suspensor cells have embryogenic potential that is suppressed by the embryo proper during normal embryogenesis." In: *PNAS* 112 (2015), pp. 12432–12437.

[157] Lodish H et al. *Viruses: Structure, Function, and Uses, Molecular Cell Biology*, 4th edition. WH Freeman, New York, 2000.

[158] Love GD et al. "Fossil steroids record the appearance of Demospongiae during the Cryogenian period." In: *Nature* 457 (2009), pp. 718–721.

[159] Lu X et al. "The retrovirus HERVH is a long noncoding RNA required for human embryonic stem cell identity." In: *Nat Struct Mol Biol* 21 (2014), pp. 423–425.

[160] Mable BK and Otto SP. "The evolution of life cycles with haploid and diploid phases." In: *BioEssays* 20 (1998), pp. 453–462.

[161] MacCallum WG. *A Text-Book of Pathology*, 2nd edition. WB Saunders Company, Philadelphia and London, 1921.

[162] Madar S, Goldstein I, and Rotter V. “Did experimental biology die? Lessons from 30 years of p53 research”. In: *Cancer Res* 69 (2009), p. 69.
[163] Magueijo J. “Connection between cosmological time and the constants of nature.” arXiv:2110.05920 (2022).
[164] Malik SB et al. “An expanded inventory of conserved meiotic genes provides evidence for sex in Trichomonas vaginalis.” In: *PLoS ONE* 3 (2008), p. 3.
[165] Manhesa G et al. “Lead isotope study of basic-ultrabasic layered complexes: Speculations about the age of the earth and primitive mantle characteristics.” In: *Earth Planet Sci Lett* 47 (1980), pp. 370–382.
[166] Margulis L. “Undulipodia, flagella and cilia.” In: *Biosystems* 12 (1980), pp. 105–108.
[167] Margulis L and Sagan D. “Order amidst animalcules: the protoctista kingdom and its undulipodiated cells.” In: *Biosystems* 18 (1985), pp. 141–147.
[168] Margulis L and Sagan D. *What is Life?* University of California Press, Oakland, 2000.
[169] Martin DIK, Cropley JE, and Suter CM. “Epigenetics in disease: leader or follower?” In: *Epigenetics* 6 (2011), pp. 843–848.
[170] Martinez DE. “Mortality patterns suggest lack of senescence in hydras.” In: *Exp Gerontol* 33 (1998), pp. 217–225.
[171] Martinez DE and Bridge D. “Hydra, the everlasting embryo, confronts aging.” In: *Int J Dev Biol* 56 (2012), pp. 479–487.
[172] Mayden RL. “Consilience and a hierarchy of species concepts: advances toward closure on the species puzzle.” In: *J Nematol* 31 (1999), pp. 95–116.
[173] Mayr E. “Two empires or three?” In: *PNAS* 95 (1998), pp. 9720–9723.
[174] McGrath J and Solter D. “Completion of mouse embryogenesis requires both the maternal and paternal genomes.” In: *Cell* 37 (1984), pp. 179–183.
[175] McKenna ES et al. “Loss of the epigenetic tumor suppressor SNF5 leads to cancer without genomic instability.” In: *Mol Cell Biol* 28 (2008), pp. 6223–6233.
[176] Mead CA. “Walking the Planck length through history.” In: *Phys Today* 54 (2001), p. 15.
[177] Meyer R. “A Long-lost spacecraft, now saved, faces its biggest test yet.” In: *The Atlantic* (June 3, 2014).
[178] Michelson AA and Morely EW. “On the relative motion of the earth and the luminiferous ether.” In: *Am J Sci* 34 (1887), pp. 333–345.
[179] Michelson AA and Morley EW. “Influences of motion of the medium on the velocity of light.” In: *Am J Sci* 31 (1886), pp. 377–386.
[180] Mihara T et al. “Taxon richness of “Megaviridae” exceeds those of bacteria and archaea in the ocean.” In: *Microbes Environ* 33 (2018), pp. 162–171.
[181] Mims FM. “Amateur science - strong tradition, bright future.” In: *Science* 284 (1999), pp. 55–56.
[182] Mitton JB and Grant MC. “Genetic variation and the natural history of Quaking Aspen.” In: *Bioscience* 46 (1996), pp. 25–31.
[183] Nadis S. “Dark matter deniers.” In: *Discov Magazine* (July 1, 2015).
[184] Narimani A, Moss A, and Scott D. “Dimensionless cosmology.” arXiv:1109.0492 (2012).
[185] Neme R and Tautz D. “Phylogenetic patterns of emergence of new genes support a model of frequent de novo evolution.” In: *BMC Genom* 14 (2013), p. 117.
[186] Newburgh R, Peidle J, and Rueckner W. “Einstein, Perrin, and the reality of atoms: 1905 revisited.” In: *Am J Phys* 74 (2006), pp. 478–481.
[187] Noether E. “Invariante Variationsprobleme.” In: *Nachr Ges Wiss Gött, Math-Phys Kl* 1918 (1918), pp. 235–257.
[188] Novarino G, Akizu N, and Gleeson JG. “Modeling human disease in humans: The Ciliopathies.” In: *Cell* 147 (2011), pp. 70–79.
[189] Nowack EC, Melkonian M, and Glockner G. “Chromatophore genome sequence of Paulinella sheds light on acquisition of photosynthesis by eukaryotes.” In: *Curr Biol* 18 (2008), pp. 410–418.
[190] Nowell RW et al. “Comparative genomics of bdelloid rotifers: Insights from desiccating and nondesiccating species.” In: *PLoS Biol* 16 (2018), p. 2004830.

[191] Okita K, Ichisaka T, and Yamanaka S. "Generation of germline-competent induced pluripotent stem cells." In: *Nature* 448 (2007), pp. 313–317.
[192] O'Raifeartaigh L. *The Dawning of Gauge Theory*. Princeton University Press, Princeton, 1997.
[193] Pandey UB and Nichols CD. "Human disease models in Drosophila melanogaster and the role of the fly in therapeutic drug discovery." In: *Pharmacol Rev* 63 (2011), pp. 411–436.
[194] Papini G. "Caianiello's maximal acceleration: Recent Developments." arXiv:quant-ph/0407115 (2004).
[195] Patel MR, Emerman M, and Malik HS. "Paleovirology: ghosts and gifts of viruses past." In: *Curr Opin Virol* 1 (2011), pp. 304–309.
[196] Patrignani C and the Particle Data Group. "The review of particle physics." In: *Chin Phys C* 40 (2016), p. 100001.
[197] Pearson K. *The Grammar of Science*. Adam and Black, London, 1900.
[198] Perrin JB. *Les Atomes*. Libraire Felix Alcan, Paris, 1913.
[199] Philippe H et al. "Acoelomorph flatworms are deuterostomes related to Xenoturbella." In: *Nature* 470 (2011), pp. 255–258.
[200] Pitzer KS. "Relativistic effects on chemical properties." In: *Acc Chem Res* 12 (1979), pp. 271–276.
[201] Posner A, Schwadron NA, and Zurbuchen TH. "Relationships of corotating rarefaction regions outside 40AU with solar observations: Heliospheric mass loading." In: *COSPAR Colloq Ser* 11 (2001), pp. 315–319.
[202] *Nomina Embryologica Veterinaria*. Prepared by the International Committee on Embryological Veterinary Nomenclature, Knoxville, Tennessee, 2003.
[203] Preuss P. "Solving the mechanism of Rett syndrome: how the first identified epigenetic disease turns on the genes that produce its symptoms." In: *Research News Berkeley Lab* (Dec. 20, 2004).
[204] Prigogine I and Stengers I. *Order Out of Chaos: Man's New Dialogue With Nature*. Bantam New Age Books, New York, NY, 1984.
[205] Pyykko P and Desclaux JP. "Relativity and the periodic system of elements." In: *Acc Chem Res* 12 (1979), pp. 276–281.
[206] Quinonez S and Innis JW. "Human HOX gene disorders." In: *Mol Genet Metab* 111 (2014), pp. 4–15.
[207] Ramesh MA, Malik SB, and Logsdon JM. "A phylogenomic inventory of meiotic genes; evidence for sex in Giardia and an early eukaryotic origin of meiosis." In: *Curr Biol* 15 (2005), pp. 185–191.
[208] Raup DM. "A kill curve for Phanerozoic marine species." In: *Paleobiology* 17 (1991), pp. 37–48.
[209] Renou MO et al. "Quantum theory based on real numbers can be experimentally falsified." In: *Nature* 600 (2021), pp. 625–629.
[210] Resnikoff S et al. "Global data on visual impairment in the year 2002." In: *Bull World Health Organ* 82 (2004), p. 82.
[211] Ronimus RS and Morgan HW. "Distribution and phylogenies of enzymes of the Embden-Meyerhof-Parnas pathway from archaea and hyperthermophilic bacteria support a gluconeogenic origin of metabolism." In: *Archaea* 1 (2002), pp. 199–221.
[212] Russell B. *The Analysis of Matter*. Kegan Paul, London, 1927.
[213] Sarkozy A et al. "Germline BRAF mutations in Noonan, LEOPARD, and cardiofaciocutaneous syndromes: molecular diversity and associated phenotypic spectrum." In: *Human Mutat* 30 (2009), pp. 695–702.
[214] Schlegel M and Hulsmann N. "Protists: a textbook example for a paraphyletic taxon." In: *Org Divers Evol* 7 (2007), pp. 166–172.
[215] Schmich J et al. "Induction of reverse development in two marine Hydrozoans." In: *Int J Dev Biol* 51 (2007), pp. 45–56.
[216] Schmitz JF and Bornberg-Bauer E. "Fact or fiction: updates on how protein-coding genes might emerge de novo from previously non-coding DNA." In: *F1000Res* 6 (2017), p. 57.
[217] Schnabel D et al. "Simultaneous deficiency of sphingolipid activator proteins 1 and 2 is caused by a mutation in the initiation codon of their common gene." In: *J Biol Chem* 267 (1992), pp. 3312–3315.
[218] Schweber SS. *QED and the Men Who Made It*. Princeton University Press, 1994.

[219] Sender R, Fuchs S, and Milo R. "Revised estimates for the number of human and bacteria cells in the body." In: *PLoS Biol* 14 (2016), p. 14.
[220] Simpson GG. "The principles of classification and a classification of mammals." In: *Bull Am Mus Nat Hist* 85 (1945), pp. 1–350.
[221] Singh TP. "Quantum theory without classical time: Octonions, and a theoretical derivation of the Fine Structure Constant 1/137." In: *Int J Mod Phys D* 30 (2021), p. 2142010.
[222] Slatko BE, Taylor MJ, and Foster JM. "The Wolbachia endosymbiont as an anti-filarial nematode target." In: *Symbiosis* 51 (2010), pp. 55–65.
[223] Smith SW. *The Scientist and Engineer's Guide to Digital Signal Processing*. California Technical Publishing, Concord, CA, 1997.
[224] Soejima H and Higashimoto K. "Epigenetic and genetic alterations of the imprinting disorder Beckwith-Wiedemann syndrome and related disorders." In: *J Hum Genet* 58 (2013), pp. 402–409.
[225] Speijera D, Lukes J, and Elias M. "Sex is a ubiquitous, ancient, and inherent attribute of eukaryotic life." In: *Proc Natl Acad Sci* 112 (2015), pp. 8827–8834.
[226] Sperling EA et al. "Oxygen, ecology, and the Cambrian radiation of animals." In: *PNAS* 110 (2013), pp. 13446–13451.
[227] Stechmann A et al. "Organelles in Blastocystis that blur the distinction between mitochondria and hydrogenosomes." In: *Curr Biol* 18 (2008), pp. 580–585.
[228] Storz JF. "Gene duplication and evolutionary innovations in hemoglobin-oxygen transport." In: *Physiology, Bethesda* 31 (2016), pp. 223–232.
[229] Suttle CA. "Marine viruses: major players in the global ecosystem." In: *Nat Rev, Microbiol* 5 (2007), pp. 801–812.
[230] Suttle CA. "Environmental microbiology: viral diversity on the global stage." In: *Nat Microbiol* 1 (2016), p. 16205.
[231] Takahashi K et al. "Induction of pluripotent stem cells from adult human fibroblasts by defined factors." In: *Cell* 131 (2007), pp. 861–872.
[232] Takahashi K, Murakami M, and Yamanaka S. "Role of the phosphoinositide 3-kinase pathway in mouse embryonic stem (ES) cells." In: *Biochem Soc Trans* 33 (2005), pp. 1522–1525.
[233] Tan T et al. "Telomere maintenance and telomerase activity are differentially regulated in asexual and sexual worms." In: *PNAS* 109 (2012), pp. 4209–4214.
[234] Tanackovic G et al. "PRPF mutations are associated with generalized defects in spliceosome formation and pre-mRNA splicing in patients with retinitis pigmentosa." In: *Hum Mol Genet* 20 (2011), pp. 2116–2130.
[235] Tanimoto Y et al. "Embryonic stem cells derived from C57BL/6J and C57BL/6N mice." In: *Comp Med* 58 (2008), pp. 347–352.
[236] Tegmark M. "The mathematical universe." In: *Found Phys* 38 (2008), pp. 101–150.
[237] Thomas K. "His doctors were stumped." In: *NY Times* (Feb. 4, 2017).
[238] Tiesinga E et al. "CODATA recommended values of the fundamental physical constants: 2018." In: *Rev Mod Phys* 93 (2021), p. 025010.
[239] Tovar J et al. "Mitochondrial remnant organelles of Giardia function in iron-sulphur protein maturation." In: *Nature* 426 (2003), pp. 172–176.
[240] Tovar J, Fischer A, and Clark CG. "The mitosome, a novel organelle related to mitochondria in the amitochondrial parasite Entamoeba histolytica." In: *Mol Microbiol* 32 (1999), pp. 1013–1021.
[241] Toyama BH and Weissman JS. "Amyloid structure: conformational diversity and consequences." In: *Annu Rev Biochem* 80 (2011), p. 10.
[242] Unlisted A. *National Academies of Sciences and Engineering and and Medicine, Examining the State of the Science of Mammalian Embryo Model Systems: Proceedings of a Workshop*. The National Academies Press, Washington, D.C., 2020.
[243] Valas RE and Bourne PE. "The origin of a derived superkingdom: how a gram-positive bacterium crossed the desert to become an archaeon." In: *Biol Direct* 6 (2011), p. 16.

[244] Valentine JW, Jablonski D, and Erwin DH. "Fossils, molecules and embryos: new perspectives on the Cambrian explosion." In: *Development* 126 (1999), pp. 851–859.

[245] Visscher PM, McEvoy B, and Yang J. "From Galton to GWAS: quantitative genetics of human height." In: *Genet Res* 92 (2010), pp. 371–379.

[246] Vogelstein B, Lane D, and Levine AJ. "Surfing the p53 network." In: *Nature* 408 (2000), pp. 307–310.

[247] Von Klitzing K, Dorda G, and Pepper M. "New method for high-accuracy determination of the fine-structure constant based on quantized Hall resistance." In: *Phys Rev Lett* 45 (1980), pp. 494–497.

[248] Von Marilaun AK. *The Natural History of Plants, translated by FW Oliver*. Blackie and Son, London, 1902.

[249] Vulliamy T et al. "Mutations in the telomerase component NHP2 cause the premature ageing syndrome dyskeratosis congenita." In: *PNAS* 105 (2008), pp. 8073–8078.

[250] Wachtershauser G. "Before enzymes and templates: theory of metabolism." In: *Microbiol Rev* 52 (1988), pp. 452–484.

[251] Wachtershauser G. "Pyrite formation, the first energy source for life: a hypothesis." In: *Syst Appl Microbiol* 10 (1988), pp. 207–210.

[252] Walker PJ et al. "Recent changes to virus taxonomy ratified by the International Committee on Taxonomy of Viruses." In: *Arch Virol* 167 (2022), pp. 2429–2440.

[253] Walker-Loud A. "Dissecting the mass of the proton." In: *Physics* 11 (2018), p. 118.

[254] Wang J et al. "Primate-specific endogenous retrovirus-driven transcription defines naive-like stem cells." In: *Nature* 516 (2014), pp. 405–409.

[255] Ward LD and Kellis M. "Interpreting noncoding genetic variation in complex traits and human disease." In: *Nat Biotechnol* 30 (2012), pp. 1095–1106.

[256] Ware SM, Aygun MG, and Hildebrandt F. "Spectrum of clinical diseases caused by disorders of primary cilia." In: *Proc Am Thorac Soc* 8 (2011), pp. 444–850.

[257] Waterham HR et al. "Autosomal recessive HEM/Greenberg skeletal dysplasia is caused by 3-beta-hydroxysterol delta(14)-reductase deficiency due to mutations in the lamin B receptor gene." In: *Am J Hum Genet* 72 (2003), pp. 1013–1017.

[258] Weatherall DJ. "Molecular pathology of single gene disorders." In: *J Clin Pathol* 40 (1987), pp. 959–970.

[259] Wetterbom A et al. "Comparative genomic analysis of human and chimpanzee indicates a key role for indels in primate evolution." In: *J Mol Evol* 63 (2006), pp. 682–690.

[260] Weyl H. *Raum, Zeit, Materie*. Springer-Verlag, Berlin, 1918.

[261] Wigner E. *The Unreasonable Effectiveness of Mathematics in the Natural Sciences*, Communications in Pure and Applied Mathematics Volume 13. John Wiley and Sons, New York, 1960.

[262] Wilczek F. "Mass without mass I: Most of matter." In: *Phys Today* 52 (1999), pp. 11–13.

[263] Wilczek F. "Mass without mass II: The medium is the mass-age." In: *Phys Today* 53 (2000), pp. 13–14.

[264] Wilkins AS and Holliday R. "The evolution of meiosis from mitosis." In: *Genetics* 181 (2009), pp. 3–12.

[265] Willis RA. *Borderland of Embryology and Pathology*. Butterworth, London, 1958.

[266] Wilson RA. "A group-theorist's perspective on symmetry groups in physics." arXiv:2009.14613 (2020).

[267] Wochner A et al. "Ribozyme-catalyzed transcription of an active ribozyme." In: *Science* 332 (2011), pp. 209–212.

[268] Woese CR. "Default taxonomy: Ernst Mayr's view of the microbial world." In: *PNAS* 95 (1998), pp. 11043–11046.

[269] Woese CR and Fox GE. "Phylogenetic structure of the prokaryotic domain: the primary kingdoms." In: *PNAS* 74 (1977), pp. 5088–5090.

[270] Wolf A et al. "Determining Lyapunov exponents from a time series." In: *Physica D* 16 (1985), pp. 285–317.

[271] Wu D et al. "A phylogeny-driven genomic encyclopaedia of Bacteria and Archaea." In: *Nature* 462 (2009), pp. 1056–1060.

[272] Yamaguchi H. "Mutations of telomerase complex genes linked to bone marrow failures." In: *J Nippon Med Sch* 74 (2007), pp. 202–209.

[273] Yu J et al. “Induced pluripotent stem cell lines derived from human somatic cells.” In: *Nature* 392 (1998), pp. 300–303.
[274] Yu J et al. “A draft sequence of the rice genome (Oryza sativa ssp indica).” In: *Science* 296 (2002), pp. 79–92.
[275] Zaher HS and Unrau PJ. “Selection of an improved RNA polymerase ribozyme with superior extension and fidelity.” In: *RNA* 13 (2007), pp. 1017–1026.
[276] Zhang J et al. “Novel RAG1 mutation in a case of severe combined immunodeficiency.” In: *Pediatrics* 116 (2005), pp. 445–449.
[277] Zhang S and Hou J. “Theoretical model for the Mpemba effect through the canonical first-order phase transition.” In: *Phys Rev E* 106 (2022), p. 034131.
[278] Zilfou JT and Lowe SW. “Tumor suppressive functions of p53.” In: *Cold Spring Harb Perspect Biol* 00 (2009), p. 001883.

Index

https://doi.org/10.1515/9783111498690-013

De Gruyter Series in Mathematics and Life Sciences

Volume 12
Olga Vasieva
Origins of Biocomplexity. The Fundamental Principles of Life in Terms of Physics
ISBN 978-3-11-133472-1, ISBN (PDF) 978-3-11-133479-0, ISBN (EPUB) 978-3-11-133483-7

Volume 11
Apul Dev, Sayantan Guha (Eds.)
Applied Engineering Mathematics. Fluid and Solid Mechanics in Life Science
ISBN 978-3-11-163853-9, ISBN (PDF) 978-3-11-163878-2, ISBN (EPUB) 978-3-11-163868-3

Volume 10
Fateh Mebarek-Oudina (Eds.)
CFD Simulation. Thermo-Fluids and Nanofluids in Engineering and Biomedicine, 2025
ISBN 978-3-11-140454-7, ISBN (PDF) 978-3-11-140509-4, ISBN (EPUB) 978-3-11-140523-0

Volume 9
Jochen Mau, Sergey Mukhin, Guanyu Wang, Shuhua Xu (Eds.)
BIOKYBERNETIKA. Mathematics for Theory and Control in the Human and in Society, 2024
ISBN 978-3-11-134197-2, ISBN (PDF) 978-3-11-134199-6, ISBN (EPUB) 978-3-11-134204-7

Volume 8/2
Igor M. Rouzine
Mathematical Modeling of Evolution. Volume 2: Fitness Landscape, Red Queen, Evolutionary Enigmas, and Applications to Virology, 2023
ISBN 978-3-11-069731-5, ISBN (PDF) 978-3-11-069738-4, ISBN (EPUB) 978-3-11-069746-9

Volume 8/1
Igor M. Rouzine
Mathematical Modeling of Evolution. Volume 1: One-Locus and Multi-Locus Theory and Recombination, 2020
ISBN 978-3-11-060789-5, ISBN (PDF) 978-3-11-061545-6, ISBN (EPUB) 978-3-11-060819-9

Volume 7
George Dassios, Athanassios S. Fokas
Electroencephalography and Magnetoencephalography, 2020
ISBN 978-3-11-054583-8, ISBN (PDF) 978-3-11-054753-5, ISBN (EPUB) 978-3-11-054578-4

Volume 6
Piotr Biler
Singularities of Solutions to Chemotaxis Systems, 2019
ISBN 978-3-11-059789-9, ISBN (PDF) 978-3-11-059953-4, ISBN (EPUB) 978-3-11-059862-9

FSC
www.fsc.org
MIX
Papier | Fördert gute Waldnutzung
FSC® C083411

Kolibri 360 GmbH
Ferdinand-Jühlke-Straße 7
99095 Erfurt, Deutschland
produktsicherheit@kolibri360.de